Mobile Technology Consumption:

Opportunities and Challenges

Barbara L. Ciaramitaro
Ferris State University, USA

Managing Director:	Lindsay Johnston
Senior Editorial Director:	Heather Probst
Book Production Manager:	Sean Woznicki
Development Manager:	Joel Gamon
Development Editor:	Joel Gamon
Acquisitions Editor:	Erika Carter
Typesetters:	Mackenzie Snader, Chris Shearer
Print Coordinator:	Jamie Snavely
Cover Design:	Nick Newcomer

Published in the United States of America by
    Information Science Reference (an imprint of IGI Global)
    701 E. Chocolate Avenue
    Hershey PA 17033
    Tel: 717-533-8845
    Fax: 717-533-8661
    E-mail: cust@igi-global.com
    Web site: http://www.igi-global.com

Library of Congress Cataloging-in-Publication Data

Mobile technology consumption: opportunities and challenges / Barbara Ciaramitaro, editor.
    p. cm.
  Includes bibliographical references and index.
  Summary: "This book explores essential questions related to the cost, benefit, individual and social impact, and security risks associated with the rapid consumption of mobile technology, covering the current state of mobile technologies and their use in various domains including education, healthcare, government, entertainment, and emerging economic sectors"-- Provided by publisher.
  ISBN 978-1-61350-150-4 (hbk.) -- ISBN 978-1-61350-151-1 (ebook) -- ISBN 978-1-61350-152-8 (print & perpetual access) 1. Mobile communication systems. 2. Mobile computing. 3. Technological innovations--Social aspects. I. Ciaramitaro, Barbara.
  TK5103.2.M63155 2012
  004--dc23
                2011017933

British Cataloguing in Publication Data
A Cataloguing in Publication record for this book is available from the British Library.

All work contributed to this book is new, previously-unpublished material. The views expressed in this book are those of the authors, but not necessarily of the publisher.

Editorial Advisory Board

Table of Contents

Preface

The growth of mobile technologies is remarkable. At a recent Mobile World Congress Conference, Eric Schmidt, CEO of Google predicted that within three years, smart phones will surpass Personal Computer sales. The number of mobile phones used worldwide has exceeded 4.6 billion with continued growth expected in the future. In fact, in the United States alone, the numbers of mobile phone users comprise over 80% of the population. There have been over 300,000 mobile applications developed over the last 3 years, and these applications have been downloaded 10.9 billion times. Whether the applications are used for communication, entertainment, socio-economic growth, crowd-sourcing social and political events, monitoring vital signs in patients, helping to drive vehicles, or delivering education, this is where the mobile technology has been transformed from a mode to a medium.

The adoption of mobile technology is growing fast in many sectors. Banking is quickly adopting mobile technology, as is education, the military, and healthcare. Of course, this fast growing world of mobile technology is exciting to watch and to participate in. However, this new world of mobility also presents many questions. What is the impact of this shift to mobile technology? Is it of benefit to consumers or ways to reduce costs on the business side – or both? What are the human factor implications of a mobile culture? Will mobile technologies expand or reduce the digital divide? What are the security and privacy risks of using mobile devices for the transfer of private information? What are the opportunities and challenges of mobile technologies? Helping to answer these questions is the goal of this book which includes contributions from mobile experts and academics around the world.

Mobile devices are no longer simple voice communication devices. They have become a medium to create voice, music, text, video, and image communications. Importantly, these various interactions can be created and shared on demand by the mobile user. In addition to communication methods, mobile devices are also a tool used to access the Internet, view television and movies, interact with GPS (Global Positioning System), play games, and read and respond to barcode and augmented reality messages. The reach and functionality of mobile devices depends on their underlying network infrastructure and the capabilities of the mobile device or handset.

Mobile technologies have become the predominant mode and medium of communication in the world. Mobile technology is comprised of several levels of technologies, devices, and applications to provide people with a powerful communication tool. At its most basic level, the reach and functionality of mobile devices depends on their underlying network infrastructure and the capabilities of the mobile device or handset. This involves the physical communication of data through a limited set of available frequencies. It involves the maturity of the underlying network and the functionality of the mobile device itself. The world of mobile applications lay on top of the technology, and this is where we see the richness and potential of mobile technology.

Chapter 1 is authored by Dr. Barbara Ciaramitaro of the United States, and introduces the core elements of mobile technologies from three perspectives. From the user perspective, the history of mobile technologies began with the use of two way radios and evolved to the current state of prolific smartphones, tablets, and other mobile devices. From the technical perspective, the history of mobile technologies originated with the limited use of radio frequencies, where the ability to establish simultaneous two-way communication (full duplex) was considered a technological feat. In the present day, mobile devices are quickly becoming IP devices that use the TCP/IP protocols to access, receive, and transmit data. From the social perspective, mobile technologies began as a rare device used by limited personnel who needed to communicate to others in real time emergencies such as police and the military. Mobile technologies are used today to provide health care services, deliver education, organize political events, market new products, provide location services, and deliver games, music and video.

Chapters 2 and 3 discuss the important use of mobile technologies in education.

In Chapter 2, Ariel Velikovsky, from Israel, discusses how the mobile phone is fast becoming an invaluable educational tool that is available the world over. Owing to the diverse applications that are available, educational content and experiences can be provided on the mobile that are simply not attainable elsewhere. Education on the mobile has the potential for becoming one of the main components of mobile functionality, and as such, represents a sizeable niche for mobile application and content developers as well as mobile operators and providers. Aspects indigenous to mobile consumption such as the use of voice and rich media can be judiciously incorporated into the mobile teaching arena, with close attention paid to best practices of effective pedagogy that is well suited and specifically tuned for mobile capabilities but that also take mobile limitations into consideration.

In Chapter 3, Dr. Douglas Blakemore, from the United States, focuses on how small electronic devices, also referred to as handhelds, are impacting education in a variety of ways including teaching methods, student life and the need for support from technical staff. This chapter discusses the importance of handhelds in education, how handhelds are being used in education, the challenges presented by handhelds for those in education, and what might happen with handhelds in the future.

In Chapter 4, Dr. Ciaramitaro discusses the exciting world of mobile marketing. Mobile technologies have dramatically changed the world's ability to communicate. They have become a medium to create voice, music, text, video, and image communications. Importantly, these various types of communication can be created and shared on demand by the mobile user. In addition to communication methods, mobile devices are also a tool used to access the Internet, view television and movies, interact with GPS (Global Positioning System), and read and respond to barcode and augmented reality messages. Each of these methods utilized by the mobile phone user becomes a tool that can be used in mobile marketing to expand beyond traditional marketing methods. The collection of accessible personal information amassed by mobile technology allows mobile marketing to target individuals at the time and place where their message will be most effective. Mobile technologies over the past 20 years have dramatically changed the way people communicate, collaborate, search for, receive, and share information. These dramatic changes have had striking impact on the world of marketing to the extent that mobile marketing has become the predominant form of customer engagement.

In Chapter 5, Dr. Jim Jones of the United States discusses how mobile technologies are used by government, military, and intelligence agencies. Government services, such as information access and certain transactions, are rapidly adopting mobile delivery mechanisms. The military is using mobile technology to share static information as well, but is also providing live data feeds and information sharing to support combat operations. Intelligence agencies are using mobile devices as a data collection platform for their own agents, and are also accessing the mobile devices of enemy agents and intelligence targets to collect data surreptitiously. Military operations face unique challenges, given that they are often conducted in regions without existing networks and against an enemy trying to actively disrupt communications. The Government, Defense, and Intelligence communities all face the challenge of securing mobile devices and data in response to regulatory and statutory requirements, as well as a dynamic and evolving threat space of identity thieves, hackers/crackers, hostile military forces, and foreign intelligence services.

Chapters 6 and 7 describe how mobile technology is being used in healthcare.

In Chapter 6, Ade Bamigboye from the United Kingdom, discusses how successfully placing mobile technology at the centre of any healthcare delivery service can enable innovations in healthcare to be distributed quickly, globally and equally. For some patients this could mean being able to gain better access to healthcare where previously there was very little or none at all. For others it could mean closer, more convenient healthcare management. For medical professionals, it is about using mobile technology to deliver and manage healthcare services from wherever they happen to be. Mobile healthcare is a complex combination of mobile operators, medical device manufacturers, care providers, software developers, funding partners and regulatory bodies. Each of these stakeholder groups is motivated to participate in the m-health discussion in different ways but need to work together in order to ensure that this technology can be deployed on a scale that enables benefits to be captured. This chapter presents the challenges that the industry must collectively overcome in order for all stakeholders to be successful.

In Chapter 7, Drs. Ciaramitaro and Skrocki from the United States talk about the revolutionary use of mobile devices in healthcare. Much of its revolutionary reach is due to the widespread adoption of mobile devices such as mobile smart phones and tablets. It is estimated that there are over five billion mobile devices in use throughout the world. In terms of demographics, in the United States, it is estimated that five out of seven Medicaid patients carry a mobile smart phone. mHealth is viewed as changing the healthcare landscape by changing the relationship between the patient, healthcare provider, and between healthcare providers. There is clearly a growing interest in, and emphasis on, mobile healthcare applications in the world today by vendors, physicians, and patients. It is predicted that the mobile health application market alone will be worth over $84 million, and that by the year 2015, more than 500 million people will be actively using mobile health care applications.

In Chapters 8 and 9 the topics of the use of more advanced mobile technologies such as augmented reality, RFID, and Bluetooth are discussed.

In Chapter 8, Jorg Kloss from Germany, presents a realistic picture about MAR (Mobile Augmented Reality), where it comes from, where it currently stands, and where it is going. The chapter distinguishes between the variants of MAR, and explains their similarities, differences, and different development stages and outlooks. While describing potential application fields and their current limitations, the opportunities and challenges of MAR become obvious, as well as how much work is still to do. The chapter intends to categorize the challenges for MAR, covering also related issues in the mobile hardware, software and operator industry, and their efforts in standardization and open interfacing. Besides the technology-driven discussion, a strong emphasis will be taken also to the essential aspect of user experience. With ubiquitous MAR, the technology will become more and more secondary, and the user and her individual context moves into the center of attention.

In Chapter 9, Dr. Greg Gogolin from the United States, examines the use of embedded mobile, RFID, and augmented reality in mobile devices. He explains that the proliferation of mobile devices such as smart phones and other handheld devices has stimulated the development of a broad range of functionality including medical, retail, and personal applications. Technology that has been leveraged to enable many of these uses includes radio frequency identification (RFID), embedded mobile, and augmented reality. RFID involves communication between a tag and a reader. Mobile RFID extends the technology by tagging the mobile device with an RFID tag to perform tasks on the device. Embedded mobile refers to preprogrammed tasks that are performed on a mobile device. Personal care and monitoring is one of the most common uses of embedded mobile. Augmented reality involves the use of computer generated or enhanced sensory input such as audio and visual components to enhance the perception of reality. This is commonly used in situations such as video games where there is feedback in the game controllers.

Chapter 10, authored by Nabil Harfoush of Canada, presents a very important discussion on the impact of mobile technologies on social and political movements. The strength of social and political movements is often correlated with the cost and risks of organizing the effort. Reaching large numbers of people to inform them of a movement's goals, and the ability to recruit supporters, has historically relied on mass media, both printed and electronic, along with traditional canvassing, public assembly, and public speaking. This has naturally favoured economic and political elites who had easier access to media channels, and who controlled in many cases the rights to public assembly and free speech. The emergence of affordable communications in general, and mobile communications in particular, is bringing radical change to this balance of power. This chapter explores some of these changes and suggests directions for future research in this area.

In Chapter 11, a team of researchers from Vicomtech Research Centre, Spain, led by Mike Zorilla, discuss the challenges of next generation multimedia on mobile devices. They explain that the multiplatform consumption of multimedia content has become a crucial factor in the way of watching multimedia. Current technologies such as mobile devices have made people desire access to information from anywhere and at anytime. The sources of the multimedia content are also very important in that consumption. They present the content from many sources distributed on the cloud and mix it with automatically generated virtual reality into any platform. This chapter analyzes the technologies to consume the next generation multimedia and proposes a new architecture to generate and present the content. The goal is to offer it as a service so the users can live the experience in any platform, without requiring any special abilities from the clients. This makes the architecture a very interesting aspect for mobile devices that normally do not have big capabilities of rendering but can benefit of this architecture.

In Chapter 12, Nalin Sharda from Australia discusses the important topic of using mobile technologies to support iMaintenance activities. iMaintenance stands for integrated, intelligent and immediate maintenance; which can be made possible by integrating various maintenance functions, and connecting these to handheld devices –such as an iPhone– using mobile communication technologies. The main innovation required for developing iMaintenance systems is to integrate the disparate systems and capabilities developed under the current eMaintenance models, and to make these immediately accessible by ubiquitous and intelligent computing technologies –such as Digital Ecosystems and Cloud Computing– connected via wireless networks and handheld devices such as the iPhone. A Digital Ecosystem is a computer-based system that can evolve with the system that it monitors and controls, and can be embedded in the system's components, thereby providing the ability to integrate new functionality without any downtime. Cloud Computing can provide access to additional software services that are not available in the local Digital Ecosystem. This chapter will show how these computing paradigms can provide mobile computing and communication facilities required to create novel iMaintenance systems.

In Chapter 13, Dr. Ciaramitaro and Velislav Pavlov, from the United States, discuss the important challenges of security in mobile technologies. Over the past few years, cyber criminals have expanded their focus from desktop PCs to mobile devices such as smart phones, PDAs, and tablet computers.

Unfortunately, even though many mobile devices approach personal computers in functionality, most mobile users are not aware of the degree of security threats in the mobile environment. There are several security threats related to mobile devices, each of which is discussed in detail in this informative chapter.

I am sure you will find the collections of chapters and perspectives on the various aspects of mobile technology consumption to be interesting and worthwhile.

Acknowledgment

This book could not happen without the efforts of a remarkable group of mobile technology experts from around the world who made up our Editorial Advisory Board and Contributing Authors. I would first like to thank the members of the Editorial Advisory Board who volunteered their time to ensure that the content of this book was worthwhile and reflected the current state of mobile technology consumption. I would then like to thank each of the individual authors from around the world who contributed countless hours of research and writing to deliver high quality content for this book.

Thanks must also be given to Editor Joel Gamon, and my hard working friends at IGI Global whose guidance and editing genius allowed a collection of disparate chapters to become an integrated book. They are able to work miracles.

Barbara L. Ciaramitaro
Ferris State University, USA

Chapter 1
Introduction to Mobile Technologies

Barbara L. Ciaramitaro
Ferris State University, USA

ABSTRACT

Mobile devices are no longer simple voice communication devices. They have become a medium to create voice, music, text, video, and image communications. Importantly, these various interactions can be created and shared on demand by the mobile user. In addition to communication methods, mobile devices are also a tool used to access the Internet, view television and movies, interact with GPS (Global Positioning System), play games, and read and respond to barcode and augmented reality messages. The reach and functionality of mobile devices depends on their underlying network infrastructure and the capabilities of the mobile device or handset. Mobile communications also rely on specific access methods which operate on top of the physical wireless architecture. There are generally three types of access methods in use: FDMA, TDMA and GSM, and CDMA. Although mobile devices are most commonly associated with mobile phones, there are many types of mobile devices. Some have broad usage such as smart phones and mobile tablet devices, and some are very specific such as telematic devices in vehicles and devices that monitor vital signs in healthcare. "It's all about the apps" is a common refrain we hear in the world of mobile technology. Not only are mobile applications the key to innovation and customer expansion, it is also a high revenue business. There have been over 300,000 mobile applications developed over the last 3 years, and these applications have been downloaded 10.9 billion times. Whether the applications are used for communication, entertainment, socio-economic growth, crowd-sourcing social and political events, monitoring vital signs in patients, helping to drive vehicles, or delivering education, this is where the mobile technology has been transformed from a mode to a medium.

DOI: 10.4018/978-1-61350-150-4.ch001

INTRODUCTION

"The mobility wave now in place may have the most profound impact of all (previous technology waves). While the previous waves were built on computers that communicate, mobility represents a complete new model: communication devices that compute." (Synder, 2009, p. x)

The profound changes that mobile technologies have introduced into our world had quite humble beginnings. It began as a simple two-way communication device and has become a powerful social tool. To gain a full understanding of the impact of mobile technologies, it is worthwhile to examine its history from several different perspectives.

- From the user perspective, the history of mobile technologies began with the use of two way radios and evolved to the current state of prolific smartphones, tablets, and other mobile devices.
- From the technical perspective, the history of mobile technologies originated with the limited use of radio frequencies; where the ability to establish simultaneous two-way communication (full duplex) was considered a technological feat. In the present day, mobile devices are quickly becoming IP devices which use the TCP/IP protocols to access, receive and transmit data. Historically, TCP/IP is viewed as a networking protocol most commonly associated with communications through the Internet. This change has required significant transformations in the underlying mobile infrastructure that is still occurring.
- From the social perspective, mobile technologies began as a rare device used by limited personnel who needed to communicate to others in real time emergencies such as police and the military. Mobile technologies are used today to provide health care services, deliver education, or-

ganize political events, market new products, provide location services, and deliver games, music and video. Mobile devices are considered to be "the most personal piece of technology that most of us will ever own" (Krum, 2010, p. 7). We usually take them with us wherever we go and are usually reachable through them. They store and record some of the most personal information about us: who we talk with; where we go; what we buy; where we bank; our family, friends and business contacts; and often employer information.

Mobile devices are no longer simple voice communication devices. They have become a medium to create voice, music, text, video, and image communications. Importantly, these various communications can be created and shared on demand by the mobile user. In addition to communication methods, mobile devices are also a tool used to access the Internet, view television and movies, interact with GPS (Global Positioning System), play games, and read and respond to barcode and augmented reality messages.

MOBILE TECHNOLOGY NETWORKS AND INFRASTRUCTURE

The reach and functionality of mobile devices depends on their underlying network infrastructure and the capabilities of the mobile device or handset. At its most fundamental level, wireless networks are based on the use of electromagnetic frequencies or spectrums. Mobile networks operate in a very small slice of available frequencies from 700MHz to 3,000MHz and are viewed as a rather scarce resource (Synder, 2009). This is the reason for the highly competitive bidding by wireless vendors when the government auctions off frequencies. Wireless communications has some unique characteristics such as frequency reuse, handoffs, and cells. Frequency reuse allows wireless vendors to maximize the number of users they can support within a set of frequencies they have licensed. Handoffs are the reason that wireless communication can traverse distances. As mobile users travel from one location to another, their call is handed off from one cell to another. If wireless users increase beyond the capacity of a given cell, a wireless vendor can divide the existing cells into smaller units to support the additional load.

Access Methods

Mobile communications also rely on specific access methods which operate on top of the physical wireless architecture. There are generally three types of access methods in use.

- FDMA (Frequency Division Multiple Access) was used in the early 1G mobile networks. It divided the allocated spectrum into different channels with each communication assigned to a specific channel. This type of access is still used in cable television.
- TDMA (Time Division Multiple Access) divides the allocated spectrum into specific time slots. Although all users communicate on the same frequency, their transmissions are synchronized to operate in different time slots. Mobile devices that use TDMA must be more sophisticated in to ensure that they can identify and operate in the appropriate time slot. TDMA and FDMA can be used together. In this case, the frequency is first divided into separate channels and then within those channels users are allocated specific time slots. "This is the case in GSM where the allocated band is divided into 124 channels using FDMA, and then each of these channels are further divided into eight time slots." (Synder, 2009, p. 144).

- CDMA spreads each transmission across the entire frequency. In order to prevent interference from multiple transmissions occurring at the same time, a unique random code is embedded into each transmission. This form of transmission is called spread-spectrum technology and was first used during World War II (Synder, 2009).

Mobile Network Generations 1G to 4G

1G

As discussed above, the ability to move from cell to cell was the key technological development that allowed the emergence of mobile phones. The earliest generation of mobile networks was based on a circuit-switching technology and relied on the transmission of analog signals sent from the phones to radio towers. This first generation is referred to as 1G and was the primary network infrastructure introduced in the 1980's and used through the first half of the 1990's. The throughput of data transmission on a 1 G network was limited to 9.6 Kbps (Kilobytes per second) (Krum, 2010). This limited throughput constrained the users of the 1 G network to only voice communications.

In 1983, the first mobile phone became available through Motorola. Its only functionally was that of voice calling. Affectionately known as "the brick", it weighed two pounds and cost nearly four thousand dollars and provided an hour of talk time before its battery would run out. The first mobile phones weighed about 2 pounds and were quite expensive. (Cassavoy, 2007) Due to the high cost, phones used with the 1 G network were rarely used by the general consumer. However, over the next few years mobile service provides began to upgrade their analog 1G network with digital networks. As a result, "Mobile phones soon went from being expensive novelty items used only by business professionals and the wealthy to being well within the reach of the masses." (Hager, 2006).

2G

The second generation of mobile networks, referred to as 2G, was launched in 1991and relied upon a digital signal to transmit rather than the analog signal used by the 1G network. The use of the digital signal improved the quality of the voice communications although the throughput remained the same at 9.6 Kbps. From the mobile technology provider's viewpoint, 2G offered an improvement in that carriers were able to transmit a higher volume of calls through their network. From the user point of view, however, the 2G network proved problematic as it relied more on a caller's proximity to a cell tower. If the caller moved out of range, the call was immediately dropped rather than gradually degrading. Text messaging became available on the 2G network although it was not quickly adopted. Although the first text message was sent in England in 1992, by 1994 users worldwide were only sending a mere.4 text messages per month. By the year 2000, the average number of text messages had increased to 35 per month. (TMC, 2007)

Mobile phone handsets in the 2G network were lighter, smaller, and less expensive which resulted in an increased consumer adoption rate. Most significantly, these handsets allowed text messaging capability in addition to voice calls. (Krum, 2010).

2.5G

The next generation of mobile networks is referred to as 2.5G. It is considered to be a bridge between 2G and 3G cellular wireless technologies and was introduced between the years 2000 and 2005. The 2.5G network provided a set of combined services including a circuit-switching domain for voice communication and a packet-switching domain for data communication. The transmission throughput in 2.5G network was 50+ Kbps.

During the era of the 2.5G network, mobile handsets with additional functionalities were introduced to the market. In 1993, BellSouth/IBM introduced the world's first mobile phone with PDA features that included phone, text, calculator, calendar, fax and email. It was small in size, weighing only 21 ounces and cost $900. And in 1996, "fashionable" mobile devices were introduced with the StarTAC mobile phone from Motorola that weighed only 3.1 ounces in a clam shell design. (Sacco, 2007)

The added functionalities available in the 2.5G network were beneficial to users and the adoption of mobile device usage increased so that by 2002 the number of mobile subscribers in the world was in excess of 1 billion users. Additionally, mobile marketing began during this era when the first mobile phone advertisement appeared as a free daily SMS text message in Finland (Membridge, 2008).

3G

It was with the release of the third generation of mobile networks, referred to as 3G that much of the capability used today came into existence. Although the first 3G network was introduced in Japan in 2001, it was not until 2005 that the use of 3G networks became widely adopted. The demand and use for mobile devices skyrocketed during this era, so that by 2006, the number of mobile subscribers exceeded 2.5 billion (Membridge, 2008). 3G actually refers to a set of standards that meet ITU (International Telecommunication Union) requirements (Union, 2005). In addition to voice, 3G is able to provide a wide variety of application services including mobile internet access, multimedia, location-based services, video calls, and mobile television. Its data transmission rates are rather robust at 384 Kbps (Krum, 2010).

In addition to the voice and text messaging capabilities of the 3G era, a significant added functionality to mobile handsets was that of the mobile web browser. The introduction of the web browser expanded the use of the mobile device so that Internet information could be searched for and accessed on demand by the user.

4G

The fourth generation of mobile networks, referred to as 4G, provides high speed data transmission and high levels of data throughput at 100 Mbit/s for high mobility communication (such as from trains and cars) and 1 Gbit/s for low mobility communication (such as pedestrians and stationary users). The 4G system provides a 100% IP based broadband network and allows for the delivery of high demand applications such as Voice over IP, gaming, and streamed multimedia (Krum, 2010). 4G networks are developed to meet QoS (quality of service) requirements in that they are able to appropriately prioritize the various types of network traffic for the most optimum results. This is necessary to support the high bandwidth capabilities offered through 4G networks including: Wireless Broadband Internet Access, MMS (Multimedia Messaging Service), Video Chat, Mobile Television, HDTV (High Definition TV), DVB (Digital Video Broadcasting), Real Time Audio, and High Speed Data Transfer (Internet, 2010). Additionally, 4G networks support reliable file transfer even if the user travels from one cell coverage area to another. Additionally, the mobile device is assigned a specific IP address that is maintained through the entire coverage area of any 4G network. This allows for easy tracking of mobile devices through the global network. (Mohr, 2002)

The standards for 4G networks are still in development at the time of this writing. There are currently several technologies competing to be the basis of 4G but many predict that the WIMAX standard will prevail (Wertime, 2008). WIMAX has been deployed in both the United States and Japan but other nations are currently using other technologies.

Infrastructure Challenges

The mobile infrastructure is currently faced with several challenges. One interesting challenge results from the increased use of mobile devices from fixed locations. Recent studies have found that users of mobile devices have significantly increased their use of those devices from static locations such as home or office. This is significant because wireless service providers must now manage Building Penetration Loss (BPL) which results from the inability of signals to travels through walls, flows and windows. "The BPL can result in 9o-99.9 percent of the energy of wireless signal being lost, just because a user was consuming a service indoors compared to a similarly located outdoor user." (Grayson, 2011, p. 7). Another challenge results from the increased use of video. Studies predict that the amount of mobile data in 2014 will be 39 times larger than mobile data transmissions in 2009. (p. 7). Referring to our previous discussion on cells, Grayson predicts that the only way mobile network will be able to support the dramatic increase in mobile traffic is through the use of smaller cells. He predict that the number of cells will need to increase "tenfold to meet the expected demand in traffic." (p. 9). The last issues results from our movement to an all IP based mobile network. As Grayson states so clearly,, "Unfortunately, the Internet is not Mobile." (p. 10). The current approach relies upon delivery of IP services is through tunneling which may result in more unreliability in the transmission of data (p. 10).

MOBILE DEVICES

Although mobile devices are most commonly associated with mobile phones, there are many types of mobile devices. Some have broad usage such as smart phones, and mobile tablet devices, and some are very specific such as telematic devices in vehicles and devices that monitor vital signs in healthcare. Although a mobile device could technically be considered any type handheld device including calculators, handheld game consoles, and CD players, we will focus our discussion of mobile devices on those that are able to transmit and receive data through the wireless network infrastructure discussed previously.

The mobile operating system is the underlying technology that controls the mobile device. The capability of a mobile operating system directly impacts the mobile device functionality. The most common mobile operating systems are Symbian from the Symbian foundation, Android from Google, iOS from Apple, RIM Blackberry OS, and Windows Mobile from Microsoft. The major improvement in mobile operating systems came with the introduction of smartphones which were supported by full featured mobile operating systems that allowed more advanced computing capability, Web browsing, and the installation of various applications and games. Most of the leading mobile operating systems now support the additional functionality provided by smart phones.

It is important to recognize that the capabilities of the various mobile web browsers differ greatly in terms of features offered and the mobile operating systems they support. The various screen sizes, handset functions, and underlying operating system of mobile devices make mobile Web browser consistency a challenge. There are currently several mobile Web browsers in use including Opera Mobile which works on the Windows Mobile and Symbian operating system; Skyfire which support Flash and works on Android, iPhone, Symbian, Windows Mobile operating systems; Safari which

work on the Apple iPhone; Google Android that works on the Google Android system; Mobile Chrome that works on the Google Android platform; Microsoft IE which works on the Microsoft Mobile platform; and Mobile Safari which works on the Apple iPhone (Nations, 2010).

Smart phones are becoming the most common mobile device in use today. In 2009, 20% of mobile phone users bought smart phones; by 2010, smart phones comprised 30% of all mobile phones purchased; it is predicted that by 2011, purchases of smart phones will exceed purchases of other mobile devices, computers and laptops (Entner, 2010). The smartphone is best portrayed as a convergent device providing a variety of mobile functionality in one handset. Smartphones provide telephone communication capabilities as well as text messaging, photo, video and music capability, location services, Bluetooth and RFID connectivity abilities, WiFi connectivity, application and gaming capability, and some 3D functionality through the use of augmented reality applications.

Other popular mobile devices include mobile music players such as the iPod and Zune. These are dedicated mobile devices that limit their functionality to primarily music, text, and videos. The emerging tablet mobile devices combine capabilities of desktop computing along with email, video and music capabilities. The tablet device is becoming a popular tool in educational environments due to its ability to clearly display texts with images (See Chapter 3). Tablet devices are also being adopted in healthcare as a mobile device supporting patient care, diagnostics and management.

Personal navigation devices are another popular mobile device. Used in conjunction with GPS (Global Positioning System), these devices are able to find location information from GPS satellites and translate that into step-by-step instructions. Electronic readers, also known as e-readers, are becoming more popular as well. Amazon, the premier online seller of books, has reported that sales of electronic books for their e-reader known as the Kindle, have surpassed their sales of hard copy books (CNN Money, 2011).

The entry of television and video into the mobile world has also provided new entertainment options. *IPTV (Internet Protocol Television)* allows traditional television shows and movies-on-demand to be sent over the Internet rather than traditional radio waves or cable. Additionally, IPTV allows the streaming of live television shows to mobile devices. However, it is important to note that although each of these mobile devices performs individual functions, they are also available in most smartphones, converged into one device.

Another category of mobile devices serves more targeted purposes. One example is the field of telematics where mobile technology is incorporated in automobiles. Telematics combines mobile communication with electronics to provide information or guidance to the vehicle driver or vehicle itself. Applications of vehicle telematics include GPS based location services, vehicle tracking, mobile television, and emergency warning systems. (SearchNetworking.com, 2011). Another example of a targeted mobile device is in the field of mobile health monitoring where patients use mobile devices to monitor their vital signs and other health conditions which is then forwarded to monitoring centers for evaluation by health care professionals. For a more detailed discussed on the use of mobile devices in healthcare, please see Chapters 6 and 7. A last interesting example of targeting mobile devices is seen in the military. Currently the military has combined the use of mobile technology and robotics to search for and neutralize incendiary devices (Singer, 2009).

MOBILE APPLICATIONS

"It's all about the apps" is a common refrain in the world of mobile technology. Not only are mobile applications the key to innovation and customer expansion, it is also a high revenue business. The Gartner Group estimated that $6.8 billion in revenue was earned in 2010 through mobile applications and predicts that by 2013, the mobile application market revenue will increase to $29.5 billion (Foresman, 2009). "The essence of the app's appeal is power…The app world is all about you…from the way we use our apps to the way they seem to spread… by word of mouth, this represents a new level of cultural and social empowerment for the individual." (della Cava, 2010, p. 1). There have been over 300,000 mobile applications developed over the last 3 years and these applications have been downloaded 10.9 billion times. The most commonly used mobile applications are news, maps, social networking and music. However, it is predicted that by 2015, more than one billion people will be using mobile financial applications to bank, invest and make purchases. (mobi Thinking, 2011).

The use of mobile devices has moved away from their original voice and messaging function. Most mobile users are now looking to their devices for entertainment. This has resulted in an explosion of mobile games. Some users will even base their selection of the mobile device on how well it handles gaming (Fung, 2010). It is predicted that revenue generated from mobile gaming will exceed $11 billion by the year 2015 (Reisinger, 2010). Social networking ranks as the fastest growing category of mobile applications with an estimated 20% of users accessing social networking sites through their mobile device (Sachoff, 2010). It is estimated that by the year 2013, there will be more than 140 million social networking subscribers on mobile devices (Research, 2010)

The government, military and intelligence communities are also using mobile devices and applications. Their uses include providing public information to citizens as well as the use of location-based services by law enforcement and the military. In Chapter 5, Jones discusses several uses of mobile technologies by the military. For example the *Raytheon Android Tactical System, or RATS,* is a mobile device with applications to provide bidirectional data flow for a soldier in the field. Example applications include streaming video, live mapping applications, and communications with other soldiers and facilities. Streaming video might include feeds from Unmanned Aerial Vehicles (UAVs) and mapping applications can geolocate friendly and hostile forces on a live map and permit annotation and sharing of map data. Another interesting application is the use of real-time language translation for soldiers in the field. For more detailed discussion of mobile technologies uses by the government, military and intelligence please refer to Chapter 5.

There are several new areas of mobile application development. One type of application relies upon Bluetooth and RFID technology. Bluetooth technology uses radio transmission bands to send signals to Bluetooth enabled devices within a close proximity of about 100 meters. When a Bluetooth enabled device enters the range of the Bluetooth hotspot, communication about the device and the user is sent to the Bluetooth server where, using its stored database of customer and device information, the Bluetooth server forwards content back to the user. RFID requires the use of RFID tags or chips which can be tracked using radio waves. For example, an RFID hotspot will recognize a passing RFID tag and send location specific messages to the device owner about a nearby store, restaurant or other business.

Another area of application development is in the area of augmented reality. Augmented reality is a mobile technology that combines location specific information with highly immersive and detail multimedia content. "AR systems integrate virtual information into a person's physical environment so that he or she will perceive that information as existing in their surroundings." (Hollerer, 2004, p. 1). For more information on Bluetooth, RFID and Augmented Reality, please refer to Chapters 9 and 10.

Similar to mobile devices, mobile applications also can be targeted to specific audiences. As mentioned earlier, mobile healthcare relies upon mobile technology applications to assist inpatient diagnosis and care, administration functions, and examination of lab, x-ray and other test reports. Other examples of mHealth devices and applications are the use of small wireless devices that are used inside the body to examine and diagnose possible health conditions. One recent development has been the use of a small capsule that is swallowed and transmits images as it travels through the stomach and small bowel. Similarly, implantable pacemakers can now transmit ongoing information for monitoring the patient's heart condition. Other implantable devices are able to assist in controlling bodily functions such as heart rhythms, provide or subdue nerve stimulation, and monitor cranial pressure (Barritt, 2010). For more detailed discussed on mobile healthcare applications, please refer to Chapters 4 and 5.

Mobile applications have also found their way into education. Their portability and small size make them ideal as multi-purpose devices. They can deliver digital reading material, multimedia lectures, and interactivity through conferences and discussions. There are also a large number of specific applications that have been developed for educational purposes in the mobile environment additions to art, science and math classes. For more detail on the use of mobile technology in education, please see Chapters 2 and 3.

Of course mobile technologies continue to advance in functionally particularly in the area of multimedia applications. For more discussion on the next generation of multimedia applications on mobile devices, please see Chapter 11 in which Zorilla et al analyze " the technologies to consume the next generation multimedia and proposes a new architecture to generate and present the content. The goal is to offer it as a service so the users can live the experience in any platform, without requiring any special abilities from the clients."

THE SOCIAL SIDE OF MOBILE TECHNOLOGIES

Although the mobile technology infrastructure and mobile applications are interesting to discuss, the social implications of mobile technologies are quite compelling. Mobile devices have changed the way the world communicates on an individual and mass level. However, there have been several concerns expressed as to whether the change is for the better. One interesting question that has been asked on several occasions is whether the introduction of mobile technologies has blurred the lines between home and work. In fact, a 2009 study has found that more than half of their respondents reported that mobile technology improved their ability to balance home and work primarily through improved ability to coordinate activities in their lives (Wajcman, 2009). Many fear that mobile technology interferes with interpersonal relationships. Contrary to the fears of many, mobile technologies have been found to tighten interpersonal connections particularly in existing social networks. This is referred by the Japanese as *telecocooning* which "refers to a zone of intimacy in which people can maintain their relationships with others who they have already encountered without being restricted by geography or time" (Oksman, 2009, p. 119)

However, mobile technologies have gone beyond individual communication. They have become communication tools for large groups of people focusing on various goals and activities. Discussed in detail in Chapter 10, Nabil Harfoush examines the various ways that mobile technologies have impacted our world and society through their communication capability. Harfoush's detailed discussions include social and political organization, economic opportunities to undeveloped communities, healthcare, data collections, tracking of outbreaks of disease and epidemics, and the delivery of education material.

One example provided by Harfoush on the socio-economic impact of mobile technology, is the example of Grameen Phone.

"One of the first programs Grameen Phone offered was the Village Phone Program, started in 1997, which used micro loans to help more than 210,000 people, mostly women living in rural areas, to acquire a mobile phone and use it to provide service to others. The program supported universal access to telecommunications service in remote rural areas: People lacking the means to own a phone could gain access to communications through the services offered by these so-called Village Phone operators, who had an opportunity to earn a living providing this essential service. Grameen Phone became remarkably successful."

An interesting use of mobile communications is in the support of crowd-sourced activities and movements. An example provided by Harfoush is described below.

"Oil Reporter, developed by Intridea for the Crisis Commons group, is an Android and iPhone application that "enables trained citizen journalists to use their mobile phones to capture and upload quantitative and qualitative data, as well as geo-tagged photos and videos to help in the recovery efforts" from the oil spill following the explosion and sinking of the Deepwater Horizon offshore drilling platform in the Gulf of Mexico in April 2010."

MOBILE TECHNOLOGY SECURITY AND MANAGEMENT CHALLENGES

Mobile devices are uniquely susceptible to security risks in that they are always on and accessible, and provide several means of communication and connectivity through text and multimedia messaging, voice, and wireless connectivity through Bluetooth and WiFi. Although they offer tremendous benefits with their convenience, functionality and immediate access to data, messaging, downloaded applications, and Web services, mobile devices are also fertile grounds for cyber criminal attacks. Over the past few years, cyber criminals have shifted their focus from desktop PC's to mobile devices. Unfortunately, even though many mobile devices approach personal computers in functionality, most mobile users are not aware of the degree of security threats in the mobile environment. "As mobile Internet usage continues its rapid growth, cyber criminals are expected to pay more attention to this sector" (Siciliano, 2010, p. 1). A troublesome trend in mobile devices is indicated by the escalating numbers of mobile malware and the increasing use of Internet capabilities in mobile devices. Together these indicate "a growing malware development community" and an "increasing source of potential attack vectors." (Jansen, 2008, pp. 3-9).

Malware is short for malicious software and refers to a collection of malevolent software tools designed to attack the pillars of information security: confidentiality, integrity, availability and authentication. In the computer world at large, some estimate that the number of malware programs exceed that of legitimate software programs (Symantec, 2008). Unfortunately, malware has become an increasingly prevalent threat to mobile devices.

A second category of malware is known as *concealment malware* and includes Trojan horses, root kits and backdoors. A *Trojan horse* is a software program designed to look like a desirable game or application but that carries with it malicious software. Its most common method of distribution is through the download of what appears to be attractive ringtones, wallpaper or games. *Root kits* are a type of software that modifies the mobile operating system so that its presence may not be detected. *Backdoors* are installed applications that allow malicious users to bypass the standard authentication processes of the phone and gain access without providing passwords or PIN numbers.

A third category of malware is specifically created for financial gain. These include spyware, keystroke loggers, dialers, and other intrusive applications. These programs are the most common ones used by organized cybercriminals to steal personal and financial information. *Spyware* software is designed to track the user's actions and provide information to the creator. They are often inadvertently downloaded by users from Web sites, attached to SMS or email messages, or hidden in games, ringtones and wallpaper downloads. *Key loggers* work by intercepting the keystrokes of a user then they are entering authentication information, credit card, banking or other personal information. The captured keystrokes are then sent to the creator of the malicious software for future exploitation. *Dialers* are malicious software that takes over control of the mobile device and either places unauthorized voice calls or sends unauthorized SMS or MMS messages.

The most common security threat associated with mobile devices is their propensity to become lost, stolen or misplaced. In fact, it is estimated that over 8 million phones are lost with 32% of them never recovered (Jansen, 2008). Unless a user specifically establishes security precautions on their mobile device, once it is in the hands of unauthorized users they have the ability to access and use any of the data and functionality provided by the mobile device.

Closely related to the loss of a phone is the security threat of unauthorized access. If a mobile device comes into the hands of an unauthorized user, their first goal will be to access the information and resources stored on the mobile device. The types of information stored on today's mobile devices go far beyond a listing of contacts and phone numbers. Mobile devices now hold personal and financial information as well as records of calls, texts, and locations visited. Unfortunately, very few mobile users implement password or PIN (Personal Identification Numbers) on their mobile device.

One common social engineering attack involves attempts to collect personal, credit card, and banking information from users. There are several variations of this scam. *Smishing* uses SMS texting as the basis of illegal requests for users to provide personal information. The smishing message will provide the user with a phone number to call or a web site to access. On the other end of the phone or web site is a cyber criminal anxious to steal whatever personal information is provided. Smishing is similar to *phishing* attacks which use the same tactics in the email environment. *Vishing* is a similar illegal scan in which voicemail messages are left for the user requesting that they call back or access a web site to provide more information.

Mobile eavesdropping and tracking can occur in two ways: direct and indirect. One of the common and direct ways for mobile eavesdropping to occur is through the use of spyware which is usually downloaded as part of a game, message or application. Once the spyware is installed, it tracks, collects, and forward information about the user their transactions to a recipient server or phone (Jansen, 2008). Most mobile devices provide the capability to connect to local wireless networks such as the WiFi connections in coffee shops, book stores, or airports. Wi-Fi refers to a set of standards that guide wireless connections in close proximity. Although Wi-Fi ad hoc connectivity is very convenient for the user, it is rife with security risks and provides the means and opportunity for an indirect eavesdropping or tracking attack (Jansen, 2008). It is common for a cyber criminal to create a rogue wireless access point in a public Wi-Fi area in order to intercept and hijack information transmitted from mobile devices (Couture, 2010).

There are two vital steps that every mobile user should implement to protect their device against malware and breaches. The first is the use of passwords and PINs. As discussed earlier, the most prevalent security risk is the lost mobile device. Requiring a password or PIN to access data stored on the mobile device will provide one level of protection. Establishing a relationship with a vendor who is able to wipe all the data from your device remotely if lost, is the next level of protection. The second essential step is the use of mobile anti-virus software. As mobile viruses become more prevalent, the use of mobile anti-virus software becomes essential. However, it is important to keep your anti-virus software updated by establishing an ongoing subscription with a mobile anti-virus software vendor. For more detailed discussion on mobile security please see Chapter 13.

Although security threats to personal users of mobile device can be disturbing, the threat to corporate enterprise systems can be devastating. The use of mobile devices such as smart phones is widespread in the business world, some of which are authorized by the enterprise, and some unauthorized. In fact, "More than 50% of enterprises have bowed to worker pressure and support personally-owned smart phone" (Lai, 2010, p. 1). Unfortunately, many corporate IT staffs continue to treat these devices as personal phones rather than portable computing devices rife with similar security concerns. The use of a smart phone or other intelligent mobile device can open a corporate network to several types of malicious attacks as discussed previously. However, the one unique aspect of enterprise mobile security is that the attacks and breaches can affect the entire corporate network rather than simply one individual's device. Enterprise networks also have the additional risk of destruction or access to confidential and proprietary data and information. "Unless actions are taken to secure this information, the mobile device represents a potentially severe security risk to the enterprise" (Good Technology, 2009, p. 1). Another concern related to mobile security is the cost of data breaches through mobile devices. "The average organizational cost of a data breach was US $3.4 million, but all countries in the study reported noticeably higher costs when the incidents involved mobile devices." (Lobel, 2010). Is this a mobile breach cost? it is essential for organizations to institute *Mobile Device Management (MDM)* policies and practices to protect the enterprise from security attacks. Mobile Device Management (MDM) includes technology, policies and procedures to manage mobile devices across the enterprise. The software and policies include the ability to remotely manage and wipe devices, password and PIN requirements, asset management and consistent application configuration settings (Joseph, 2006). Over-the-Air (OTA) ability is considered an essential component of MDM which allows the central IT department to remotely configure, lock or wipe a device Although MDM policies and practices have been in place for many years, the current emphasis of MDM is on security (Winthrop, 2010). For further discussion on iMaintenance, please see Chapter 12.

CONCLUSION

Mobile technologies have become the predominant mode and medium of communication in the world. Mobile technology is comprised of several levels of technologies, devices, and applications to provide people with a powerful communication tool. At its most basic level, the reach and functionality of mobile devices depends on their underlying network infrastructure and the capabilities of the mobile device or handset. This involves the physical communication of data through a limited set of available frequencies. It involves the maturity of the underlying network and the functionality of the mobile device itself.

The world of mobile applications lay on top of the technology, and this is where we see the richness and potential of mobile technology. Whether the applications are used for communication, entertainment, socio-economic growth, crowd-sourcing social and political events, monitoring vital signs in patients, helping to drive vehicles, or delivering education, this is where the mobile technology has been transformed from a mode to a medium.

"Communication technologies are not reversible. One cannot undiscover media developments; one cannot reject the concept of mobility. It has become an integral part of our social values." (Gumpert, 2007, p. 19)

REFERENCES

Ahonen, T. (2010). *An inconceivable truth: MMS is a global success at 30B dollars.* Retrieved on December 16, 2010 from http://communities-dominate.blogs.com/brands/2010/06/an-inconceivable-truth-mms-is-a-global-success-at-30b-dollars.html

Becker, M. A. (2010). *Mobile marketing for dummies.* Wiley Publishing, Inc.

Bowser, M. (2009). *IVR - A marketer's dream.* Retrieved on December 18, 2010, from http://www.mobilemarketingmagazine.co.uk/content/ivr-marketers-dream-no-really

Butcher, D. (2010). *7 key trends mobil marketers need to know.* Retrieved on December 15, 2010 from http://www.mobilemarketer.com/cms/news/research/7342.html

Butcher, D. (2010). *Snickers ties first branded mobile game to in-store marketing.* Retrieved on December 17, 2010 from http://www.mobilemarketer.com/cms/news/gaming/5468.html

Cassavoy, L. (2007). *In pictures: A history of cell phones*. Retrieved on December 16, 2010, from http://www.pcworld.com/article/131450/in_pictures_a_history_of_cell_phones.html

Cassela, D. (2009). *What is augmented reality?* Retrieved on December 16, 2010, from http://www.digitaltrends.com/mobile/what-is-augmented-reality-iphone-apps-games-flash-yelp-android-ar-software-and-more/2/

CBS News. (2010). *Number of cell phones worldwide hits 4.6B*. Retrieved on December 15, 2010, from http://www.cbsnews.com/stories/2010/02/15/business/main6209772.shtml

Cellphone Advertising. (2007). *Product placement in mobile phone advertisement*. Retrieved on December 17, 2010 from http://www.cellphone-advertising.com/product-placement-in-mobile-phone-advertising/

Channel Insider. (2010). *Salesforce chatter social networking goes mobile*. Retrieved on December 16, 2010, from http://www.channelinsider.com/c/a/Cloud-Computing/Salesforce-Chatter-Social-Networking-Goes-Mobile-443229/

CTIA. (2011). *Basics of CSC FAQs*. Retrieved on December 16, 2010 from http://www.ctia.org/business_resources/short_code/index.cfm/AID/10341

della Cava, M. (2010). *It's an app world, and it could swallow all computing*. Retrieved on February 16, 2011, from http://www.usatoday.com/tech/products/2010-03-31-1Aappworld31_CV_N.htm

Durrell, J. (2010). *Mobile game marketing*. Retrieved on December 16, 2010 from http://mmaglobal.com/articles/mobile-game-marketing-greystripe

Entner, R. (2010). *Smartphones to overtake feature phones in U.S. by 2011*. Retrieved on December 19, 2010 from http://blog.nielsen.com/nielsenwire/consumer/smartphones-to-overtake-feature-phones-in-u-s-by-2011/

Foresman, C. (2009). *Apple responsible for 99.4% of mobile app sales in 2009 (Updated)*. Retrieved on February 16, 2011, from http://arstechnica.com/apple/news/2010/01/apple-responsible-for-994-of-mobile-app-sales-in-2009.ars

Free Trade Commission. (2004). *Sham site is a scam: There is no national do not e-mail registry*. Retrieved on December 18, 2010 from http://www.ftc.gov/opa/2004/02/spamcam.shtm

Free Trade Commission. (2010). *FTC testifies on do not track legislation*. Retrieved on December 18, 2010 from http://www.ftc.gov/opa/2010/12/dnttestimony.shtm

Fung, L. (2010). Marketing mobile games. Retrieved on December 17, 2010, from http://www.selfgrowth.com/articles/Marketing_Mobile_Games.html

Grayson, M. S. (2011). *Building the mobile internet*. Indianoplis, IN: Cisco Press.

GSMA. (2010). *European framework for safer mobile use by younger teenagers and children*. Retrieved on December 18, 2010 from http://www.eubusiness.com/topics/telecoms/gsma.10-06-09/

Gumpert, G. A. (2007). Mobile communication in the twenty-first century or everybody, everywhere, at any time. In Kleinman, S. (Ed.), *Displacing space* (pp. 7–20). New York, NY: Peter Lang Publishing.

Hager, F. (2006). Mobile communications. Retrieved on December 18, 2010 from http://www.fredhager.com/index.asp?CategoryID=67&SubCategoryID=587&ContentID=1047

Havenstein, M. (2008). *LinkedIn social networking goes mobile*. Retrieved on December 16, 2010 from http://www.cio.com/article/187401/LinkedIn_Social_Networking_Goes_Mobile

Hollerer, T. A. (2004). Mobile augmented reality. In Karimi, H., & Hammad, A. (Eds.), *Telegeoinformatics: Location-based computing and services*. Taylor and Francis Books Ltd.

Interactive Blend. (2010). *QR codes: The future of marketing*. Retrieved on December 16, 2010 from http://interactiveblend.com/blog/interactive/qr-codes/

International Telecommunications Union. (2005). Cellular standards for the third generation: The ITU's IMT-200 family. Retrieved on December 15, 2010, from http://www.itu.int/osg/spu/imt-2000/technology.html#Cellular%20Standards%20for%20the%20Third%20Generation

Jansen, W. A. (2008). *Guidelines on cell phone and PDA security. National Institute of Standards and Technology*. US Department of Commerce.

Kee, T. (2010). *4 ways that 4G will impact mobile marketing in 2011*. Retrieved on December 16, 2010 from http://econsultancy.com/us/blog/6965-4g-or-not-4g-four-ways-it-will-impact-mobile-marketing-in-2011

Krum, C. (2010). *Mobile marketing: Finding your customers no matter where they are*. Indianopolis, IN: Que.

Membridge. (2008). *History of cell phones*. Retrieved on December 16, 2010 from http://www.historyofcellphones.net/

Miller, B. (2009). *RFID technology being added to mobile marketing campaigns*. Retrieved on December 16, 2010, from http://blog.armoryideas.com/2009/06/11/rfid-technology-being-added-to-mobile-marketing-campaigns/

MJelly. (2009). *7 viral marketing tactics for mobile internet services*. Retrieved on December 18, 2010 from http://blog.mjelly.com/2009/01/viral-marketing-on-mobile.html

mobi Thinking. (2011). *Global mobile statistics 2011*. Retrieved on February 16, 2011 from http://mobithinking.com/mobile-marketing-tools/latest-mobile-stats

Mobile Augmented Reality. (2010). *The absolute latest in Android and iPhone augmented reality*. Retrieved on December 16, 2010, from http://www.mobileaugmentedreality.info/

Mobile Market Watch. (2010). *Research: Mobile proximity marketing to reach $750M by 2011 and nearly $6b by 2015*. Retrieved on December 16, 2010, from http://www.mobile-marketingwatch.com/research-mobile-proximity-marketing-to-reach-750m-by-2011-and-nearly-6b-by-2015-10252/

Mobile Marketing Association. (2008). *Code of conduct*. Retrieved on December 18, 2010 from http://www.mmaglobal.com/codeofconduct.pdf

Mobile Marketing Association. (2010). *Consumer best practices*. Retrieved on December 18, 2010 from http://www.mmaglobal.com/codeofconduct.pdf

Mohr, W. (2002). *Mobile communications beyond 3G in the global context*. Retrieved on December 15, 2010 from http://www.cu.ipv6tf.org/pdf/werner_mohr.pdf

Money, C. N. N. (2011). *Amazon sales pop as Kindle books overtake paperbacks*. Retrieved on February 16, 2011, from http://money.cnn.com/2011/01/27/technology/amazon_earnings/index.htm

Murphy, D. (2008). *It's as easy as IPTV*. Retrieved on December 18, 2010, from http://www.mobile-marketingmagazine.co.uk/content/its-easy-iptv

Nations, D. (2010). *A list of mobile Web browsers*. Retrieved on December 16, 2010, from http://webtrends.about.com/od/mobileweb20/tp/list_of_mobile_web_browsers.htm

Neustar. (2010). *CSC implementation: A mobile marketing plan*. Retrieved on December 16, 2010 from http://www.scribd.com/doc/21139025/CSC-Implementation-a-Mobile-Marketing-Plan

Nielsen. (2010). *Nielsen unveils retail 2015 forecast.* Retrieved on December 15, 2010, from http://www.nielsen.com/us/en/insights/press-room/2010/nielsen_unveils_retail.html

Oksman, V. (2009). Media content in mobiles. In Goggin, G. A. (Ed.), *Mobile technologies: From telecommunciations to media* (pp. 118–130). New York, NY: Routledge.

Pollard, S. (2008). *Mobile email marketing tips.* Retrieved on December 16, 2010 from http://www.lyris.com/resources/email-marketing/articles/mobile-email-marketing-tips/

Reisinger, D. (2010). *Mobile game revenue to top $11 billion by 2015.* Retrieved on Decemver 17, 2010 from http://news.cnet.com/8301-13506_3-20024103-17.html

Research, A. B. I. (2010). *Online social networking goes mobile: 140 million users by 2013.* Retrieved on December 16, 2010 from http://www.abiresearch.com/press/2998-Online+Social+Networking+Goes+Mobile%3A+140+Million+Users+by+2013

Sacco, A. (2007). *A brief history of the mobile phone (1973-2007).* Retrieved on December 16, 2010, from http://advice.cio.com/al_sacco/a_brief_history_of_the_mobile_phone_1973_2007?page=0%2C0

Sachoff, M. (2010). *Mobile social networking grows 240%.* Retrieved on December 16, 2010, from http://www.webpronews.com/topnews/2010/06/02/mobile-social-networking-grows-240

SearchNetworking.com. (2011). *Telematics.* Retrieved on February 16, 2011, from http://searchnetworking.techtarget.com/definition/telematics

Singer, P. (2009). *Wired for war: The robotics revolution and conflict in the 21st century.* Penguin.

Synder, S. (2009). *The new world of wireless.* Wharton School Publishing.

TMC. (2007). *The history of SMS messaging.* Retrieved on December 15, 2010, from http://www.tmcsms.com/sms-history.aspx

W3. (2010). HTML5 differences from HTML 4. Retrieved on December 18, 2010 from http://www.w3.org/TR/html5-diff/

Wajcman, J. B. (2009). Intimite connections: The impact of the mobile phone on work/life boundaries. In Goggin, G. A. (Ed.), *Mobile technologies: From telecommunications to media* (pp. 9–22). New York, NY: Routledge.

Wallace, L. (2009). *Blink-182 rocks augmented reality show in Doritos bag.* Retrieved on December 16, 2010 from http://www.wired.com/underwire/2009/07/blink-182-rocks-augmented-reality-show-in-doritos-bag/

Warren, C. (2010). *Mobile social networking usage soars.* Retrieved on December 16, 2010 from http://mashable.com/2010/03/03/comscore-mobile-stats/

Wauters, R. (2009). There's money in mobile dating. Retrieved on December 16, 2010, from http://techcrunch.com/2009/01/19/juniper-research-theres-money-in-mobile-dating-services/

Wertime, K. A. (2008). *DigiMarketing: The essentail guide to new media and digital marketing.* John Wiley& Sons.

Wireless Internet. (2010). *What's this about 4G?* Retrieved on December 16, 2010 from http://www.wirelessinternet.org/4G-network.php

Chapter 2
Mobile Education

Ariel Velikovsky
SpeakingPal Ltd., Israel

Shaunie Shammass
SpeakingPal Ltd., Israel

ABSTRACT

The mobile phone is fast becoming an invaluable educational tool that is available the world over. Owing to the diverse applications that are available, educational content and experiences can be provided on the mobile that are simply not attainable elsewhere. Education on the mobile has the potential for becoming one of the main components of mobile functionality, and as such, represents a sizeable niche for mobile application and content developers as well as mobile operators and providers. Aspects indigenous to mobile consumption such as the use of voice and rich media can be judiciously incorporated into the mobile teaching arena, with close attention paid to best practices of effective pedagogy that is well suited and specifically tuned for mobile capabilities but that also take mobile limitations into consideration.

INTRODUCTION

Some ten years ago, a mobile phone in the classroom was considered to be a disturbance to a teacher trying to keep the lesson flow. Today we are in the process of witnessing a dramatic change. The mobile phone is more and more frequently being considered as a tool that may actually be helpful in performing various educational tasks. Mobile technology offers students and teachers new ways to communicate with each other, as well as novel ways to interact with learning resources when outside of the classroom to enrich the learning experience. Using a mobile phone for learning is fun, allows utilizing time in a better way, and connects the communicative nature of mobile services with the learning experience. Does it replace other ways of learning? No, probably not. However, it makes the whole learning process more efficient and enjoyable. The technological and cultural infrastructure for these changes is already here and the real challenge is to identify and implement the right learning models to benefit both students and teachers.

DOI: 10.4018/978-1-61350-150-4.ch002

Wide implementation of mobile education is not here yet. There are still significant challenges of scale and sustainability, and of blending mobile learning technologies with other established and emerging educational technologies (Guy, 2009). Some of the issues that surround the difficulties in implementation are mobile service costs, the need to change attitudes, and institutions' policy against using electronic devices (De Lorenzo, 2010). These and other problems may be resolved as mobile education frameworks become better defined.

One objective this chapter is to advance the consideration of mobile education as one of the most important areas in mobile technology consumption. Proliferation of mobile devices together with technology development of higher bandwidth networks and better smartphones make the mobile phone a powerful platform for educational applications. The idea of using mobile devices in the learning process had captivated educators and mobile industry players long before the conditions were set for ubiquitous mobile penetration. Today, we are beginning to see significant adoption of these technologies in further and higher education, in schools, in the community, and in training and updating (Kukulska-Hulmes & Traxler, 2005). This points to a growing trend towards a more widespread use of mobile education.

Another objective of this chapter is to address the compatibility and symbiosis between usual and widespread functions of the mobile with learning objectives. Too often the types of learning activities are not conducive to the technology and vice versa. For instance, simple porting of content intended for other media such as printed material, wide-screen computers, television or film, does not afford the maximal usage of the intrinsic capabilities of the mobile device. Learning material coupled with appropriate platform development should ideally be created specifically for the mobile in mind, or at least undergo substantive alteration for optimal benefit.

BACKGROUND

Mobility has now become a staying factor in our societies, and the mobile phone industry has responded to a genuine need – to allow seamless communication between people even though they are moving about constantly in different places, conditions and environments. This basic communication, coupled with the proliferation of features and content that are enabled through mobile phone access, allow for ever-increasing enrichment of everyday experiences.

To this end, the mobile industry is burgeoning. It is estimated that there is over 3.3 billion cell phones in use globally. The Horizon Report states that mobile technology is quickly becoming mainstream in both developed and developing countries alike (Johnson et. al., 2010). Cell phone usage currently outstrips television and internet usage in many parts of the world (Kolb, 2008). Furthermore, there is an explosive worldwide growth in use of mobile web access (Perez, 2009).

The factor of mobility is just now starting to be considered in the realm of education. Mobile learning will allow everyone to learn anything at anytime, irrespective of location, status or culture, empowering both students and teachers with educational freedom, and not only at specific times in specified locations (Ally, 2009). The use of mobile phone technology also alters the nature of learning by focusing on personal and communal activities to find and process information, rather than possess or know static information (Traxler, 2009).

Kukulska-Hulme and Traxler (2005) comment on the impact that mobile devices have on the blurring between formal and informal learning as well as the weakening discrepancy between work versus leisure. They note that the newer features of mobile phones offer spontaneous access to almost limitless educational resources in a relatively low-cost and already widely used, accessible manner. They also warn that judicious coupling of the new technology with appropriate design and learner support is both complex and challenging.

Development of mobile learning has been largely driven by a combination of opportunity, necessity, innovation and perceived disappointments with the e-learning experience in a complex interplay between technologists and educators (Guy, 2009). However, new directions in mobile education will be forged by a symbiotic relationship between education and technology, each having a proportionately equal say in the ensuing development. To this end, private enterprise, and government sponsored programs and educational institutions are in a key position to find ways to leverage new technologies with appropriate content and delivery methods that focus on the ever-important user experience (Kossen, 2010). The key to success is to refrain from using technology for the sake of technology, and ensure that sound pedagogical considerations are developed in tandem. Future trends will focus on typical consumer-driven concerns, such as enhancing user experience by providing engaging content along with new delivery and design features.

MOBILE EDUCATION

Setting the Stage for Mobile Learning

Mobile technology has progressed significantly over the last few decades. An average phone is a full-fledged mini-computer in terms of computational power and network access. Bigger displays provide better user experience and enable more types of mobile applications to run on mobile devices. Open platforms, support for advanced mobile browsers and a more developed ecosystem for support, distribution, discovery and billing mechanisms for third- party applications are additional important enablers that facilitate mobile innovation.

As educational applications vary from very simple ones such as offline dictionaries and casual educational games to more complex network dependent client-server applications with enhanced user interface and downloadable content, the required technology requirements vary as well. The mobile environment is highly fragmented and this fragmentation should be taken into account when planning an educational application. Certain technologies, such as short messaging (SMS) and WAP browsing are quite ubiquitous and may be used by most mobile users. Other technologies such as, for example, high-speed data access or video calls are still limited by choice of mobile carrier, geography and a range of available mobile devices. The choice of a specific technology for implementation of an application will be influenced by the above factors along with marketing data regarding mobile service consumption by the target consumer segment.

Mobile learning is not necessarily just the conjunction of 'mobile' and 'learning' (Guy, 2009). While often approached from the technological point of view, mobile learning is also about changing learning process paradigms. What may work perfectly well in traditional education or even e-learning, may not fit the dynamic mobile environment or be comfortable within specific mobile constraints. Creating educational frameworks that appropriately reflect the mobile nature of m-learning will prepare ground for further m-learning growth and absorption by educators and students alike.

Learning on the Go: The Promise

Greater access to on-the-go functionality has shaped expectations of 'here and now' compliance. As more and more services are provided by the mobile that can be accessed anytime, anywhere, the more people are inclined to demand that other services be accessible in the same way. Only a few years back, the notion that such services such as purchase and payment or scheduling and management, could be accomplished from outside confined spaces or specific locations was considered impossible. Today, these and many more applications are being offered as standard functions. Remote access to material, media, and social interconnectivity, all within easy reach via the modern mobile phone are now becoming accepted as normal and usual operations.

To this end, it seems reasonable that education will also be a service demanded to be available beyond the confines of the traditional classroom. Learning as a structured event within blocked time limits and in specific and anchored locations will give way to a more freer and spontaneous notion of learning.

As the widespread use of the internet has shown, the most insatiable demand asked of connectivity is for community and social interaction. The mobile is an extension of this, being a device that is, at its heart, based on the core demand for connecting people remotely so that they need not occupy the same space or even time in order to communicate with one another. Newer and more powerful generations of phones combine the positive aspects of being connected via the internet with the positive aspects of being connected via a mobile communication device. As web and other server access are now attainable through the mobile phone, community functions offered in 'the cloud' are also now available on-the-go. This enables a plethora of heretofore unimaginable opportunities for social interaction and development, which can be judiciously harnessed for educational purposes.

Never has this leap in capabilities been as important as in the developing countries. "Leap-frogging" the traditional stages of requiring expensive and long-cycle infrastructure development, developing countries can now attempt to be on a more equal footing with that of developed countries as penetration of cellular phones increases. If remote schools were at one time inaccessible and substandard, new access to the repository of knowledge once only available to the elite can now help to level the playing field. Connectivity to the internet may no longer be dependent on preconditions imposed by cable or satellite access.

Rather, it is dependent on the one infrastructure that is being developed in any case, that of cellular communication.

The educational repercussions of this are that access to once elitist learning material, mentors, teachers, and methods are no longer confined to the elite. If, for example, I happen to live in a country where there are few native English speakers with whom to speak to in order to practice the language, this at one time would have severely hampered my ability to learn English. This may no longer be the case. And if, in this country, the internet connectivity is prohibitively expensive, or not available, or not easy to develop, the use of the mobile may allow access to these resources in a much less expensive and easier way (Conway-Smith, 2010). This represents a type of 'double leap-frogging', bypassing one stage of requiring costly landline infrastructure and a second stage of needing difficult-to-obtain internet access at one and the same time.

Mobile Technology for Education

The world of mobile educational applications is in a nascent ascendency. The gamut of mobile educational practices encompasses the whole range of mobile capabilities. They are also evident on every kind of mobile device. Even simpler, low-end mobile phones have been employed for educational projects, such as in Bangladesh (Power, & Shrestha, 2010) or Mexico (PR Newswire, 2010). This chapter outlines the types of mobile technology used for educational projects, including well-worn solutions native on the mobile such as use of voice and SMS, as well as newer functionalities available such as use of audio, video, camera and video-camera capabilities to provide podcasts, web-browsing and internet access.

Native solutions include use of voice or SMS communications that are already on the mobile phone. Use of SMS and other communication functions have grown exponentially. In the realm of education, such tools have generally been used for management and administrative purposes in a teacher-centric and conservative manner (Herrington, et. al., 2009). Functioning as a remote and mobile calendar, timetable, reference and library provider and the like, use of the mobile in concrete educational programs has focused on the functional aspects of mobility. SMS is used for imparting information between teacher to student, school to parents, and the school administration to both student and teachers (ZapitSMS, 2010). It functions to provide in-class communication such as notification of grades, homework assignment and assessment, exam scheduling with content overview, etc. In addition, SMS is also used to provide school administrative communications such notifications of school closures, scheduling of intramural sports, cultural events or other activities, notification of illness or emergencies, or monitoring of absenteeism and even bullyism (De Waard, 2010).

Students and teachers find SMS notifications useful and report that it ensures smoother classroom logistics. (Terbilang, 2008; Vassal, et. al., 2006). Educational goals are also enhanced, such as assignment completion and student achievement (Crisp, 2009). Podcasts have been used for educational purposes, both for audio (for example, Cebeci, & Tekdal, 2006) and video (Copley, 2007). Podcasting has been an effective and desirable medium for educational content distribution (Gribbins, 2007). Owing to the plethora of podcasts now available, the attainment of quality and appropriate content is one drawback. Collaborative sites such as (EPN, 2010) allow for easier and more targeted searching capabilities in order to find appropriate material.

Mobile presentations and enhanced podcasts over the mobile have been recently employed with increasing effectiveness and student acceptance (Evans, 2008). Mobile podcasting enables educational content to be delivered that is unbound to specific times or locations and can thus be customized to allow students to follow their own preferred or idiosyncratic studying behaviors – they can watch the podcast as often or as little as they want, in the morning if they are 'morning people' or late at night if they prefer to study then. Research shows that students who are allowed to use podcasting material in their studies can outperform those that are denied such access (McKinney, et. al., 2009). Podcasts can be used to enhance existing in-class activities and lectures, or as a stand-alone educational delivery system, although there is current debate if the stand-alone model can actually replace a blended solution (McKinney, et. al., 2009).

Mobile web browsing is another popular activity that can be easily harnessed for educational purposes. Access to web-based information and functionality through web-browsing on the mobile can aid in such diverse areas as literacy, numeracy, sciences, social studies, and language learning (De Lorenzo, 2009). However, usability studies such as that of Shrestra (2007) indicate that it is imperative for the user interface to be appropriate to the mobile screen and take into account other such limitations of the mobile for utmost effectiveness.

Dedicated (i.e., embedded, native) mobile applications for education are growing rapidly. As an example, just a few months after the opening of the Apple AppStore in mid 2008, there were over 1600 learning applications in the store (Ambient Insight, 2008). Developers understand that education related products must be tailored to the mobile device to ensure the best user experience. Furthermore, research has shown the efficacy of such dedicated developments (Brown, 2009).

Finally, mobile phones or video recorders can be used for data collection, collaborative reporting, easy content dissemination and delivery, and other functions. There is current debate on the double-edged sword of allowing ultimate freedom of mobile usage for capturing information for educational content gathering purposes vs. invasion of privacy and other issues within the school walls (Ramirez, 2009).

In summary, the use of the mobile for education is slowly coming into its own. However, with the advent of a new technological medium comes the caveat that proper and well thought out guidelines must be employed to enhance best practices. The fact that people have all of the functions of the mobile at their fingertips does not necessary indicate that they can use the material effectively in an educational context (Corbeil, & Valdes-Corbeil, 2007). This is the upcoming challenge for mobile educators and educational developers alike.

Review of Existing Applications

Existing applications for the mobile cover a wide range of topics and include several design types. One common element among all such types of applications is the inclusion of some type of 'micro-learning', or chunking of the educational content and pedagogy into small portion sizes for consumption while on the go. These smaller educational portions enable more efficient use of idle time, capitalizing on short time intervals and filling them with bite-size and appropriately small learning units. The fragmentation of the learning day into many small sections is merely a reflection of the general fragmentation of our modern, busy, daily lives.

Mobile applications are now available for diverse training such as: test preparation, skill-based learning like languages or mathematics and a multitude of other educational topics. They cover levels from preK-12 through to university, institutional as well as stand-alone solutions, first-time training to review and refresher courses – the range is wide and growing. This chapter is only a cursory overview of the existing mobile applications.

Test preparation applications are typically based around the concept of flash cards, with accompanying multiple choice or true/false types of

questions. They may be timed or not, and have various navigation and learning mechanisms for enhancing the learning experience, such as search capabilities, control over inclusion or selection of items, ordering of teaching elements, etc. Many train for simpler aspects of test preparation, such as providing memory games to help students remember simple facts, aiding vocabulary retention, assessing mathematical or other content-specific correct responses, etc. Test preparation is a widespread industry that has a large worldwide demand. In many cases, the mobile applications simply mimic web-based content, so that they are adapted to physical and other limitations of the mobile, but not to the functionality of the mobile. Future trends will be to develop true test preparatory material that includes a wider range of mobile capabilities and options.

A desirable application of mobile test preparation is to be able to actually test the person remotely on the phone for real certification and compliance-testing. This is still problematic since it is not easy to control for verification of the test-takers. Perhaps technology can ultimately provide an answer to this, by looking into true speaker verification using biometrics (e.g. speaker voice recognition, fingerprints, etc.) or behavior patterns that can be adequately monitored and classified for more exact profiling.

Language training has been one area where there are several types of mobile applications currently available, as outlined in Hockly (2010) and Ballantyne (2010). They are largely an import of web-based material and run the gamut from vocabulary training and development to grammar tests, listening and reading comprehension and even some types of elementary writing practice. The ability to record your voice on the mobile allows for oral practice that is both engaging and effective (Jobi, 2009). Some are more cleverly adapted than others, and most use typical language training activities such as multiple choice questions, true/false, cloze, matching pictures, matching columns, etc. There are programs for training general language skills in one's native language, including the upgrading of literacy and development of early childhood language, as well as applications for foreign or second language training. Almost all of the major world languages are covered.

Some applications also come with glossaries, phrasebooks, dictionaries, learning tips, examples, games, and other learning aids. Some utilize the community-based aspect of mobile phones by allowing for instant sharing of ideas, stories, media, and opinions or ratings, culminating in a rich and wide array of possible peer-to-peer and teacher-to-peer interactions. Collaborative efforts along with appropriate feedback provide language training that can be both creative and engaging.

Coupled with a wide range of ever-growing content available on the mobile, including e-books, e-zines, websites, wikis, blogs, and other online resource materials, the language learner is never at a loss for authentic content. In fact, the real job is to pick and choose material, and establish criteria for better learning versus substandard content for learning purposes. Specially created language training content has a pedagogical legitimacy that may tradeoff genuineness and naturalness for learning authenticity and relevancy, while existing material, edited or not, may have the problem of requiring decisions as to whether such resources are legitimate for educational use.

Even if recycling older web or printed material, when placed in the new medium of the mobile, a new user experience is created that rejuvenates and revitalizes these older forms. In a sense, porting to the mobile may be the best way of utilizing available material that lies within the 'long tail' of the publication industry by re-inventing tried and true methodologies with a more youthful and up-to-date look. However, the caveat of this usage is that simple porting is almost never enough – the real upgrading comes at a price of providing additional and judicious finishing touches to create a fulfilling user experience on the mobile.

This includes resizing to the mobile screen, text manipulation, application of appropriate sound and video effects, etc.

Grading and testing is another area where mobile phones are just beginning to be used in the area of language training. Teachers can provide feedback via SMS or other means regarding homework assignments or test scores on a more one-to-one basis, or to groups of students. Of course, this puts another strain on teachers to respond remotely and within a given time to all students. A new trend is to outsource such grading functions, currently performed on remote computers (Rampell, 2010), but this has the potential of drifting to the mobile as well.

In most cases, mobile language learning applications are seen as an adjunct to more traditional language learning services. However, there is some courseware available designed to function as a stand-alone course for personal use and learning. The issue of blended m-learning vs. consumer based product development has not reached the education market as such. The current offerings seem to be a rather ad hoc assembly of types of language learning training, each occupying their own niche in the mobile language training domain.

Mobile applications in other skill training, such as mathematics, are also widely available (Davis, 2009). The use of the small screen for animations, graphs, equations, etc. coupled with explanations and trendy user interfaces makes math more appealing and less frightening to users. Instant feedback on questions and skill development is also important. The ability to solve different types of problems over and over again, without the need for a human teacher to provide the correct responses each time, provides students with more ample practice time and a wider range of problem types that can be tackled. The efficiency of learning, therefore, is upgraded. Deubel (2010) has outlined the types of technological integration used for educational purposes, focusing on mathematics training. The survey includes use of several technological innovations such as the use of Web 2.0 for creation of user generated content, access to math-rich innovative teaching materials and techniques and community building. It is only a matter of time before all of these are transported successfully to the mobile.

Other domains also richly incorporate technology-driven content and methods. There is not a subject taught that cannot be found on the cloud, and, with mobile intervention, these will all soon be available, if not already so, on mobile devices. The mobile's entertainment value for attracting learners coupled with its staying power provided by creating an engaging user experience are a powerful duo that can be harnessed for educational purposes irrespective of the subject matter at hand. In addition, the creation of mobile-centric collaborative educational efforts for sciences, arts, history, geography or any other teaching area, will only serve to increase mobile education consumption in the near future.

Students' Experience with Mobile Education

Students are now beginning to welcome the mobile as a new educational phenomenon. The device is an integral part of their world, but so is education. The integration and blurring of their non-learning environment with their learning universe is permeating the student scene. Mobiles are everywhere, used by everyone, and specifically, used excessively by youth the world over. The need for constant community and togetherness, together with an undercurrent of social need and status differentiation makes the mobile their constant companion.

Students like the mobile for administrative functions, and the easy-to-use connections that enable smooth teacher-student and peer-to-peer communication. They know the base technology well since they use the mobile on a daily basis. While there are some discrepancies in the level of technological savvy between different students,

and between students and teachers, this is quickly becoming less and less of an issue as more practical experience is garnered by all, and as easier user interfaces and intuitive flows are employed. If the user experience is not affected by prior mobile experience, the distinction between technological wizards and technophobic 'dummies' becomes blurred – and a type of equalization of education is allowed to occur.

Chen (2009) reports on the use of freely provided iPhones at a private US campus in Texas along with integration into the university curriculum. Students plan and turn in homework, find their way around campus, watch their lectures on mobile podcasts and get their changes in class schedules along with their grades, all through the mobile device. In addition, anonymous feedback and polling applications allow them to communicate their opinions or answer questions without being self-consciousness or feeling foolish. This allows the shy and quieter students to contribute and feel counted as much as the open, gregarious ones, and levels the playing field among students that have varying personality traits. Chen (2009) also reports on a similar though less inclusive project implemented at Stanford.

There are several examples of a blended approach that combine the best parts of classroom instruction with those of the mobile. Wang, et. al. (2009) reports on the use of a blended program for teaching English that podcast classroom instruction for unlimited and unconstrained usage, along with support of SMS and polling. Wang reports that student involvement increased, and that the project turned students into active learning participants rather than passive listening recipients. Ogata, et. al. (2008) outlines a project conducted in Japan that supports students in using their mobile devices on field trips for gathering and collating information. Here, teacher assignments were provided to students, who had to fulfill provided tasks using all aspects of their mobile devices. Students were also allowed to collaborate and share information on the mobile.

The use of the mobile for collaborative efforts is gaining some momentum, ranging from mobile blogging capabilities (Huang, 2009) to collaborative games using mobile devices (Moore, et. al., 2009). Some mobile collaboration projects are highly classroom structured (Winters, & Mor, 2008), while others allow heretofore unimaginably flexible interactions such as those that provide opportunity for collaborative music-making separated by both time and distance, all through the mobile phone (Johansson, & Hagman, 2009; Zhou, 2010). This represents a type of blurring between formal and informal education, a trend that will be accelerated with increased mobile usage.

Students report that advantages for having education on the mobile includes: connectivity to the educational setting at all times, natural usage of a familiar object, informality and friendliness of the interaction, engaging and more playful learning activities, convenient for their needs and customized to their personalities and learning behaviors (Kukulska-Hulme & Pettit, 2009). When asked about what they considered to be new and innovative about their experience of learning with mobile devices, students responded that the following qualities of the mobile were most advantageous: availability, flexibility, portability, low cost, ease of checking as often as they wished and finally, the sense of being in control.

Implementation Difficulties

In this section, some of the difficulties of mobile education implementation will be addressed. Some relate to problems related to the individual, such general attitudes of the teachers or the fear of improper usage, while others are more societal based, such as inadequate technology structure or consumer economic considerations.

The general attitude of educators is crucial for implementing educational projects and programs in the classroom. MacCallum and Jeffrey (2009) report that teachers respond with a rather mixed reaction when asked about their attitudes regarding use of mobile education. Most are highly receptive to simpler integrations of mobile functions such as SMS notifications to students, mobile blogging or podcasting or quiz management. However, barriers to adoption of mobile technology into classroom practices include cost of mobile phones and services, access to devices and technical support, lack of time for adequate preparation of lesson plans specific to the mobile, and limitations of the mobile itself for educational purposes (e.g., small screen, difficulty in typing input). Device dependency was also listed as a problem resulting in perceived uneven opportunities for students within a given class.

MacCallum and Jeffrey (2009) report that teachers who have more experience in using e-learning material have more favorable attitudes towards integrating mobile education into their teaching practice. This is consistent with other research that indicates a link between technological acceptance and ICT experience (Hew, & Brush 2007).

The distractions associated with mobile phone usage in an educational setting are also of concern. Ringing of mobile phones in public and at inappropriate times can be a nuisance and an annoying distraction to others. Other mobile-enabled temptations, such as the desire to play games or be involved in non-classroom activities on the web, is also problematic. Campbell (2006) outlines some of the negative attitudes associated with such distractions, which are often accompanied by institutional restrictions placed on the use of mobiles or even their outright ban. Certainly there is room for timely limitations on the use of mobiles in the classroom, but this does not mean that there is no room in the classroom for the more positive and educational aspects of mobile training.

Of course, many are concerned with improper use of the mobile during class time. Copying or cheating during exams, or copying verbatim of works available on the web are some of the issues that immediately come to mind. The control of student behavior and limitations on the use of the mobile towards purely educational goals is an aspect of discipline that garners deep concern. Teachers may be wary of their ability to control and monitor judicious, wholesome and moral mobile behaviors among the student population, and may see this as yet another burden to tackle in addition to the myriad of roles they already have on their agenda.

Of course, improper use of web material can be said for using computers in the class, or for that matter, in the home, and the improper use of mobile web applications can likewise take place in the school or home alike. It is not the technology, per say, that is problematic, but the social norms and control mechanisms that must be developed in order for the problem to be marginalized. With adequate teacher training and judicious student monitoring, the benefits of education on the mobile can far offset the negatives, just as in any type of new technological offering.

There are a number of technical difficulties in implementing mobile learning applications. For example, for dedicated applications, the fragmented nature of mobile devices is one of the major factors impacting the complexity of providing a solution that covers a wide range of mobile phones. The development of open mobile platforms, which seemed to start converging some five years ago, opened up the doors to new players such as Apple's iOS (originally iPhone OS) and Google's Android. The mobile application platform's variety is further complicated by differences in screen sizes and resolutions, multimedia support, and access to the native phone capabilities.

Unstable wireless data connectivity may be a challenge in creating good user experience in a client-server mobile application. The possibility of frequent disconnections due to poor network coverage or overloaded network needs to be properly handled. Yet another issue, often ignored by mobile solution developers, is the battery life. This is still one of the biggest showstoppers in mobile industry. A big phone screen, computational re-

sources required for multimedia processing and heavy network usage may quickly drain the battery, thus making the service usable only in a lab.

Finally, an additional difficulty in implementation of mobile learning is the mobile infrastructure costs. Depending on choice of the mobile device and usage pattern for a specific application, the costs may be a prohibitive factor in some economies. Powerful capabilities of modern mobile phones come at a higher cost of the devices to a consumer, comparable to and even higher than those of personal computers. Mobile subscription plans and wireless data connectivity packages add to these expenses. The cost impact may be greater in the secondary school, where either the institution or the parents must have the means to afford such additional technological requirements.

Suggestions

To help overcome any resistance posed by teachers' wariness of the new mobile technology, it is recommended that they undergo targeted teacher training in the use of mobile devices in the classroom for their professional development. Such training must first desensitize those who are somewhat leery of the idea of mobile content delivery, and must then go on to help them develop a clearer understanding of the huge potential that the mobile has to offer. Teachers can also be provided with methods for maintaining proper 'mobile discipline' in order to restrain students from abusing the mobile in an educational context. Further to introductory pedagogical training, on-going support in the form of web blogs, forums, and the like can aid teachers so that they are better able to cope with the dizzying world of constantly changing circumstances and opportunities.

On the technological side, it is important to monitor the latest industry developments to be able to provide stable, cost effective solutions that would be applicable to a wide range of mobile devices. For example, one technology worth close monitoring in our opinion is HTML5. When there is a sufficient install base of supporting devices and assuming that there is no significant fragmentation in terms of the mobile browsers, technical implementation of new and advanced mobile learning applications may become a much simpler task.

Dealing with the mobile service costs may prove another significant challenge. The costs of the mobile devices will probably not change significantly in the observable future. However, the costs of data connection, often quite high, may be avoided by means of local connectivity solutions. For example, as more and more mobile phones support WiFi, this may be the right choice for services that require data communication. In this case, an education body may merely have to provide a few WiFi access points to allow students to have access to free learning content.

FUTURE RESEARCH DIRECTIONS

As the sheer volume of mobile educational applications begins to mushroom, there will be a need for rating the effectiveness of solutions within each type of teaching domain. Meaningful comparisons that include user feedback combined with clear-cut effectiveness ratings such as increase in retention or types of additional skills mastered will be required. Enhanced enthusiasm and participation of learners can also be monitored and compared to more traditional methods of educational content delivery.

To this end, creation of mobile-centric educational material is a needed direction, whereby utilization of functions best done on the mobile is optimized. This requires the cooperation and melding of both pedagogical and technological teams in order to create the best and most effective types of educational experiences available.

Mobile technology infiltrates our everyday experiences at an ever increasing rate. Education is moving out of a location-based, teacher-centric universe, and into the informal learning arena.

One of the challenges for both students and their "educators of the future" will be blending the advantages of mobile learning with traditional methods of education. Currently, there seems to be a gap between formal mobile learning framework definitions available from a few academic and commercial bodies and the actual mobile device usage for learning practices. More research is required to understand how to create more dynamic and flexible frameworks dictated by and oriented to the current mobile usage-patterns by students. Getting the most of the communicative, personal and content-aware nature of mobile devices for educational needs may boost the impact on mobile learning performance.

CONCLUSION

In conclusion, mobile education is here to stay. Educational applications are penetrating the private-study consumer market and gradually being absorbed into the institutional and training markets as well. Just as mobile phones are now accepted as indispensible, so too will such educational applications. However, in order to be effective, mobile educational material must be delivered in seamless mobile platforms with clear pedagogical goals and objectives along with transparent methods of tracking student progress and learning efficacy.

REFERENCES

Ally, M. (2009). *Mobile learning transforming the delivery of education and training*. Edmonton, Canada: AU Press.

Ambient Insights. (2008). *Marketing report*. The US market for eight learning technology products and services: 2008-2013 forecast and analysis.

Ballantyne, N. (2010). Mobile learning: What is it and what's it got to do with me? *IATEFL Call Review, Spring/Summer,* 8-11.

Brown, L. (2009). Using mobile learning to teach reading to ninth-grade students. *Journal for Computing Teachers, Fall,* 1-17. Retrieved September 1, 2010, from http://www.iste.org/jct

Campbell, S. W. (2006). Perceptions of mobile phones in college classrooms: Ringing, cheating, and classroom policies. *Communication Education, 55*(3), 280–294. doi:10.1080/03634520600748573

Cebeci, Z., & Tekdal, M. (2006). Using podcasts as audio learning objects. *Interdisciplinary Journal of Knowledge and Learning Objects, 2,* 47–57.

Chen, B. X. (2009). *How the iPhone could reboot education.* Retrieved September 1, 2010, from http://www.wired.com/gadgetlab/2009/12/iphone-university-abilene/

Conway-Smith, E. (2010). *Teaching with cell phones.* Retrieved September 1, 2010 from http://www.globalpost.com/dispatch/education/100720/south-africa-teaching-cell-phones?page=0,0

Copley, J. (2007). Audio and video podcasts of lectures for campus-based students: production and evaluation of student use. *Innovations in Education and Teaching International, 44*(4), 387–399. doi:10.1080/14703290701602805

Corbeil, J. R., & Valdes-Corbeil, M. E. (2007). Are you ready for mobile learning? *EDUCAUSE Quarterly, 30*(2). Retrieved September 1, 2010, from http://www.learning-centric.net/mobile.cc/relatedinfo.pdf.

Crisp, M. S. (2010). *Modernizing school communication systems: Using text messaging to improve student academic performance.* Unpublished doctoral dissertation, Oregon State University, Oregon.

Davis, G. (2009). *Teaching mathematics with technology. Computational media: The universal acid of mathematics teaching,* vol. 4. Retrieved Sept 28, 2010 from http://republicofmath.wordpress.com/2009/12/29/computational-media-the-universal-acid-of-mathematics-teaching-4/

De Lorenzo, R. (2009). *The powerful combination of mobile devices and learning apps.* Retrieved on June 6, 2010, from http://themobilelearner.wordpress.com/2009/11/15/ combining-mobile-devices-with-apps/

De Lorenzo, R. (2010). *Five difficulties in mobile learning implementation.* Retrieved on June 6, 2010, from http://themobilelearner.wordpress.com/2010/06/11/on-five-difficulties-in-mobile-learning-implementation/

De Waard, I. (2010). *Free mobile service getting to grips with bulling in schools: Bullyproof.* Retrieved September 1, 2010 from http://ignatiawebs.blogspot.com/2010/02/free-mobile-sms-service-getting-to.html

Deubel, P. (2010). *Technology integration: Essential questions.* Retrieved October 1, 2010, from http://www.ct4me.net/technology_integr.htm

EPN. (2010). *Website.* Retrieved October 1, 2010, from http://epnweb.org/

Evans, C. (2008). The effectiveness of m-learning in the form of podcast revision lectures in higher education. *Computers & Education, 50*(2), 491–498. doi:10.1016/j.compedu.2007.09.016

Gribbins, M. (2007). The perceived usefulness of podcasting in higher education: A survey of students' attitudes and intention to use. *Proceedings of the Second Midwest United States Association for Information Systems,* Springfield, IL May 18-19.

Guy, R. (2009). *The evolution of mobile teaching and learning.* Santa Rosa, CA: Informing Science.

Hew, K. F., & Brush, T. (2007). Integrating technology into K-12 teaching and learning: Current knowledge gaps and recommendations for future research. *Educational Technology Research and Development, 55*, 223–252. doi:10.1007/s11423-006-9022-5

Hockly, N. (2010). Mobile learning: What is it and what's it got to do with me? *In IATEFL. California Law Review*, (Spring/Summer): 5–8.

Huang, Y.-M., Jeng, Y.-L., & Huang, T.-C. (2009). An educational mobile blogging system for supporting collaborative learning. *Journal of Educational Technology & Society, 12*(2), 163–175.

Jobi, P. (2009). Cell phones in the (language) classroom: Recasting the debate. *EDUCAUSE Quarterly Magazine, 32*(4). Retrieved September 1, 2010, from http://www.educause.edu/EDUCAUSE+Quarterly/EDUCAUSEQuarterlyMagazineVolum/CellPhonesintheLanguage-Classro/192995

Johansson, F., & Hagman, P. (2009). *MiniBands - A collaborative mobile music concept*. Unpublished Master's Thesis, Umea, Sweden.

Johnson, L., Levine, A., Smith, R., & Stone, S. (2010). *The 2010 horizon report*. Austin, TX: The New Media Consortium. Retrieved September 1, 2010, from http://wp.nmc.org/horizon2010

Kolb, L. (2008). *Mobile learning is global: Examples of teachers using cell phones for learning!* Retrieved June 6, 2010, from http://www.cellphonesinlearning.com/

Kossen, J. S. (2001). *When e-learning becomes m-learning*. Retrieved June 6, 2010, from http://www.palmpowerenterprise.com/issuesprint/issue200106/elearning.html

Kukulska-Hulmes, A., & Pettit, J. (2009). Practitioners as innovators: Emergent practice in personal mobile teaching, learning, work, and leisure. In M. Ally (Ed.), *Mobile learning transforming the delivery of education and training*. Edmonton, Canada: AU Press. Retrieved from http://www.aupress.ca/books/120155/ebook/99Z_Mohamed_Ally_2009-MobileLearning.pdf

Kukulska-Hulmes, A., & Traxler, J. (2005). *Mobile learning: A handbook for educators and trainers*. London, UK: Routledge.

MacCallum, K., & Jeffrey, L. (2009). Identifying discriminating variables that determine mobile learning adoption by educators: An initial study. In *Proceedings of ASCILITE 2009: Same Places, Different Spaces*. Retrieved October 1, 2010, from http://www.ascilite.org.au/conferences/auckland09/procs/maccallum.pdf

McKinney, D., Dyke, J., & Luber, E. S. (2009). iTunes university and the classroom: Can podcasts replace professors? *Computers & Education, 52*, 617–623. doi:10.1016/j.compedu.2008.11.004

Moore, A., Goulding, J., Brown, E., & Swan, J. (2009). AnswerTree – A hyperplace-based game for collaborative mobile learning. *Proceedings of mLearn*, Orlando, Florida.

Newswire, P. R. (2010). *Blackboard launches mobile education platform in Mexico*. Retrieved October 1, 2010, from http://www.prnewswire.com/news-releases/blackboard-launches-mobile-education-platform-in-mexico-84415947.html

Ogata, H., Hui, G. L., Yin, C., Ueda, T., Oishi, Y., & Yano, Y. (2008). LOCH: Supporting mobile language learning outside classrooms. *International Journal of Mobile Learning and Organisation, 2*(3), 271–282. doi:10.1504/IJMLO.2008.020319

Perez, S. (2009). Opera reports explosive mobile web growth worldwide. Retrieved June 6, 2010, from http://www.readwriteweb.com

Power, T., & Shrestha, P. (2010). *Mobile technologies for (English) language learning: An exploration in the context of Bangladesh*. In: IADIS International Conference: Mobile Learning, Porto, Portugal.

Ramirez, E. (2009). *Why teachers want to ban cellphone cameras from classrooms*. Retrieved September 1, 2010, from http://www.usnews.com/blogs/on-education/2009/03/23/why-teachers-want-to-ban-cellphone-cameras-from-classrooms.html

Rampell, C. (2010, April 6). Outsourced grades. *New York Times*. Retrieved September 28, 2010, from http://economix.blogs.nytimes.com/2010/04/06/outsourced-grades/

Shresta, S. (2007). Mobile web browsing: Usability study. In *Mobility '07: Proceedings of the 4th International Conference on Mobile Technology, Applications, and Systems and the 1st International Symposium on Computer Human Interaction in Mobile Technology*, (pp. 187-194).

Terbilang, S. (2008). *Studying through SMS*. Retrieved September 1, 2010, from http://mjrevenge.blogspot.com/2009/12/studying-through-sms.html

Traxler, J. (2009). Current state of mobile learning. In Ally, M. (Ed.), *Mobile learning transforming the delivery of education and training*. Edmonton, Canada: AU Press.

Van Rooyen, A. (2010). Effective integration of SMS communication into a distance education accounting module. *Meditari Accountancy Research*, *18*(1), 47–57. doi:10.1108/10222529201000004

Vassel, C. (2006, August). *Mobile learning: Using SMS to enhance education provision*. The 7th Annual Conference of the Higher Education Academy for Information and Computer Sciences, Dublin, Ireland.

Wang, M., Wang, M., Shen, R., Novak, D., & Pan, H. (2009). The impact of mobile learning on students' learning behaviours and performance: Report from a large blended classroom. *British Journal of Educational Technology*, *40*(4), 673–695. doi:10.1111/j.1467-8535.2008.00846.x

Winters, N., & Mor, Y. (2008). *CoMo: Supporting collaborative group work using mobile phones in distance education*. Retrieved September 1, 2010, from http://www.lkl.ac.uk/como/CoMo-Final-Report.pdf

Zapit, S. M. S. (2010). *SMS for education*. Retrieved October 1, 2010, from http://www.zapitsms.com/SMS-For-Education

Zhou, Y., Percival, G., Wang, X., Wang, Y., & Zhao, S. (2010). *MOGCLASS: A collaborative system of mobile devices for classroom music education*. Retrieved September 1, 2010, from www.comp.nus.edu.sg/~zhaosd/paper/mmshc05843-zhou.pdf

ADDITIONAL READING

Beale, R., & Lonsdale, P. (2004). Mobile context aware systems: The intelligence to support tasks and effectively utilise resources. Mobile Human-Computer Interaction (Mobile HCI 2004) [Springer-Verlag.]. *Lecture Notes in Computer Science*, *3160*, 240–251. doi:10.1007/978-3-540-28637-0_21

Beddall-Hill, N. L., & Raper, J. (2009). Mobile devices as 'boundary objects'? Special Handheld Learning Conference issue. Journal *of the Research Center for Educational Technology, 6(1)*, 28-46, Retrieved Oct 1, 2010 from http://www.rcetj.org/index.php/rcetj/article/view/84.

Brown, E. (2010). Education in the wild: contextual and location-based mobile learning in action. *A report from the STELLAR Alpine Rendez-Vous workshop series*, Nottingham, UK.

Brown, E., Börner, D., Sharples, M., Glahn, C., Jong, T. D., & Specht, M. (2010). Location-based and contextual mobile learning. *A STELLAR Small-Scale Study*. Retrieved Oct 1, 2010 from http://www.stellarnet.eu/d/1/2/images/2/23/Sss6.pdf.

Bruner, J. (1966). *Toward a theory of instruction*. Cambridge, MA: Harvard University Press.

Clough, G. (2010). Geolearners: Informal learning with mobile and social technologies. *IEEE Transactions on Learning Technologies*, *3*(1), 33–44. doi:10.1109/TLT.2009.39

Cook, J., Pachler, N., & Bradley, C. (2008). Bridging the gap? Mobile phones at the interface between informal and formal learning. Journal of the Research Center for Educational Technology, Spring. Retreived October 1, 2010 from: http://www.rcetj.org/index.php /rcetj/article/view/34.

Corlett, D., Sharples, M., Chan, T., & Bull, S. (2004). A mobile learning organiser for university students. *Proceedings of the 2nd International Workshop on Wireless and Mobile Technologies in Education* (pp 35-42). JungLi,Taiwan: IEEE Computer Society.

DuVall, B., Powell, M. R., Hodge, E., & Ellis, M. (2007). Text messaging to improve social presence in onlinelearning. *EDUCAUSE Quarterly*, *3*, 24–28.

Hartnell-Young, E., & Heym, N. (2008). *How mobile phones help learning in secondary schools*. Coventry: Becta.

Hayes, P., Pathak, R., Rovcanin, L., & Joyce, D. (2004). *Mobile technology in education: A multimedia application*. Tralee, Ireland: EdTech.

Holzinger,A., Nischelwitzer,A., & Meisenberger, M. (2005). Mobile phones as a challenge for m-learning: examples for mobile interactive learning objects (MILOs). In D. Tavangarian (Ed.), *3rd IEEE PerCom*, (pp. 307–311).

Lonsdale, P., Baber, C., & Sharples, M. (2004). A context awareness architecture for facilitating mobile learning. In Attewell, J., & Savill-Smith, C. (Eds.), *Learning with mobile devices: Research and development* (pp. 79–85). London: Learning and Skills Development Agency.

Lyons, L. (2009). Designing opportunistic user interfaces to support a collaborative museum exhibit. *Proc. CSCL 2009, ISLS*, pp. 375-384.

Martin, J. (2008). Restructuring activity and place: Augmented reality games on handhelds. *Paper presented at the International Conference of the Learning Sciences (ICLS)*. Retrieved October 1, 2010 from http://www.fi.uu.nl/en/icls2008/416/paper416.pdf.

Niasmith, L., Lonsdale, P., Vavoula, G., & Sharples, M. (2004). *Report 11: Literature review inmobile technologies and learning. Futurelab series*. Bristol: Futurelab.

O'Malley, C., Vavoula, G., Glew, J. P., Taylor, J., Sharples, M., & Lefrere, P. (2003).*Guidelines for Learning/Teaching/Tutoringin a Mobile Environment*. Retrieved Oct 1, 2010 from http://www.mobilearn.org/results/results.htm.

Pachler, N., & Bachmair, B. 7 Cook, J. (2010). *Mobile learning: Structures, agency, practices*. New York: Springer.

Patten, B., Arnedillo-Sánchez, I., & Tangney, B. (2006). Designing collaborative, constructionist and contextual applications for handheld devices. *Computers & Education*, *46*(3), 294–308. doi:10.1016/j.compedu.2005.11.011

Rosas, R., Nussbaum, M., Cumsille, P., Marianov, V., Correa, M., & Flores, P. (2003). Beyond Nintendo: design and assessment of educational video games for first and second grade students. *Computers & Education*, *40*(1), 71–94. doi:10.1016/S0360-1315(02)00099-4

Roschelle, J. (2003). Unlocking the learning value of wireless mobile devices. *Journal of Computer Assisted Learning, 19*(3), 260–272. doi:10.1046/j.0266-4909.2003.00028.x

Sharples, M., Taylor, J., & Vavoula, G. (2007). A theory of learning for the mobile age. In Andrews, R., & Haythornthwaite, C. (Eds.), *The Sage Handbook of E-learning Research* (pp. 221–247). London: Sage.

Strom, P. S., & Strom, R, D. (2002). Personal digital assistants and pagers: a model for parent collaboration in school discipline. *Journal of Family Studies, 8*(2), 226–238. doi:10.5172/jfs.8.2.226

Vavoula, G., Sharples, M., O'Malley, C., & Taylor, J. (2004). A study of mobile learning as part of everyday learning. In Attewell, J., & Savill-Smith, C. (Eds.), *Mobile learning anytime everywhere. A book of papers from MLEARN 2004* (pp. 211–212). London: Learning and Skills Development Agency.

Wishart, J. (2009). Ethical considerations in implementing mobile learning in the workplace. *International Journal of Mobile and Blended Learning, 1/2,* 76–92. doi:10.4018/jmbl.2009040105

Zurita, G., Nussbaum, M., & Sharples, M. (2003). *Encouraging face-to-face collaborative learning through the use of hand-held computers in the classroom. Proceedings of Mobile HCI 2003* (pp. 193–208). Udine, Italy: Springer-Verlag.

32

Chapter 3
Mobile Handheld Devices in Education

Douglas Blakemore
Ferris State University, USA

David Svacha
Ferris State University, USA

ABSTRACT

Small electronic devices, also referred to as handhelds, are impacting education in a variety of ways including teaching methods, student life, and the need for support from technical staff. This chapter discusses the importance of handhelds in education, how handhelds are being used in education, the challenges presented by handhelds for those in education and what might happen with handhelds in the future.

INTRODUCTION

Small electronic devices have been around in one form or another for decades but now with the introduction of the iPhone, the iPad, the Android phone and even devices such as the iPod Touch some real computing power has been put into small packages. Small electronic devices such as these, also referred to as "handhelds," are beginning to have an impact on education. This chapter will discuss what is included in the handheld category and will look at the impact of

handhelds in education from the standpoint of the instructor, the student, the technical staff, and the educational institution.

BACKGROUND

Before discussing the status of handhelds in education, it is important to discuss the various kinds of small devices available because each type of device gives different options for use by students, educators and school administrators. At present, there appear to be a few different broad categories of handhelds along with some sub-categories.

DOI: 10.4018/978-1-61350-150-4.ch003

Copyright © 2012, IGI Global. Copying or distributing in print or electronic forms without written permission of IGI Global is prohibited.

One category is the smart phone. The iPhone and the Android phone can be categorized as *smart phones*. There are several other brands that fit into this category but these two by far make up the bulk of the market currently. A smart phone starts with the phone concept and adds in the ability to run other applications (normally called just "apps") such as web browsers and a host of other types of programs.

Another category would be those devices that are similar to smart phones but do not have the capability to make or receive phone calls. In this category would be devices such as the iPad™, the iPod Touch™ and a host of other similar devices. This category is in the process of being refined into two different groups. The first, and more traditional group, is the pocket-sized MP3 player. The smallest of these can simply play music while the larger, more sophisticated ones, can do almost everything a smart phone can do except handle phone calls. The second group is a bit larger and more powerful. This group includes devices such as the Amazon Kindle, the Barnes and Noble Nook and, of course the iPad which sparked a revolution in thinking about the personal computer and small devices.

Apple introduced the iPad in April of 2010. Computer World ran an article in March 2010 talking about the host of iPad™ clones based on Linux, Windows and other operating systems that were in the process of being developed (Vaughan-Nichols, 2010). According to CNET News in a September 7, 2010 article, there are currently at least a half dozen or more iPad clones either on the market or soon to be on the market from manufacturers such as Dell, Toshiba, HP and Samsung. To blur the categories a bit, some of these iPad-like devices will be able to also be used as communication devices making calls and sending text messages. Virtually all of these devices will be able to run applications, connect to Bluetooth devices such as headsets and keyboards and most will be able

to connect to 3G networks. Some, such as the Samsung Galaxy Pad will come only with a 3G subscription and will be sold through the various telephone carriers' outlets (Ogg, 2010).

CURRENT STATUS OF HANDHELDS IN EDUCATION

From the Teaching Perspective

Advantages

Small devices can add a significant number of advantages to the classroom. The most obvious is size. Most devices are no bigger than a deck of playing cards. Even those that are much larger than a deck of cards are very thin and easy to store. Applications are being developed almost daily for these devices. Even though there are different operating systems, many, if not most developers are creating their apps on multiple platforms making them available for both the Apple and the Android line of small devices. Book management applications are a good example of this cross-platform development. Currently, in addition to reading books and magazines in print, it is possible to read them with specially designed hardware/software such as the Apple iPad, Amazon's Kindle and Barnes and Noble's Nook. All three of these devices have some real advantages for the instructor and for the student. The Kindle has probably one of the most impressive statistics for battery life of just about any device; it boasts a full month of use on one charge with the wi-fi turned off ("Kindle", 2011). It accomplishes this by the use of a special e-ink technology that is able to keep a display on the screen even though the computer is in a kind of sleep mode. For between $139 and $379, depending on options and screen size, it is possible to have a device that can hold up to 3500 books at the same time ("Kindle", 2011).

The Barnes and Noble Nook Color is a very similar device that can hold up to 6000 books at a time plus it has the ability for expanded storage and can double as an MP3 player. Nook Color is currently selling for just under $250 ("Nook Color," 2011). The color screen, while very nice for many things does come at a cost to battery life. The battery for the Nook Color is advertised at up to 8 hours ("Nook Color", 2011). The standard Nook is a closer relative to the Kindle. It is also based on e-ink technology and lasts up to 10 days on a charge ("Nook", 2011). Not as impressive as the battery life of the Kindle but more than enough for most class projects. The Nook has a built in web browser and email application to make it more than just a book reader.

The Cadillac of the three is probably the iPad from Apple. The starting price for the iPad is a hefty $499 for the Wi-Fi only version ("iPad", 2011). One of the interesting things about the iPad is that since it is not developed by a book company, it has the ability to be all things to all people so to speak. The iPad has an excellent book reading program called iBook that can be downloaded for free. Users of iBook can purchase virtually millions of books through the iTunes store online. Additionally there is a huge collection of out-of-copyright books and classics that can be downloaded for free ("iBooks", 2011). This feature is also available on the Kindle and the Nook. Since the iPad is not necessarily seen as a competitor, both the Kindle application and the Nook application can be downloaded for the iPad for free effectively turning the iPad into a Kindle or a Nook device whenever needed; this gives the option of using whichever one has the best selection and/or price.

Apple has some other advantages for the classroom. One of these is iTunes University. Podcasting has been around for several years now and there are hundreds of thousands of podcasts available online on almost every topic in the world. Couple this with sites like YouTube and other video publishing outlets and there is a wealth of information available. iTunes University, or iTunes U for short is a podcasting service Apple has developed specifically for Universities. Lectures, videos, presentations and notes can be made available to students or can also be made available to the general public. In order to set up an iTunes U account with Apple, the university must provide its own Apple server and store the content locally ("iTunes U", 2011). Once set up, it is possible to use either an Apple computer or a Windows computer using a web-based application to create content to be uploaded to the iTunes U server. Once there, the server "brands" the creation with the university's own header information and lead-in messages. Since iTunes is available on just about every computer, the student can subscribe to the lectures using his desktop computer, laptop computer and/or small devices such as the iPod, iPad or the iPhone.

Another product that is now becoming popular on small devices is web-conferencing software. This software originally required a full sized computer, but now it is possible to catch live lectures while traveling about. DimDim, Adobe Connect and Webex all have small device applications developed for connecting to conferences and lectures anywhere it is possible to either have a good 3G signal or Wi-Fi signal. This is very handy for use by students who cannot attend in person. It also opens up the possibility of bringing in guest lecturers from very distant places in a matter of seconds without the time and cost of travel arrangements.

One only needs to spend a little time browsing the applications available for devices such as iPads to see that there is a host of specific applications developed for art instruction, various science classes, math classes and many other topics. An excellent example is the SPARKvue™ app by PASCO scientific. According to their iTunes description:

SPARKvue brings real-time measurement, data visualization, and analysis to science education everywhere. Students can use the new PASPORT AirLink 2 Bluetooth interface to connect to over 70 PASCO sensors, measuring a wide range of phenomena, including pH, temperature, force, carbon dioxide level, and many more. SPARKvue is designed for scientific inquiry in biology, chemistry, earth science, environmental science, physics, and physical science. SPARKvue can also record data from the iPhone or iPod Touch's internal accelerometer, with no additional hardware needed. ("SPARKvue", 2010)

Disadvantages

While there are many advantages to using handhelds in the classroom there are also disadvantages. Small electronic devices may be provided by educational institutions for use in the classroom or students may bring their own into the classroom. In either case instructors must be aware of the disadvantages and challenges that arise from having handhelds in the classroom in order to teach most effectively. One of the biggest disadvantages to handhelds from a teaching perspective occurs when students bring their own devices into the classroom for personal use. Ask almost any teacher at the K-12 or college level and they would agree, having smart phones, iPods and MP3 players can be a real distraction in the classroom. Virtually all of the devices have some means of communications capability. While in the classroom, a student can use their smart phone and/or other small device to do things such as send/receive text messages to friends or other classmates and check their email or social networking accounts such as Facebook. This problem is not new; it is just a new twist. Instructors have had this type of problem for years in the computer labs. Even before smart phones and other small devices instructors have had annoyances in the classroom from computers, cell phones and MP3 players.

Another disadvantage of having small devices in the classroom is the potential for cheating. Whether purchased by the school for a particular class or by the student either for the class specifically or just their own equipment, virtually all modern small devices have the ability to communicate. The email application comes on almost every device. Even if it did not, most devices have the ability to connect to the web and use a web based email, text or instant message application. This of course gives students an opportunity to network with each other or with others who could help the student with information on tests, quizzes and/or in class projects. An established policy for in-class use of handhelds may be advisable to help alleviate this potential problem.

The two disadvantages previously discussed must be recognized regardless of whether or not the teacher wants to use small devices for instructional purposes in the classroom. If these devices are introduced into the classroom in order to use them as an instructional aid, another level of problems is added that must be overcome. First of all, not all small devices use the same operating system. As in desktop computers, it is common that what work on one device frequently will not work on another device. As mentioned before, the largest market share of small electronic devices is with Apple using their proprietary operating system(s) and the Android phone which uses various versions of the Linux operating system. While both of these operating systems have a common background, they are far from being completely compatible. Add in the significant amount of phones using operating systems from Microsoft and Google and the problem of being compatible becomes even more complex. Additionally, even within the same operating system, there are version requirements that must be met in order to run various applications. With this in mind, if an instructor wants to use a particular device in class, it is well advised that either the instructor specify make, model and operating system versions in the course requirements or that the small devices be provided in the classroom. Additionally, if they are provided, when it comes time to upgrade or buy new, it would also be well advised to replace or upgrade the entire set of devices at one time since not all devices will be able to handle future upgrades. For example, the first generation, second generation and third generation of the iPod device look very much alike and have the same name and function. However, only the most recent version of the iPod Touch can accept Apple's newest operating system upgrade. If the devices were purchased a few at a time, then there is a significant chance that a particular application would only be available in the newest devices. If the class required that application, there would be a significant number of devices that would be effectively useless to that class.

The size of these devices adds an additional potential problem. Even considering devices as large as the iPad some web pages can be difficult to read. Zooming in on any particular part of a page can compensate for this but can make reading large amounts of text difficult due to the fact that much of the page being viewed is beyond either the vertical or horizontal limits of the device; the user would need to scroll to see everything. This is why there is another significant change happening in the programming field – making web pages and applications "device aware." Originally, web pages were deployed and were always taken as is. As the popularity of different web browsers such as Internet Explorer (IE), Firefox and Google's Chrome increased, programmers needed to make various versions that would allow the web page(s) to display properly depending on the brand of browsers. Now the same sort of thing is happening but rather than just the brand of browser, it also includes the size and type of the device displaying the information. This is a significantly more radical adjustment than just accommodating for different display parameters all on the basic same size and shape of a screen.

A web page for a normal size computer can display a lot of information. However, due to the size of the small device's screen, the web page needs to be broken up into a series of links to much smaller pages that will display well on the device. Add to this, the complication that in devices such as the iPad and iPod touch, the screen can be rotated making the aspect ratio of whatever is being viewed on the screen appear differently. Most applications handle this in one of two ways. The application developers lock the aspect ratio so that it cannot be rotated forcing the user to always view the screen in one position, such as only vertically or only horizontally. Another option is to display a certain amount of information in the vertical mode but when turned horizontally, in addition to rotating the data on the screen, it is re-formatted and usually either more information is added or removed from the screen based on what will display properly for that particular display. One example of this is the iPhone. When viewing information in the email application vertically, the font is small giving the ability to see more information on each line of text. However, when the device is turned horizontally, the screen reformats for the aspect change and automatically slightly increases the font making it easier to read and taking advantage of the wider screen space. On the iPad email application, not only does the font change but also, what is displayed on the screen is different. Holding the iPad vertically all the user sees is the particular email message that is being displayed. Rotating the iPad horizontally, the application takes advantage of the wider display and opens up the Mailboxes list so that you can browse through various other email messages and/or entire email accounts while still looking at the particular email message that was displayed when the device was held vertically.

From the Students' Perspective

The advantages to students of using small devices and smart phones are almost endless. To begin with, no longer do students have to get to a computer to check their email or instant messages. Almost all of the current small devices mentioned above have the ability to view, compose and send email, browse the Internet, and run a wide variety of applications. Depending on the size of device, various other applications are available.

The smaller devices such as the smart phones and iPod touch can run most of the applications that are also available on the iPad and the iPad clones coming on the market in 2011. These range from applications to find your way to school using GPS applications to finding your way around school with university developed applications that include GPS positioning combined with maps of the-campus. For student life, there are applications for connecting to web services such as Facebook, Twitter, and MySpace. One of the most unique things about the use of mobile devices is the ability to perform several tasks on the same device. Formerly handhelds could do many of these tasks but it took a different device to accomplish each task. Now that several tasks can be accomplished with one device the result can be a significant savings. For example, a GPS device such as a TomTom or a Garmin GPS device was being sold for between $100 to $400 ("GPS", 2011) depending on the model and features. Now, many of the Android phones have GPS apps on them for free when the student purchases the phone. If the student has the iPhone, the price is about $0.99 for the GPS application with an optional $20 per year service fee if voice navigation is preferred ("Motion X GPS", 2011). These applications appear to be as feature rich and equally as accurate as the individual devices costing much more.

When it comes to devices as large as the iPad and their soon-to-arrive competitors, there is even more functionality to make the devices appealing to students. While not designed as a replacement for a laptop or desktop computer, these devices can go a long way towards handling many of the functions. At present, the iPad is the first on the market with this type of technology and therefore will be the only one evaluated. By the time of the publishing of this text, other devices having similar or better functionality will no doubt be available.

At first glance, the iPad looks like an oversized iPod Touch. However, because of its size and power, the iPad can handle several more jobs that will be very attractive to students. At present, Microsoft Office is not available on the iPad and it does not look likely that it will be made available anytime in the near future but that should not slow down most students. Apple sells the iWork application for all of its platforms including the iPad. While not natively compatible with Microsoft, iWork can read and write in Microsoft Word, Excel and Powerpoint formats. On the iPad, each application is purchased individually so the student can choose which applications he or she wants: Pages, their word processor, Numbers, their spreadsheet and Keynote, their presentation program. Each is available through iTunes for $9.99, which is well within most students' ability to purchase. While they lack many of the features of their desktop relatives, they can provide an excellent platform for taking notes, working on basic formulas in the spreadsheet or creating and presenting presentations with Keynote. The iPad can also connect to external projectors using a VGA adapter sold separately.

One of the features sorely lacking on the iPad is a basic file structure – at least one that is available to the user. Therefore any file management tasks seem to be handled by each individual application that uses those files. For example, when looking at documents in the Pages program, it is impossible to see documents made with anything else. Moving documents into the iWork applications and out of them back on to a desktop/laptop computer is not overly difficult but it is not the smooth easy task that one would expect it to be. Also, there is no organization of files at present time within the applications unless the applications themselves provide it and, at the time of this writing, iWork programs do not provide any.

One of the features that is becoming popular with businesses and students is the ability to connect to cloud-based computing on the Internet. iWork has some of these features built in but there are other applications that have come on the market that possibly have even more flexibility. For example, for note taking the student can download the free Evernote application. Once it is set up both on the iPad and on a desktop/laptop computer the student can take notes in classes or meetings; as long as they are connected to a Wi-Fi or 3G network all the notes are automatically backed up to Evernote's server and transferred to the student's main computer("Evernote", 2011). The same goes for anything that the student puts into Evernote on his main computer; it automatically gets backed up and transferred to the iPad the next time the application is opened. The application is able to manage text, audio, images and web pages inserted into it and can handle keeping information together in Evernote folders. As of this writing, the folders must be created on the computer, not the iPad application.

There are other applications similar to Evernote available and with similar features. Virtually all of these applications take advantage of cloud-based computing. This enhances the functionality of devices such as the iPad. Office² HD offers very handy tools for the student. For $7.99 the student is able to get a Microsoft Office word processor and spreadsheet application which includes being able to connect to online document management programs such as Google Docs, DropBox, and MobileMe ("Office 2 hd", 2011). Rather than worrying about being able to transfer the documents, the student can read from and write directly to the online storage system of choice. If one is not available due to the lack of Wi-Fi or 3G connectivity, the application can store files and folders on the iPad itself as easily as it can save them to one of the cloud applications.

Add to the productivity applications such as the ones that have been mentioned above the ability to hold dozens of television shows, full length movies, thousands of songs, hundreds of pictures and applications to handle virtually almost any interest, hobby or topic imaginable and it is easy to see how these devices can become so popular. The student can have all of this and yet the amount of space required in the student's backpack is not any larger than the size of a typical tablet of paper. As mentioned above, books are now being made available online as well. Since the user of a device such as the iPad can purchase books for the Kindle and the Nook as well as the iBook application itself through iTunes, the student can browse through multiple options for getting the best price on whatever e-books are available. These three are at the top of the popularity list currently but are not anywhere near the extent of what is available online for e-readers.

Typically e-books are less expensive than printed books and it is possible to store an amazing number of books on one device; however there are some drawbacks to e-books. One of the first things that is noticeable is the fact that most e-book readers such as iBook re-paginate their documents to accommodate the size and format of the screen. For example a book that has 50 pages in a printed format may show up in the iBook application as having 85 pages when viewing the book horizontally. To complicate things even more, turn the iPad vertically and the book is now re-formatted to be read in 42 pages. If an instructor wants to assign specific material from a text, the student or instructor may need to specify the content found such as chapter and sub-chapter headings in the assigned reading rather than just referencing page numbers. As another option for handing this situation, the student may want to investigate if the book(s) are available in pdf format or other format that would retain the proper page numbers. Another issue that students could contend with if not in pdf format is images, graphs and charts, which may or may not display properly – or at all, in some e-reader formats.

Another issue with e-books is related to ownership and the ability to re-sell. Typically when students are finished with a course they look forward to recouping some of their costs and reselling the book either to someone else or back to the bookstore. With e-books this can be very difficult. Typically, e-books are locked down so that it is not possible to copy them to another person's account although several of the e-book services are now incorporating the feature of being able to loan a book to another person. After a period of time, usually about 2 weeks, the book automatically returns to the original owner. One possible way to avoid buying and reselling is to rent electronic versions of textbooks and other e-books required for class. This is a recent development. Some the companies allow off line reading while others require the student to be logged in to their account to read the text. This option is typically cheaper than just buying an e-book but the student needs to be careful to rent it for the full period of the class.

From the Technical Support on Campus Perspective

The results of discussions with technical support staff from various colleges and universities showed that very few had much in the line of comments about the issues of how small devices are impacting their schools. Most that were willing to talk at all stated that they were concerned that their infrastructure was not ready for all the devices that students are bringing to campus.

Gone are the days when managing student computers could be accomplished merely by their network cable and special applications installed on the student's computers for managing access, content and sharing. Now technical support not only has to provide connectivity to the students' computers in their dorm and in the library, but it also has to provide connectivity to their cell phones, MP3 players, iPads, and an ever expanding list of small electronic devices.

Obviously, schools are concerned about connectivity and being able to handle the infrastructure. Options being considered include adding more routers, wireless access points and expanded dhcp services for handling IP address leasing. Expansions place additional pressure on already tight budgets. Approval for funding for this type of expansion may depend upon being certain that all stakeholders understand the technical support requirements created by small electronic devices and their role in education. Every piece of equipment purchased for the college or university requires someone to account for it, manage it and at times repair it. Each different type of device purchased regardless of size adds that much more of a complication to the management system for handling equipment.

From the University Life Perspective

Some educational institutions are making great strides in utilizing the power of small electronic devices for a wide number of uses. Applications are being developed and adopted to enhance student life and market to future potential students. The applications are coming from three different sources. First, some schools are getting their applications compliments of student projects and internships. Secondly, for universities that have the programming staff available some are being developed internally. Thirdly, there are companies that specialize in developing university applications for small devices. The following is a brief examination of a sample of more than 50 college and university applications available for download through iTunes (see Table 1).

In addition to developing specialized applications for the small devices such as the Android and the iPhone, universities are re-working their websites to be small device aware. As mentioned previously, websites can be viewed by smart phones, e-readers and a host of other devices; all

Table 1. Mobile applications of university web sites

University	News	Map	Media	Calendar	Directory	Blog	Classes	Athletics	Events	Images	Library	iTunes U	Help	Emergency	Developed by Blackboard	Developed by Students/ School	Developed by other company	Unknown Developer
Univ of Texas	X	X	X	X	X	X	X	X	X	X								X
Stanford	X	X	X		X		X	X	X	X	X	X	X	X	X			
Texas Tech	X	X	X		X		X	X	X	X	X	X	X	X	X			
Grand Valley	X	X	X		X	X		X	X	X						X		
Univ of Mich	X	X			X				X			X				X		
Northwestern	X	X	X		X			X	X	X	X				X			
Carnegie Mellon	What options are available depends on who the user identifies himself as, student, future student, graduate student, parent, alumni, etc.																	
Florida State	X	X	X	X	X			X	X	X	X		X	X	X			
Duke Univ.	X	X	X		X		X	X	X	X	X	X	X		X			
Central Michigan Univ.	X		X		X	X		X	X	X	X						X	
Capella Univ.	X	X				X	X		X				X					X
Indiana Univ.	X	X		X	X			X	X			X	X	X				
Texas A&M	X	X	X		X		X	X			X				X			
MIT	X	X			X		X		X							X		
Perdue Univ.	X	X	X		X	X	X		X					X		X		
Louisiana State Univ	X	X						X							X			
Oregon Univ.	X	X	X		X	X	X		X					X				

of which may have different screen sizes and capabilities and many having different web browser programs in which to view the web pages. All of these issues make developing a web page that is easy to view and easy to navigate a significant challenge.

In addition to the above grid, there are several other options on many of the applications. Letters of welcome from the president of the university were offered by both University of Texas and the University of Florida. This was the very first thing that came up before showing any of the menu options. Carnegie Mellon starts by having the user select a category from a pre-set list. The list they give is student, staff/faculty, alumni, potential undergrad student, potential grad student, parent, other. Depending on what choice is made, a customized menu for that type of user is displayed. Capella's application is designed solely for the user so their opening screen is a user login screen and no information is displayed other than a link to their web site to visit if the user is not a student. Not listed in the grid but still fairly popular is the bus schedule included in the University of Michigan, Indiana University, Texas A&M and MIT.

Several universities have features that are unique to only the individual university. For example, University of Texas includes an app to display waitlist for classes that are full, an eGradebook, finals schedules and even extras such as puzzles and songs. They also have a historic campus so in addition to their normal map, they have a map on their app for landmark locations around campus. Stanford, in anticipation of potential students downloading their app, offer a virtual tour of campus in addition to their normal map. They also have an app that will display the students current balance. Duke University also does something similar through an app called the DukeCard. Grand Valley State University (GVSU) has much of their app linking to external documents and web links so they extend their app significantly by using the web application. Central Michigan University also includes an app

for weather. Central Michigan has, in the author's opinion the most creative display which was created for them by the Straxis company. The display is a rotatable screen between the menu choices that can be made in addition to a more standard menu at the bottom of the screen. At first attempt, the menu options tended to pan around faster than intended but with a very short amount of practice, it is very easy to scroll between choices without problems. Central Michigan also includes a poll option which is not included on any of the other applications evaluated. The options that an educational institution includes seem to be limited only by the developer's vision of what is important to students or other stakeholders of the institution.

CONCLUSION

Since the introduction of the iPad in the Spring of 2010, there has been a flurry of new handheld devices being developed similar to the iPad. It appears that this trend will continue into the future. Although similar in look and feel, they each have or will have some unique differences such as size, options for buttons and controls, file structures, screen resolution etc. This would indicate that manufacturers are willing to invest in this type of technology and they are betting that there is a significant future in this type of device. What is not fully known yet is what choices will be most sought after by the consumer. For example:

• What is the most popular size and combination of height to width?
• What will the most popular on-screen controls be like?
• How close to a laptop/desktop's functioning file structure should it have?

Only time and maturity of the market will be able to answer the question of whether or not Apple remains the small tablet king or not. Is this going to be a settled technology and the way people work in the future or is this a transition technology to yet something different?

Mac Life magazine in their cover story for their January 2011 magazine makes some significant predictions about "Apple's Next Big Thing" (Curthoys, Phillips, Aguilera, Ochs, 2011). In their article they foresee stunning screen resolution, motion sensing of hands without having to touch the screen and synchronizing to a computer or iPhone without cables via Bluetooth. In 2013 they are predicting that the screen of the iPad will take on a 3D look and feel. Finally, by 2017 they are predicting that the iPad, or whatever it will be called then will be as thin as a sheet of paper with only a slightly thicker portion at one end holding the controls and interfaces.

There are others with vastly different perceptions of what the future will hold. According to Kim Zetter in the online version of Wired Magazine (Zetter, 2009), MIT's fluid lab group developed a wearable device that scans faces, images, bar codes and ISBN numbers to identify people and objects. Once identified, the device projects information about the person or object on whatever surface is in front of it such as a wall, another person, a sheet of paper or even a hand or arm.

The question is not so much about technology such as what is possible to be built. It is now possible to create almost anything – in almost any size imaginable. The question is functionality and interface. What will students want in their backpacks when they go to class? How will they want to take notes, write papers, buy tickets and interact with their friends in the social media applications available in the future? Companies will create an amazing armada of digital devices in coming years; then students and all of society will vote with their wallets as to which ones will be accepted and which ones will not. The educational field will need to stay informed. Instructors will need to adapt teaching methods to accommodate new devices and capabilities. Administrators and technical support staff will need to evaluate additional infrastructure and support requirements to meet new demands. Those focusing on applications for university life will want to evaluate how changing technology affects applications. The age of small electronic devices is here. The challenge to educators is to make the most of it by finding the most effective and cost efficient ways to integrate handhelds into the educational system.

REFERENCES

Curthoys, P. C., Phillips, J. P., Aguilera, R. A., & Ochs, S. O. (2011, January). Apple's next big thing. *Mac Life*, *5*(1), 22–31.

Evernote. (2011). *Learn more*. Retrieved from http://www.evernote.com/about/learn_more/

GPS. (2011). *Amazon search*. Retrieved from http://www.amazon.com/s/ref=nb_sb_noss?url=search-alias%3Daps&field-keywords=GPS&x=0&y=0

iBooks. (2011). *iPad features*. Retrieved from http://www.apple.com/ipad/features/ibooks.html

iPad. (2011). *Apple shop: iPad*. Retrieved from http://store.apple.com/us/browse/home/shop_ipad/family/ipad?mco=OTY2ODA0NQ

iTunes U. (2011). *What is iTunes U?* Retrieved from http://www.apple.com/education/itunes-u/what-is.html

Kindle. (2011). *Amazon search*. Retrieved from http://www.amazon.com/dp/B002Y27P3M/?tag=googhydr-20&hvadid=6912512736&ref=pd_sl_98pmgzpmhj_e

Motion, X. GPS drive. (2011). *iTunes app*. Retrieved from http://itunes.apple.com/us/app/motionx-gps-drive/id328095974?mt=8

Nook. (2011). *Features and tech specs*. Retrieved from http://www.barnesandnoble.com/nook/features/techspecs/index.asp?cds2Pid=35611

Nook Color. (2011). *Features and tech specs.* Retrieved January 26, 2011, from http://www.barnesandnoble.com/nookcolor/features/tech-specs/index.asp?cds2Pid=35607

Office 2 HD. (2011). *iTunes app.* Retrieved from http://itunes.apple.com/us/app/id364361728?mt=8

Ogg, E. O. (2010, September 7). *iPad competitors lining up.* Retrieved from http://news.cnet.com/8301-31021_3-20015610-260.html

Sparkvue. (2010). *iTunes app.* Retrieved from http://itunes.apple.com/us/app/sparkvue/id361907181?mt=8

Vaughan-Nichols, S. V. (2010, March 12). *Here comes Linux's iPad clones.* Retrieved from http://blogs.computerworld.com/15742/here_comes_linuxs_iPad_clones

Zetter, K. (2009, February 5). *TED: MIT students turn internet into a sixth human sense.* Retrieved from http://www.wired.com/epicenter/2009/02/ted-digital-six/

ADDITIONAL READING

5 Android Apps, I. Can't Live Without (and Why) http://chronicle.com/blogs/profhacker/5-android-apps-i-cant-live-without-and-why/26604

Education World http://www.educationworld.com/a_tech/tech083.shtml

Educause www.educause.edu

End-user programming to support classroom activities on small devices http://www.computer.org/portal/web/csdl/doi/10.1109/VL-HCC.2008.4639106

Good Things Come in Small Packages http://campustechnology.com/articles/2009/02/01/mobility.aspx

Hand Held Devices in the Classroom http://eduscapes.com/tap/topic78.htm

Handheld computers in Education http://www.littleferry.k12.nj.us/FeaturedLinks/Handhelds/Handhelds.html

Handheld Devices in the Classroom http://eet.sdsu.edu/eetwiki/index.php/Handheld_devices_in_the_classroom

Handhelds in Education http://www.education.wichita.edu/m3/mobility/handhelds/education.htm

Handhelds in Education http://www.tribeam.com/educator.html

iPhone in the Classroom – One Teacher's Story http://ithinked.com/archives/2007/11/the-iphone-in-the-classroom-one-teachers-story-dr-richard-beck/

iPod Touch in the Classroom http://www.cmich.edu/ehs/x27737.xml

Kathy Schrock http://kathyschrock.net/power/

Learning in Hand http://learninginhand.com/

New Electronic Devices Could Interest Schools. http://www.eschoolnews.com/2010/01/07/new-electronic-devices-could-interest-schools/

Tech Teachers http://www.techteachers.com/handhelds.htm

Tiny Electronic Devices go to College http://www.peterli.com/cpm/resources/articles/archive.php?article_id=1080

Types of Technology Used with Interactive Learning http://www.ehow.com/list_7368118_types-technology-used-interactive-learning.html#ixzz19kB7CsRa

KEY TERMS AND DEFINITIONS

Adobe Connect: An application developed by Adobe for holding online conferencing.

Apps: An abbreviation for the word application, a program designed to run on a particular device. Usually it is used in reference to small devices but it can also be on any computing device.

Android: A mobile device operating system developed by Android Inc. The Android company is now owned by Google.

Bluetooth: a wireless technology for transmitting data over short distances. Bluetooth uses short wavelength radio transmissions that are generally limited to about 10 meters or less, although there are ways to increase this distance.

Cloud-Based Computing: Computers, applications and disk-storage space all available on the Internet. It is location independent and applications, processes and file storage can be used regardless of not only where a person is but what computer or small device they are using as long as it is connected to the Internet.

DHCP Service: A networking services that hands out Internet addresses temporarily to computers and other devices wanting to access the Internet.

DimDim: An online service for hosting web conferencing.

DropBox: A web-based file hosting service – a type of cloud computer application.

eGradebook: An application developed for small devices for students to manage their class grades.

e-ink: A special, proprietary type of electronic paper developed by E Ink Corporation.

eReader: An electronic device developed for reading books, papers, articles, and other electronically delivered media.

Evernote: A cloud based service for managing an electronic notebook. The service is available on most desktop/laptop computers and small devices. What is created in one device is synchronized between all other devices that have the Evernote application and are logged in to the service with the same username/password.

Facebook: An online service for social gatherings and sharing of information and online games.

Galaxy Tablet: A small device similar to, but smaller than the iPad tablet. It is developed by Samsung and uses the Android operating system.

Google Docs: A cloud computing service by Google for online file storage and Microsoft Office compatible word processing, spreadsheet and presentation programs.

GPS: Global Positioning System. An application or device for using external points of reference such as satellites in orbit to determine location and to chart travel.

Handhelds: Any computing device that is small enough to fit into one hand as opposed to desktop and laptop computers which are usually much larger. Frequently, but not always incorporates the functions of a cell phone.

iBook: An application developed for the iPad and the iPod Touch for purchasing and reading electronically printed material.

IP Address: The Internet Protocol IP address a device receives from a DHCP server in order to function on the Internet.

iPhone: A device developed by Apple Computers, Inc. for cell phone use and for working online.

iPad: A tablet computer developed by Apple Computers, Inc. It normally communicates with a host computer such as a desktop or laptop computer through a special cable. It also has the ability to connect to the Internet. All iPads come with built-in ability to connect to wi-fi signals. Many of them also come equipped to connect to the Internet through a 3G cell phone signal with an accompanying service through AT&T.

iTunes: An application that connects to Apple Computers online store for purchasing media such as music, videos and for subscribing to podcasts.

iTunes University: An application within iTunes specifically developed for hosting podcasts from colleges and universities. The school is required to set up their own podcasting server which hosts the media for local consumption at the campus and uploads those podcasts that are allowed for public consumption to the iTunes store.

iWork: An application developed by Apple Computers, Inc. which contains a word processor called Pages, a spreadsheet program called Numbers and a presentation program called Keynote.

Keynote: A presentation program in the iWork suite that is available for Apple computers and Apple iPads.

Kindle: The Kindle is an eReader device developed and marketed by Amazon for purchasing and reading books and other printed material in an electronic device. Kindle is also an application developed by Amazon for doing the same thing but instead of on their proprietary Kindle device, the application can run on small devices such as the iPad.

Mobile Me: A cloud computing service by Apple Computers for hosting web pages, video and still image media, email applications and other tools.

MP3 Player: A device or an application developed for playing audio media that is formatted in the MP3 format. The device or application can frequently also play other media formats as well.

Myspace: An online social gathering service for individuals to meet.

Nook: The Nook is an eReader device that is developed and marketed by the Barnes and Noble book company for purchasing and reading books and other printed material in an electronic device. The Nook is available in both e-ink and in a color screen format. The Nook is also available as an application available for download on devices such as the iPad.

Numbers: Numbers is the spreadsheet program in the iWork suite developed by Apple Computers which is available for both desktop/laptop computers and for the iPad.

Office[2] HD: A cloud computing application developed for small devices such as the iPad that offers word processing and spreadsheet applications. It also has the ability to organize files and to attach to cloud computing services such as me.com, dropbox and Google Docs.

Pages: Pages is the word processing program in the iWork suite developed by Apple Computers which is available for both desktop/laptop computers and for the iPad.

Smartphones: A cell phone that can handle many computing applications as well such as email, web browsing and playing of various types of media.

Sparkvue: Software developed by the Pasco corporation (www.pasco.com) for use on desktop/laptop computers and small devices for use in viewing scientific experiments.

TomTom: A GPS device that is normally in a vehicle. It is used to display current location as well as plot course to destinations selected.

VGA Adapter: A device designed to connect a computer or small device to a desktop monitor or projector.

Web Conferencing: The ability to connect with other people remotely and share information. Web conferencing usually has the ability to have both audio and video capability as well as various forms of text and whiteboard communication.

Webex: A web conferencing program that can run on most computers and many small devices for attending meetings and conferences remotely.

Wi-Fi: A networking radio signal that gives computers and small devices the ability to connect to the Internet without the need for plugging in a networking cable.

Chapter 4
Mobile Marketing

Barbara L. Ciaramitaro
Ferris State University, USA

ABSTRACT

Mobile technologies have dramatically changed the world's ability to communicate. The number of mobile phones used worldwide has exceeded 4.6 billion with continued growth expected in the future. In fact, in the United States alone, the numbers of mobile phone users comprise over 80% of the population. Mobile phones and tablets (mobile devices) are not simply voice communication devices. They have become a medium to create voice, music, text, video, and image communications. Importantly, these various types of communication can be created and shared on demand by the mobile user. In addition to communication methods, mobile devices are also a tool used to access the Internet, view television and movies, interact with GPS (Global Positioning System), and read and respond to barcode and augmented reality messages. Each of these methods utilized by the mobile phone user becomes a tool that can be used in mobile marketing to expand beyond traditional marketing methods. Mobile devices are considered to be "the most personal piece of technology that most of us will ever own" (Krum, 2010, p. 7). We usually take them with us wherever we go and are usually reachable through them. However, mobile devices also provide the ability to access the most personal information about us. Mobile devices know who we communicate with and how often. They know our schedule – both business and personal. They often know all of our email addresses and frequently accessed web sites. They know what videos, music, television shows, and movies we like. They know about us through pictures and text messages sent and received. They know where we go, how often, and how long we stay through location tracking technology. This collection of accessible personal information allows mobile marketing to target individuals at the time and place where their message will be most effective. Mobile technologies over the past 20 years have dramatically changed the way people communicate, collaborate, search for, receive, and share information. These dramatic changes have had striking impact on the world of marketing to the extent that mobile marketing has become the predominant form of customer engagement.

DOI: 10.4018/978-1-61350-150-4.ch004

INTRODUCTION

"Whenever our world changes, so must the practice of marketing." (Becker, 2010, p. 9).

Mobile technologies have dramatically changed the world's ability to communicate. In recent years, the number of mobile phones used worldwide has exceeded 4.6 billion with continued growth expected in the future (News, 2010). In fact, in the United States alone, mobile phone users are soon expected to comprise over 80% of the population. (Butcher, 2010). The growth of a subset of mobile phones, referred to as smart phones, is also increasing at a rapid rate. In 2009, 20% of mobile phone users bought smart phones. By 2010, smart phones comprised 30% of all mobile phones purchased. It is predicted that by 2011, purchases of smart phones will exceed purchases of other mobile devices, computers and laptops (Entner, 2010) Smart phones provide users with added functionality such as email, web browsing and the ability to download games and other applications. Smart phones also provide an array of marketing opportunities. According to a recently published Nielson Study (Nielsen, 2010), by 2015 smart phones "will be the primary enabler of consumer shopping engagements" (p. 1).

It is also important to take into account other mobile technologies such as the growing use of tablet mobile devices. Led by the Apple iPad, which shipped 13 million devices in 2010 alone, tablet mobile devices are clearly changing the face of mobility and computing. It is predicted that by the year 2015, tablets will represent 23% of all PC sales (Butcher). As the widespread adoption of mobile technologies becomes the platform for several new approaches to marketing available only in the mobile environment, the impact of mobile technologies on marketing is significant.

"Fundamentally, mobile phones are now media" (Wertime, 2008, p. 4) Mobile phones and tablets (mobile devices) are not simply voice communication devices. They have become a medium to create voice, music, text, video, and image communications. Importantly, these various communications can be created and shared on demand by the mobile user. In addition to communication methods, mobile devices are also a tool used to access the Internet, view television and movies, interact with GPS (Global Positioning System), play games, and read and respond to barcode and augmented reality messages. Each of these methods utilized by the mobile phone user becomes a tool that can be used in mobile marketing to expand beyond traditional marketing methods.

Mobile devices are considered to be "the most personal piece of technology that most of us will ever own" (Krum, 2010, p. 7). We usually take them with us wherever we go and are usually reachable through them. However, mobile devices also provide the ability to access the most personal information about us. Mobile devices know who we communicate with and how often. They know our schedule – both business and personal. They often know all of our email addresses and frequently accessed web sites. They know what videos, music, television shows and movies we like. They know about us through pictures and text messages sent and received. They know where we go, how often, and how long we stay through location tracking technology. This collection of accessible personal information allows mobile marketing to target individuals at the time and place where their message will be most effective. It also presents one of the greatest dilemmas in mobile marking concerning the balance between collecting marketing data and respecting personal privacy.

BRIEF HISTORY OF MOBILE TECHNOLOGIES AND ITS IMPACT ON MOBILE MARKETING

Network Technologies and Handsets

The reach and functionality of mobile devices depends on their underlying network infrastructure and the capabilities of the mobile device or handset. Limited bandwidth results in limited functionality. Limited reach results in limited access and communication capability. The ability to effectively market to mobile device users requires an understanding of the strengths and limitation of their network infrastructure and the design of a marketing plan that accommodates those strengths and weaknesses. Over the past 15 years, the mobile network infrastructure has evolved. However, it is important to recognize that on a global basis, not all regions and countries have evolved similarly and many underdeveloped areas continue to rely on less current network technologies.

1G

The earliest generation of mobile networks was based on a circuit-switching technology and relied on the transmission of analog signals sent from the phones to radio towers. This first generation is referred to as 1G and was the primary network infrastructure introduced in the 1980's and used through the first half of the 1990's. The throughput of data transmission on a 1 G network was limited to 9.6 Kbps (Kilobytes per second) (Krum, 2010). This limited throughput constrained the users of the 1 G network to only voice communications.

In 1983, the first mobile phone became available through Motorola. Its only functionally was that of voice calling. Affectionately known as "the brick", it weighed two pounds and cost nearly four thousand dollars and provided an hour of talk time before its battery would run out. The first mobile phones weighed about 2 pounds and

were quite expensive. (Cassavoy, 2007) Due to the high cost, phones used with the 1 G network were rarely used by the general consumer. As a result of its limited availability, limited throughput of the network, and the restricted functionality of the early handsets, mobile marketing in the 1G network was virtually non-existent. However, over the next few years mobile service provides began to upgrade their analog 1G networks with digital networks.

As a result, "Mobile phones soon went from being expensive novelty items used only by business professionals and the wealthy to being well within the reach of the masses." (Hager, 2006).

2G

The second generation of mobile networks, referred to as 2G, was launched in 1991 and relied upon a digital signal to transmit rather than the analog signal used by the 1G network. The use of the digital signal improved the quality of the voice communications although the throughput remained the same at 9.6 Kbps. From the mobile technology provider's viewpoint, 2G offered an improvement in that carriers were able to transmit a higher volume of calls through their network. From the user point of view, however, the 2G network proved problematic as it relied on a caller's proximity to a cell tower. If the caller moved out of range, the call was immediately dropped rather than gradually degrading. Text messaging became available on the 2G network although it was not quickly adopted. Although the first text message was sent in England in 1992, by 1994 users worldwide were only sending a mere .4 text messages per month. By the year 2000, the average number of text messages had increased to 35 per month. (TMC, 2007)

Mobile phone handsets in the 2G network were lighter, smaller, and less expensive which resulted in an increased consumer adoption rate. Most significantly, these handsets allowed text messaging capability in addition to voice calls. (Krum, 2010) However, due to the slow adoption of text messaging, mobile marketing remained virtually non-existent in 2G network environments.

Texting

For a mobile marketer, it is important to recognize the generational and global differences in the adoption of text messaging. In the United States, texting is most widely used by users between the ages of 13 and 21 (O'Rourke, 2007). "Fully 72% of all teens -- or 88% of teen cell phone users -- are text-messagers." (Lenhart, 2010, p. 1) However we are seeing growth in the use of mobile devices by an even younger generation. In 2009, 66% of children in the United States between the ages 8 to 18 had a mobile phone with their texting as their primary method of communicating (eMarketer, 2009). There are also differences in the use of texting on a global basis. In Japan, the age group of high text message users broadens to the ages between 13 and 30. In China, there are less generational differences in the use of texting as mobile phones are used by all generations as a replacement for personal computers. In Britain more than half of the children between the ages of 5 and 9 now own a cell phone (Naish, 2009)

2.5G

The next generation of mobile networks is referred to as 2.5G. It is considered to be a bridge between 2G and 3G cellular wireless technologies and was introduced between the years 2000 and 2005. The 2.5G network provided a set of combined services including a circuit-switching domain for voice communication and a packet-switching domain for data communication. The transmission throughput in 2.5G network was 50+ Kbps. As was mentioned previously, although text messaging was possible it was used minimally and did not provide a reliable avenue for mobile marketing.

During the era of the 2.5G network, mobile handsets with additional functionalities were introduced to the market. In 1993, BellSouth/IBM introduced the world's first mobile phone with PDA features that included phone, text, calculator, calendar, fax and email. It was small in size, weighing only 21 ounces and cost $900. And in 1996, "fashionable" mobile devices were introduced with the StarTAC mobile phone from Motorola that weighed only 3.1 ounces in a clam shell design. (Sacco, 2007)

The added functionalities available in the 2.5G network were beneficial to users and the adoption of mobile device usage increased so that by 2002 the number of mobile subscribers in the world was in excess of 1 billion users. Additionally, mobile marketing began during this era when the first mobile phone advertisement appeared as a free daily SMS text message in Finland (Membridge, 2008).

3G

It was with the release of the third generation of mobile networks, referred to as 3G that much of the capability used for mobile marketing came into existence. Although the first 3G network was introduced in Japan in 2001, it was not until 2005 that the use of 3G networks became widely adopted. The demand and use for mobile devices skyrocketed during this era, so that by 2006, the number of mobile subscribers exceeded 2.5 billion (Membridge, 2008). 3G actually refers to a set of standards that meet ITU (International Telecommunication Union) requirements (Union, 2005). In addition to voice, 3G is able to provide a wide variety of application services including mobile internet access, multimedia, location-based services, video calls, and mobile television. Its data transmission rates are rather robust at 384 Kbps (Krum, 2010).

In addition to the voice and text messaging capabilities of the 3G era, a significant added functionality to mobile handsets was that of the mobile web browser. The introduction of the web browser expanded the use of the mobile device so that Internet information could be searched for and accessed on demand by the user.

It was during the 3G era that the use of mobile marketing exploded. The Internet browser functionality on 3G handsets allowed mobile marketing efforts to expand their focus from merely using text messaging to providing interactive Internet links. One interesting use of this functionality in the mobile marketing area was the use of special bar codes that could be read by cell phone cameras. After reading the barcode, data such as music, videos or coupons were made available to the user from the Internet. Another interesting use was the ability to send a text message to a special number and receive links to special offerings.

The addition of GPS (global positioning system) data provided marketers with information as to the location of the mobile device and provided the ability to send messages or links for specials or services relevant to the exact location of the user. This combination of location based services and other functionality also provided the foundation for augmented reality marketing efforts. Augmented reality refers to interaction of graphics, audio and other immersive tools displayed in real-time. The most common use of augmented reality in mobile marketing is implemented through the use of two dimensional bar codes where the user reads the bar code through the mobile device camera, and an augmented reality application is opened on their mobile device. (Cassela D., 2009) One very popular augmented reality marketing effort directed at young mobile users was the effort launched by Doritos in 2009. Doritos printed a special AR (augmented reality) barcode on limited edition packages of Doritos chips. Mobile users would take a picture of the bar code and then access a special web site where they would flash the AR picture at their webcams and get access to a performance of a popular rock group (Wallace, 2009).

Lastly, the ability of 3G handsets to download applications and games has opened up a new world for mobile marketing. Applications specific to companies' marketing efforts are provided and product placement in mobile games has become common place. One example of mobile marketing through games is the game launched by Coca-Cola in India which offered a rock climbing adventure with weekly prize giveaways. The results were significant in that more than 350,000 games were downloaded by users each week (Durrell, 2011).

4G

The fourth generation of mobile networks, referred to as 4G, provides high speed data transmission and high levels of data throughput at 100 Mbit/s for high mobility communication (such as from trains and cars) and 1 Gbit/s for low mobility communication (such as pedestrians and stationary users). The 4G system provides a 100% IP based broadband network and allows for the delivery of high demand applications such as Voice over IP, gaming, and streamed multimedia (Krum, 2010). 4G networks are developed to meet QoS (quality of service) requirements in that they are able to appropriately prioritize the various types of network traffic for the most optimum results. This is necessary to support the high bandwidth capabilities offered through 4G networks including: Wireless Broadband Internet Access, MMS (Multimedia Messaging Service), Video Chat, Mobile Television, HDTV (High Definition TV), DVB (Digital Video Broadcasting), Real Time Audio, and High Speed Data Transfer ((Internet, 2010). Additionally, 4G networks support reliable file transfer even if the user travels from one cell coverage area to another. Additionally, the mobile device is assigned a specific IP address that is maintained through the entire coverage area of any 4G network. This allows for easy tracking of mobile devices through the global network. (Mohr, 2002)

The standards for 4G networks are still in development at the time of this writing. There are currently several technologies competing to be the basis of 4G but many predict that the WIMAX standard will prevail (Wertime, 2008). WIMAX has been deployed in both the United States and Japan but other nations are currently using other technologies.

From a marketing perspective, it is interesting to note that most consumers do not recognize any significant difference in the terms "3G" or "4G" networks. Many consumers remain unfamiliar with these terms even though mobile providers are trying to inform consumers of their differences through various marketing efforts. Consumers are more responsive to terms such as high-speed and high-throughput rather than 3G and 4G. In fact, several studies have found that references to 3G and 4G cause consumer confusion. It is predicted that it will take three to four years to achieve a high level of consumer adoption of 4G networks. (Kee, 2010).

However, the increase in speed and throughput provided by 4G networks will impact the mobile user experience and will therefore impact mobile marketing. It is predicted that the mobile user will spend significantly more time browsing and accessing Internet content. As a result, mobile marketers will need to focus more on their mobile web sites than the current focus on mobile applications. Additionally, users are expected to flock to multimedia streaming services such as Pandora (www.pandora.com) and Slacker (www.slacker.com), both of which provide streaming music services. (Kee, 2010)

As can be seen from the above chronology, mobile technologies over the past 20 years have dramatically changed the way people communicate, collaborate, search for, receive, and share information. These dramatic changes have had striking impact on the world of marketing to the extent that mobile marketing has become a major form of customer engagement. In the next section we will discuss some of the unique mobile technologies that have supported the growth of mobile marketing.

MOBILE MARKETING TECHNOLOGIES

Mobile marketing is distinguished from traditional marketing efforts through its use of technologies and capabilities available only on mobile devices. Several of these technologies are worth more detailed discussion.

Mobile E-Mail

"Email is perfect for sending highly personalized, targeted, private, and interest-specific messages to a large number of people at once." (Becker, 2010, p. 152) However, producing effective mobile email campaigns requires more attention and configuration than traditional email efforts. It is important to note that most mobile devices will not display email messages in the same way as personal computers. The screen size, operating system, browser and other technical issues make the mobile email platform unique and requires special attention by companies interested in implementing a mobile email campaign. It is very different to send a targeted email message to a device with a 13 to 15 inch viewing screen than that with a 2 to 4 inch viewing screen available on most mobile devices. Mobile email content must be streamlined and focused to include only the essential content. It is also important to recognize that most users of mobile devices use them to triage their email. This means that they use available short bursts of time to quickly scan their email determining which should be opened right away, which can be deleted, and which require additional attention when they arrive at their computer. Therefore, any mobile message requesting an immediate response must be obvious to the user and require minimal action to respond. (Pollard, 2008)

There are several technical issues that must be addressed in mobile email marketing efforts. The first issue concerns the various mobile devices used by your target audience. Display screen sizes vary with mobile devices which impact their ability to display email messages. It is important to structure mobile email messages in such a fashion that they are readable on the majority of mobile device screens. Additionally, different mobile devices render email messages differently along with the Internet links embedded in the message. For example, it can prove to be quite frustrating for a user to try and activate an embedded link when they are using a touch screen device such as those prevalent on Android and iPhones. Related to this issue is the ability of the phone to handle HTML email formats. The viewing screens on mobile devices vary significantly which impact the ability to for all users to consistently and accurately view the HTML email message. One alternative is to send text email messages. But, remember that although most text messages have 60 to 80 characters displayed on a computer screen, mobile devices are limited to 20 to 40 characters in 10 to 15 lines per screen (Pollard, 2008).

SMS (Short Message Service) or Text Messaging and CSC Codes

"Text messaging is the cornerstone of mobile marketing" (Becker, 2010, p. 79) SMS or text messaging was the first non-voice application introduced on mobile devices and continues to be the most heavily used mobile technology as 95% of mobile devices are text messaging enabled. SMS mobile marketing has proven to be successful in that research has found 24% mobile users respond to mobile marketing based text messages they receive (Neustar, 2010). "While there are other mobile marketing tools out there, they don't have the reach of SMS and don't work with all carriers and all handsets. The SMS medium elicits high audience response rates because of the lack of media clutter." (Ibid, p. 8)

Text message-based mobile marketing campaigns are often tied to the use of CSC (common short codes). Common Short Codes (CSCs) are short numeric 6 to 8 digit codes to which text messages can be sent from a mobile phone. Companies obtain CSC codes for their marketing campaign through the Common Short Code Administration (http://www.usshortcodes.com/). One important benefit of the use of CSC codes is that it provides common access through all mobile carriers. This is an important consideration when planning a widespread mobile campaign (CTIA, 2011). CSC codes are often displayed on billboards and print media or told to users through web sites, radio or television where users are invited to send a text message to the CSC. When mobile users send a message to the CSC code, it is routed to the appropriate company which responds to the user with a confirmation or follow-up message. Once the confirmation or message is acknowledge by the user, they become enrolled in a marketing program that usually involves voting, polling, coupons or contests (Ibid). As will be discussed later in this chapter, it is essential that the customer "opts in" to text based and other mobile marketing campaigns in order to prevent a violation of consumer privacy.

MMS (Multimedia Messaging Service)

MMS expands on the capabilities of text messaging through the use of images, sound and video. However MMS is used far less than SMS as reflected in the current mobile messaging mix with 95% of messages sent and received via SMS and 5% sent and received via MMS (Ahonen, 2010). Many would look at these numbers and consider MMS a failure. However, there are currently 1.7 Billion users of MMS across the globe. This number totals more than the number of personal computers or televisions sets on the planet (Ibid). It is also important to recognize that adoption of MMS is different in various countries around the world. China leads MMS adoption primarily due to the fact that there is very low personal computer ownership in China, and for the majority of the mobile users their mobile device provides their only experience with multimedia (Ibid).

While not as widespread as SMS, MMS has developed its own place in mobile marketing. MMS has been adopted by many companies as a way to deliver their news, entertainment and advertising. Mobile marketers must recognize that MMS should not be expected to replicate or replace SMS. Most SMS messaging is a person-to-person communication that is performed many times during the day. MMS, on the other hand, is a terrific media platform for broad distribution of marketing messages. Also, MMS messaging is not needed or used as many times per day as text messaging. As with all mobile technologies, marketing professionals must select the right tools for their campaigns. MMS will provide very effective methods for a media rich consumer experience.

QR (Quick Response) Codes

A QR Code is a two dimensional bar code that is readable by QR scanners on mobile devices with a camera. A QR code consists of black modules arranged in a square pattern on white background in which text, URL or other information can be encoded. A regular store price bar code holds about 20 digits of information while QR codes can hold up to 7,000 digits (Blend, 2010). When a camera on a mobile device scans the QR code, the code is immediately interpreted and the user is directed to a website in the mobile browser. "QR codes can effectively turn any 'flat' space such as a billboard or a print ad, into a direct response mechanism that can automatically link a viewer or reader to the web" (Wertime, 2008, p. 6).

QR codes have been implemented on a massive basis in Japan where they are used in magazines, billboards, business cards, shop windows, and T-shirts. In the US, Google is sponsoring QR codes through its Favorite Places program (www. google.com/help/maps/favoriteplaces/) where

they are providing over 100,000 businesses with QR stickers for their windows. (Blend, 2010). The adoption of the use of QR codes is slow due to the fact that many phones do not come equipped with QR scanners as a default capability however new smart phones such as the 4G Apple iPhone, Android phones, and Blackberries are now coming equipped with QR scanners installed as part of their camera phone capability.

Bluetooth, RFID and Location-Based Marketing

Bluetooth and RFID (Radio Frequency Identification) are used for a type of mobile marketing referred to as *proximity* marketing. Bluetooth technology uses radio transmission bands to send signals to Bluetooth enabled devices within a close proximity of about 100 meters. Bluetooth marketing is generally used to target shoppers in a retail location as they pass stores or restaurants. When a Bluetooth enabled device enters the range of the Bluetooth hotspot, communication about the device and the user is sent to the Bluetooth server. Using its stored database of customer and device information, the Bluetooth server forwards content back to the user. The returned content has been optimized for that user and their handset (Krum, 2010). Use of Bluetooth proximity marketing requires that the mobile user allow his device to be Bluetooth discoverable. Many users do not enable this feature due to either privacy concerns or battery drainage although the battery capabilities of mobile devices are continually improving. This added requirement on the part of the user limits the usage of Bluetooth based proximity marketing.

RFID requires the use of RFID tags or chips which can be tracked using radio waves. RFID marketing works similarly to Bluetooth in that a RFID hotspot will recognize a passing RFID tag and send location specific messages to the device owner about a nearby store, restaurant or other business. One significant difference in RFID campaigns is the use of RFID tags that mobile users affix to their phones in order to opt-in to the marketing program. Customers who want to receive special offers from the mobile marketing program request an RFID chip, which is affixed to their phone like a sticker. Coupons and specials are sent to the customers on a regular basis. In order to redeem the coupons or specials, the customer goes to the store where the RFID chips are scanned at a terminal at the POS. From a privacy point of view it is important to note that although no name or personal information is shared with the company, the user's buying behavior is tracked through an ID number assigned to the RFID tag attached to the phone. This data is immensely important to the company but can also be beneficial to the user as future offers can be tailored more personally to the user. (Miller, 2009)

As of 2010, two hundred million dollars has been spent on mobile proximity marketing. Although this investment is currently limited primarily to major retail national companies, continued investment in proximity marketing is expected to expand to small and midsize companies. It is predicted that by 2015, proximity-based marketing will account for 29% of all mobile marketing spending (Watch, 2010).

Location-Based Marketing uses information gathered from the GPS capabilities of the mobile device to locate the geographical position of the user. This information is then used to identify services available near the location of the user such as the location of a nearby ATM machine. It is also used to send location specific coupons or advertising to users based on their current location. Applications that utilize this capability include weather services and navigation services. Most location based marketing messages are sent via SMS and include coupons or discounts to nearby advertising restaurants, cafes, movie theatres and other businesses.

Augmented Reality

Augmented reality is a mobile technology that combines location specific information with highly immersive and detail multimedia content. "AR systems integrate virtual information into a person's physical environment so that he or she will perceive that information as existing in their surroundings." (Hollerer, 2004, p. 1) One example of AR was developed by Adidas in which an Adidas shirt is superimposed on a person who is standing in front on an augmented reality display or mirror. (Becker, 2010). Another use of AR is the application Google Goggles (www.google.com/mobile/goggles/) in which a picture can be used to conduct a search of the Internet. Other interesting AR applications include Ghost Capture (http://itunes.apple.com/us/app/ghost-capture/id349479650?mt=8) which allows a user to add a realistic ghost to photos; Argo (http://www.realareal.com/argo-augmented-reality-app) allows you to aim your phone at any building or direction and will show you nearby restaurants, ATMS and stores; and Cheap Gas (http://itunes.apple.com/us/app/cheap-gas/id290765007?mt=8) which uses your mobile device GPS to locate your current position and then present you with a list of gas stations close to you sorted by distance or price (Reality, 2010).

Mobile Operating Systems and Browsers

The mobile operating system is the underlying technology that controls the mobile device. The capability of a mobile operating system directly impacts mobile Web browsers as well as other mobile device functionality. One significant difference that varies with mobile operating systems, particularly in older models, is the ability to handle JavaScript and Flash based data, although most newer systems will accommodate both. The most common mobile operating systems are Symbian from the Symbian foundation, Android from Google, iOS from Apple, RIM Blackberry OS, and Windows Mobile from Microsoft. The major improvement in mobile operating systems came with the introduction of smart phones which were supported by full featured mobile operating systems that allowed more advanced computing capability, Web browsing, and the installation of various applications and games. Most of the leading mobile operating systems now support the additional functionality provided by smart phones.

One challenging aspect of mobile marketing is accommodating the various mobile Web browsers in your marketing campaign. It is important to recognize that the capabilities of the various Mobile web browsers differ greatly in terms of features offered and the mobile operating systems they support. The various screen sizes, handset functions, and underlying operating system of mobile devices make mobile Web browser consistency a challenge. There are currently several mobile Web browsers in use including Opera Mobile which works on the Windows Mobile and Symbian operating system; Skyfire which support Flash and works on Android, iPhone, Symbian, Windows Mobile operating systems; Safari which work on the Apple iPhone; Google Android that works on the Google Android system; Mobile Chrome that works on the Google Android platform; Microsoft IE with works on the Microsoft Mobile platform; and Mobile Safari which works on the Apple iPhone (Nations, 2010).

Mobile Search and Websites

"As of mid-2007, search is the fastest growing part of Internet marketing." (Wertime, 2008, p. 96). The advantage of using search marketing is that by the very act of performing the search, users are indicating an interest in the product or service. Mobile search marketing utilizes the same marketing tools used in traditional Internet marketing: paid placement and SEO (Search Engine Optimization). The goal is to have a company's product or service rank high on the search results

or appear on the right of a search result screen. This is particularly important in the mobile arena where screen size and navigation are limited.

Paid placement refers to the practice of paying the search providers money to raise the prominence of a company's position in search results. A common provider of paid placement is AdWords, which is a service offered by Google. The second approach to achieving high relevance in search results is through the use of keywords and tags placed within your web site and its HTML code. This is known as SEO (Search Engine Optimization), as these keywords and tags are indexed by the search engine and, if used effectively, result in higher ranking in search results. It is important to note that keywords and tags are not the only information used by search engines to rank search results. Search engines also use other information in their algorithms including the history of the site in terms of current updates, known as content freshness, and the number of links to the site from other high ranking sites.

However, having mobile users directed to your site will not do any good if your site is not optimized for mobile viewing. One way to distinguish your mobile web site from your traditional one is to create a new domain with the.mobi extension. This allows for the searching of only web sites that have been designed specifically for mobile use. Other approaches include building a sub-domain or folder within your traditional web site that is designed and optimized for mobile users. Google, Twitter and Facebook are good examples of companies that have built mobile versions of their web sites to better suit the needs of their mobile users.

Mobile web pages should be designed for small screen size and minimal ability to navigate. The ability for the user to view images and other media will depend on their device. It is essential that mobile web sites are able to render their information on the various mobile devices. One approach is the use of *transcoding and mobile-browser detection* which adapts the mobile web site code to display on various mobile devices. HTML code is added to the web pages which identify, first, that a mobile device is accessing the page, and second, which specific mobile device is in use (Krum, 2010). The use of transcoding and mobile-browser detection allows the mobile marketer to help ensure that his web message is received in a readable format by the requesting user.

The emerging HTML5 standard is expected to have a positive effect on mobile marketing. HTML5 will include specifications to manage audio and video as well as geolocation services. API's will become part of the specification through the *embed* function which will dramatically improve the use of plug-ins such as Adobe Flash and JavaScript easier to handle across all devices and operating systems (W3, 2010). One mobile development platform that has endorsed HTML is an open-source engine called Webkit (http://webkit.org/). It was first developed for use on Apple devices but has since become integrated with both the Symbian and Android mobile operating systems. Development continues on the Webkit platform as it is an ongoing open source project.

Social Networking

Social networking now ranks as the fastest growing category of mobile content with an estimated 20% of users accessing social networking sites through their mobile device (Sachoff, 2010). It is estimated that by the year 2013, there will be more than 140 million social networking subscribers on mobile devices (Research, 2010). Twitter and Facebook lead the mobile social networking sites in usage. From the start, Twitter's integration with SMS messaging made it a strong mobile player. Facebook's high mobile usage speaks to the success of its mobile strategy. In fact mobile users spend more time on Facebook than personal computer users. (Warren, 2010). In addition to public social networking applications, the growth of mobile applications for internal company use is also increasing. Salesforce (www.salesforce.com) offers a

mobile version of its software tool, *Chatter*, which is used for social networking within companies. It allows employees to create and share profiles, update their status and subscribe to news feeds (Insider, 2010). LinkedIn, a professional networking site, has also made its application available to mobile users after tracking over 100,000 users a month accessing their traditional Web site from their mobile phone (Havenstein, 2008).

One type of social networking that is gaining in popularity is mobile dating. This capability is enhanced by the new video chatting capability available on 3G networks. It is predicted that the mobile dating and chat room market will grow to nearly $1.4 billion by the year 2013 (Wauters, 2009). Users of the dating services provide information about themselves in profiles that are stored on the phone and are assigned a dating ID or username. Users can then search for other IDs based on age, gender and sexual preference. Some mobile dating sites used proximity services such as Bluetooth to identify users when another user of the same dating service is nearby. Mobile dating is gaining popularity over online Internet dating particularly in Europe and Asia.

Games and Applications

The use of mobile devices has moved away from their original voice and messaging function. Most mobile users are now looking to their devices for entertainment. This has resulted in an explosion of mobile games. Some users will even base their selection of the mobile device on how well it handles gaming (Fung, 2010). It is predicted that revenue generated from mobile gaming will exceed $11 billion by the year 2015 ((Reisinger, 2010).

"Games are an ideal way to integrate mobile into an advertising campaign .... Games fill the whole screen and offer advertising without competition for user attention. Gaming is also attractive since it is a leisure time activity when users are more receptive to marketing messages.

"(Durrell, 2010, p. 1). Mobile game marketing usually occurs in one of two ways. In some cases the company develops an entire game around its product similar to the Coca-Cola Thumbs-Up Everest game launched in India. Mars Inc has developed a mobile marketing campaign around the Snickers candy bar that incorporate a mobile game with more traditional in-store marketing (Butcher, Snickers ties first branded mobile game to in-store marketing, 2010). Other companies use product placement within mobile games to develop brand and product recognition. One example of effective product placement would be in a sports game where players would wear jerseys of a specific manufacturer, the players drink a certain sports drink, or the scoreboard would include the logo of a company. "Ads like these are sure to catch our attention and yet they don't disturb us in any way." (Advertising, 2007)

Due to the lack of size and processing capability, mobile games are generally small in scope and rely more on game play over graphics. Most mobile games are single player although several multi-player mobile games have been released that rely on the use of proximity technologies such as Bluetooth. Mobile devices are also being used to access mobile gaming environments such as Microsoft's Xbox Live games so that users can continue their game play from console to mobile device. The most common form of mobile game delivery is through download from application stores.

One category of mobile games use location identification devices such as GPS and are referred to as location-based games. These mobile games integrate the player's geographical position and movement into the game play. One well known location-based game is *geocaching* in which players search for buried treasure based on longitude and latitude parameters. The most common treasures are logbooks in which the geochacher who finds the treasure records their name and date. Treasures can also include small toys or trinkets. Geocaches are currently located in over a hundred countries with an estimated 1.2 million active geocaches (www.geocaching.com/seek/).

It is important for a mobile marketer to realize that there are millions of mobile games available for download and that it is essential to effectively distinguish your game through its title and description. The most effective way to expand the use of mobile games is through word of mouth. Therefore it is important to have the mobile game recognized and talked about on channels such as mobile game review and blog sites in order to increase its adoption.

MOBILE MARKETING STRATEGIES

Mobile marketing is often used in combination with other marketing campaigns on television, print or the Web. This is referred to as *integrated marketing*. "One of the most important parts of an integrated, participatory mobile marketing campaign is to ensure that no part of your marketing effort stands alone." (Krum, 2010, p. 226). One way to inform consumers of the availability of mobile campaigns is through traditional web sites or print ads where information about the mobile site is offered. Integrating the use of CSC codes in this effort allows the user to send a text message to a certain number and to receive a link to the mobile website in return. Another approach is through the use of QR codes on billboards or other print formats. By taking a picture of the QR code, the user will receive a response that can be as simple as text or as sophisticated as a link to an augmented reality application.

Integrated marketing has taken on a new dimension through it use in mobile gaming. As mentioned earlier, mobile gaming is in high demand by mobile users, with the continued growth expected in the future. Mobile users play games generally when they have free time and are relaxed. This is an ideal time for marketing. One approach to integrated mobile marketing with games is to develop an entire game built around product recognition. Other approaches include product placement within the game environment. Although neither of these approaches will net an immediate purchase, they are effective in increasing brand awareness and loyalty.

Integration of mobile marketing with social network sites is another effective approach to increase brand awareness and loyalty. Through the use of *pages,* companies are able to communicate with potential customers and offer specials such as coupons or contests. It was anticipated that ads on social network sites would dramatically increase sales. This promise has not been fulfilled (Teo, 2010). However, social networking sites remain one of the best ways to building brand awareness and loyalty which are the first steps in successful marketing campaigns.

Personalization is the key to mobile marketing. Mobile devices provide marketers with information about users' identity, preferences and location. This combination of information allows mobile marketing campaigns to target users on an individual basis even using time of day as a differentiating factor. It is predicted that smart agents, which are small artificial intelligence programs, will soon be used on mobile devices to identify mobile sites and offerings of interest to the mobile user based on their prior use (Wertime, 2008).

Another upcoming strategy is to integrate voice messages with mobile marketing. This can be delivered through the more traditional *IVR (Interactive Voice Response)* technology or designed to send individual unique voice messages to users based on their personal data and preferences. "Compared to the 160-character limitations of SMS, IVR is a phenomenal way of delivering targeted content." (Bowser, 2009). Some believe that the use of voice in mobile marketing provides a higher quality customer experience and will result in higher response rates (Wertime, 2008). Of course, IVR and voice technology can also be successfully integrated into CSC campaigns where users receive a toll free number to call when they text to the CSC number.

The entry of television and video into the mobile world has also provided new marketing opportunities. *IPTV (Internet Protocol Television)* allows traditional television shows and movies-on-demand to be sent over the Internet rather than traditional radio waves or cable. Sites such as Hulu (www.hulu.com) allow users to view television shows and videos on demand which include a brief one to two minute advertisement. Additionally, IPTV allows the streaming of live television shows to mobile devices. A functionality of IPTV of interest to mobile marketers is the ability of IPTV providers to track which users watch which television shows or movies (Murphy, 2008). Mobile marketers can also make the IPTV user experience more interactive by adding review and voting abilities (Wertime, 2008).

Mobile marketing relies a great deal on customers sharing information with other customers. This is referred to as *viral marketing.* "The beauty of viral marketing is that the customer both passes on the message and identifies the target market." (Wertime, 2008, p. 141). One approach to increasing viral marketing is through integration with existing social network sites such as Facebook. Social network sites allow users share information with others such as company pages or offerings that they like. Additionally, the use of microblogging sites such as Twitter also allow the spread of viral information as users *tweet* and *retweet* information of interest to other Twitter users (MJelly, 2009). Another approach is to attach an identifier to the link of a downloaded game, ringtone or other message. When users share these applications with others, the source company is identified through the link.

MOBILE MARKETING CHALLENGES

There are several significant challenges facing mobile marketing. The first is the constant proliferation of new mobile phones, devices and operating systems. The mobile device market is very dynamic with new smart phones and devices being introduced on a regular basis, each offering new services and functionality. In order to effectively market to a wide audience, a mobile marketing campaign has to reach as many potential customers as possible. The constant flux in mobile devices makes that goal challenging. Additionally, there continue to be various mobile operating systems in use with no evidence of any upcoming convergence. The emerging HTML5 standard may solve some mobile web site issues, but the challenge of effectively rendering information on a constantly changing landscape of mobile devices remains significant. Mobile marketers must find the balance between presenting highly engaging multimedia content such as augmented reality against communicating with the widest audience, many of who are still reliant on text based communication.

Mobile E-Commerce

Another challenge facing mobile marketers is the integration of payment systems in their marketing campaigns. Mobile payment systems have grown in use in Asia, Africa and Europe but the United States lags behind in the use of this technology. Small payments, referred to as *micropayments,* are generally under $5 and used for applications such as ringtones. Charges for micropayments usually occur as additions to the monthly phone bill. It is the handling of larger purchases and payments, referred to as *macropayments,* which pose most issues for mobile marketers and consumers. There are two approaches to macropayments currently in use. The first approach allows mobile users to create pre-paid accounts with companies that can be used for purchases through the mobile device. The second approach is for consumers to enter their banking or credit card information on their mobile device and use it for purchases.

Prepaid services can be delivered to the mobile device in several ways. In some countries, prepaid services such as bus or subway passes are down-

loaded to the device and tied to the attached RFID chip (Krum, 2010). Similarly, prepaid accounts to be used on vending machines have also become popular around the world. Purchases of tickets or other services usually result in the download of a bar code which is scanned at the point of entry. Retail purchases can also be made in the same way with the purchase redeemed through the use of bar codes. It is important to recognize that one limiting constraint of the use of downloaded bar codes is the lack of scanning technology at stores and other businesses. Airlines are also beginning to use mobile boarding passes for their customers. When customers use the online check-in feature, they receive an email with a two dimensional bar coded boarding pass that is scanned at the airline when they arrive for their flight ((Pawlowski, 2010). Banks are also offering services that allow their customers to make payments associated with their accounts through their mobile phone. Mercantile Bank of Michigan has introduced their *MercMobile Personal Payment* service. Through this service, a customer sends a text message to the bank indicating the email address or other mobile identification of the recipient of the payment. The payment is then generated from the bank and the customer receives a verification email (Michigan, 2010).

One area of concern for mobile users is that in all cases of micro and macropayments, the financial and user identification information remains with the mobile device and, if lost, would provide access to this personal data and purchased services. One approach to manage this concern is the use of "remote kill" features on phones. This would allow users to remotely wipe out all data on their mobile devices should it be stolen or lost (Krum, 2010).

Privacy

Another challenge to mobile marketing is the risk to consumer privacy though invasive mobile marketing campaigns that collect vast amounts of data about mobile user behavior. Mobile device providers can provide almost unlimited amounts of information about mobile users including the services they use, applications they have downloaded, where they have been and when, and which web sites they have visited. The use of mobile *cookies,* which are short pieces of code downloaded to the mobile device, allow companies to track the individual practices of potential customers (Krum, 2010).

In order to respect mobile user privacy, none of this information should be used in mobile marketing campaigns without specific user agreement. Mobile users should be required to opt-in to a marketing campaign. During the opt-in process, they should be made aware of what information will be captured through their participation, and how it will be used. It is also important to inform the user if the data collected with be sold to other third parties. Mobile users should also have the right to opt-out at any time which would signal the end of any related data collection. There is currently proposed legislation to allow a Do Not Track list for users who do not want their online activities tracked (Commission, 2010)

One particular area of concern in mobile marketing is protecting children and teens from unsuitable content. This can be a difficult challenge to meet as often the only information requested on questionable sites is a date of birth. The European Framework for Safer Mobile Use for Children and Teenagers provide a model for mobile carriers and marketers. Features of this framework include additional checks on user ages at times of purchase or download, the ability of parents to block mobile phone content, and efforts to label content available through the mobile device as adult or otherwise (GSMA, 2010).

Spam

Spam generally involves the indiscriminate transmission of email or other messages to a wide audience, regardless of their interest in the message content. There are several laws and regulations around the world that attempt to regulate the sending of spam messages. In the United States SPAM is regulated through the CAN-SPAM Act enacted in 2003. Although this law covers traditional email messages, it does not cover SMS text or other types of mobile messaging (Krum, 2010). In Europe, the Privacy and Electronic Communication Regulation requires that marketers specifically obtain opt-in permission from mobile users before they send a marketing message (Ibid).

In order to retain a positive reputation with potential customers, it is important that mobile marketers do not allow their communications to be seen as spam. One way to avoid this is to only send marketing messages to those mobile users who have requested information or opted in through a prior message. A Do Not Email or Do Not Spam Registry was considered by the FTC (Federal Trade Commission) several years ago, but to date no action has been taken by the Legislature. (Commission, 2004).

MOBILE MARKETING CODE OF CONDUCT

Much of mobile marketing involves using the newest mobile technology in novel and innovative ways. As a result, specific guidance on mobile marketing practices is hard to find. However, the Mobile Marketing Association (www.MIMAglobal.com) has established a code of conduct that can be applied universally to all mobile marketing efforts. This code can be used to balance the rights of the user with the goals of a mobile marketing campaign. The key elements of the Mobile Marketing Association Code of Conduct are as follows (Association, 2008):

- *Choice and Consent.* Mobile marketers respect the rights of users to determine and control the messages they receive. Users must explicitly opt-in to a marketing campaign through an email, text or voice message. Users must also have the rights to stop receiving messages when they so desire. Users must be supplied with an opt-out option that will immediately terminate the receipt of messages.

- *Customization.* Mobile marketers must take every step necessary to ensure that data collected for the purpose of a marketing effort is "handled responsibly, sensitively, and in compliance with application laws." (p. 1).
- *Constraint.* Mobile marketers should only deliver messages and communications to users who have expressly stated interest in receiving them. This element also requires that content delivered to the user should be of value. Examples include coupons, service reminders, discount, entertainment news or other requested information.
- *Security.* Mobile marketers are expected to take every action necessary to protect user information collected during marketing efforts from unauthorized access, use or disclosure.
- *Enforcement and Accountability.* Mobile marketers are expected to self-police their actions against this code of conduct.

CONCLUSION

It is clear that mobile technologies have dramatically changed the landscape of communication and entertainment. The number of mobile devices in the world currently outnumbers personal computers and this trend is expected to continue. The mobile device is unique in that it is intensely personal and travels with the user throughout their days. It is used for both personal and business communication email and text messaging. It has become a primary device for entertainment including games, television and video. The technical capabilities of the mobile device also allow for unique immersive experiences such as those offered by augmented reality applications. As the world of mobile technology has expanded, so has the world of mobile marketing. Utilizing the various technologies available in mobile devices, mobile marketers are able to establish very personal marketing messages to users based on their preferences, demographics and location information. The information presented to users can be as simple as text messages and email or include various multimedia components including sound, video and images. The challenges are many to the mobile marketer as the technologies are still in constant flux. However great the challenges are, mobile marketing presents a unique opportunity to market in ways not yet possible. It is an exciting time for mobile marketing!

REFERENCES

Ahonen, T. (2010). An inconceivable truth: MMS is a global success at 30B dollars. Retrieved on December 16, 2010, from http://communities-dominate.blogs.com/brands/2010/06/an-inconceivable-truth-mms-is-a-global-success-at-30b-dollars.html

Becker, M. A. (2010). *Mobile marketing for dummies.* Wiley Publishing, Inc.

Bowser, M. (2009). *IVR - A marketer's dream.* Retrieved on December 18, 2010, from http://www.mobilemarketingmagazine.co.uk/content/ivr-marketers-dream-no-really

Butcher, D. (2010). *7 key trends mobile marketers need to know.* Retrieved on December 15, 2010 from http://www.mobilemarketer.com/cms/news/research/7342.html

Butcher, D. (2010). *Snickers ties first branded mobile game to in-store marketing.* Retrieved on December 17, 2010 from http://www.mobilemarketer.com/cms/news/gaming/5468.html

Cassavoy, L. (2007). *In pictures: A history of cell phones*. Retrieved on December 16, 2010, from http://www.pcworld.com/article/131450/in_pictures_a_history_of_cell_phones.html

Cassela, D. (2009). *What is augmented reality?* Retrieved on December 16, 2010, from http://www.digitaltrends.com/mobile/what-is-augmented-reality-iphone-apps-games-flash-yelp-android-ar-software-and-more/2/

CBS News. (2010). *Number of cell phones worldwide hits 4.6B*. Retrieved on December 15, 2010, from http://www.cbsnews.com/stories/2010/02/15/business/main6209772.shtml

Cellphone Advertising. (2007). *Product placement in mobile phone advertisement*. Retrieved on December 17, 2010, from http://www.cellphone-advertising.com/product-placement-in-mobile-phone-advertising/

Channel Insider. (2010). Salesforce chatter social networking goes mobile. Retrieved on December 16, 2010 from http://www.channelinsider.com/c/a/Cloud-Computing/Salesforce-Chatter-Social-Networking-Goes-Mobile-443229/

CTIA. (2011). *Basics of CSC FAQs*. Retrieved on December 16, 2010 from http://www.ctia.org/business_resources/short_code/index.cfm/AID/10341.

Durrell, J. (2010). *Mobile game marketing*. Retrieved on December 17, 2010 from http://mmaglobal.com/articles/mobile-game-marketing-greystripe

Durrell, J. (2011). *Mobile game marketing*. Retrieved on December 16, 2010, from http://mmaglobal.com/articles/mobile-game-marketing-greystripe

eMarketer. (2009). *Two-thirds of kids and teens now mobile*. Retrieved on December 19, 2010 froom http://www3.emarketer.com/Article.aspx?R=1007780

Entner, R. (2010). *Smartphones to overtake feature phones in U.S. by 2011*. Retrieved on December 19, 2010, from http://blog.nielsen.com/nielsen-wire/consumer/smartphones-to-overtake-feature-phones-in-u-s-by-2011/

Federal Trade Commission. (2004). Sham site is a scam: There is no national do not e-mail registry. Retrieved on December 18, 2010, from http://www.ftc.gov/opa/2004/02/spamcam.shtm

Federal Trade Commission. (2010). *FTC testifies on do not track legislation*. Retrieved on December 18, 2010, from http://www.ftc.gov/opa/2010/12/dnttestimony.shtm

Fung, L. (2010). *Marketing mobile games*. Retrieved on December 17, 2010, from http://www.selfgrowth.com/articles/Marketing_Mobile_Games.html

GSMA. (2010). *European framework for safer mobile use by younger teenagers and children*. Retrieved on December 18, 2010 from http://www.eubusiness.com/topics/telecoms/gsma.10-06-09/

Hager, F. (2006). *Mobile communications*. Retrieved on December 18, 2010, from http://www.fredhager.com/index.asp?CategoryID=67&SubCategoryID=587&ContentID=1047

Havenstein, M. (2008). LinkedIn social networking goes mobile. Retrieved on December 16, 2010 from http://www.cio.com/article/187401/LinkedIn_Social_Networking_Goes_Mobile

Hollerer, T. A. (2004). Mobile augmented reality. In Karimi, H., & Hammad, A. (Eds.), *Telegeoinformatics: Location-based computing and services*. Taylor and Francis Books Ltd.

Interactive Blend. (2010). *QR codes: The future of marketing*. Retrieved on December 16, 2010 from http://interactiveblend.com/blog/interactive/qr-codes/

International Telecommunications Union. (2005). Cellular standards for the third generation: The ITU's IMT-200 family. Retrieved on December 15, 2010, from http://www.itu.int/osg/spu/imt-2000/technology.html#Cellular%20Standards%20for%20the%20Third%20Generation

Kee, T. (2010). *4 ways that 4G will impact mobile marketing in 2011*. Retrieved on December 16, 2010, from http://econsultancy.com/us/blog/6965-4g-or-not-4g-four-ways-it-will-impact-mobile-marketing-in-2011

Krum, C. (2010). *Mobile marketing: Finding your customers no matter where they are*. Indianapolis, IN: Que.

Lenhart, A. (2010). *Teens, cell phones and texting*. Retrieved on December 19, 2010, from http://pewresearch.org/pubs/1572/teens-cell-phones-text-messages

Membridge. (2008). *History of cell phones*. Retrieved on December 16, 2010 from http://www.historyofcellphones.net/

Merc Bank Michigan. (2010). *MercMobile® personal payments*. Retrieved on December 19, 2010, from https://www.mercbank.com/personal/electronic/p2p.asp

Miller, B. (2009). *RFID technology being added to mobile marketing campaigns*. Retrieved on December 16, 2010 from http://blog.armoryideas.com/2009/06/11/rfid-technology-being-added-to-mobile-marketing-campaigns/

MJelly. (2009). *7 viral marketing tactics for mobile internet services*. Retrieved on December 18, 2010, from http://blog.mjelly.com/2009/01/viral-marketing-on-mobile.html

Mobile Augmented Reality. (2010). *The absolute latest in Android and iPhone augmented reality*. Retrieved on December 16, 2010, from http://www.mobileaugmentedreality.info/

Mobile Marketing Association. (2008). *Code of conduct*. Retrieved on December 18, 2010 from http://www.mmaglobal.com/codeofconduct.pdf

Mobile Marketing Association. (2010). *Consumer best practices*. Retrieved on December 18, 2010, from http://www.mmaglobal.com/codeofconduct.pdf

Mobile Marketing Watch. (2010). *Research: Mobile proximity marketing to reach $750M by 2011 and nearly $6B by 2015*. Retrieved on December 16, 2010, from http://www.mobile-marketingwatch.com/research-mobile-proximity-marketing-to-reach-750m-by-2011-and-nearly-6b-by-2015-10252/

Mohr, W. (2002). *Mobile communications beyond 3G in the global context*. Retrieved on December 15, 2010, from http://www.cu.ipv6tf.org/pdf/werner_mohr.pdf

Murphy, D. (2008). *It's as easy as IPTV*. Retrieved on December 18, 2010, from http://www.mobile-marketingmagazine.co.uk/content/its-easy-iptv

Naish, J. (2009). *Mobile phones for children: A boon or a peril?* Retrieved on December 19, 2010, from http://women.timesonline.co.uk/tol/life_and_style/women/families/article6556283.ece

Nations, D. (2010). *A list of mobile web browsers*. Retrieved on December 16, 2010, from http://webtrends.about.com/od/mobileweb20/tp/list_of_mobile_web_browsers.htm

Neustar. (2010). *CSC implementation: A mobile marketing plan*. Retrieved on December 16, 2010, from http://www.scribd.com/doc/21139025/CSC-Implementation-a-Mobile-Marketing-Plan

Nielsen. (2010). *Nielsen unveils retail 2015 forecast*. Retrieved on December 15, 2010, from http://www.nielsen.com/us/en/insights/press-room/2010/nielsen_unveils_retail.html

O'Rourke, J. (2007). *US SMS penetration by age group*. Retrieved on December 19, 2010, from http://psmsus.blogspot.com/2007/01/us-sms-penetration-by-age-group.html

Pawlowski, A. (2010). *Paperless boarding takes off at United*. Retrieved on December 19, 2010, from http://articles.cnn.com/2010-03-15/travel/mobile.boarding.passes_1_boarding-airport-gates-passes?_s=PM:TRAVEL

Pollard, S. (2008). *Mobile email marketing tips*. Retrieved on December 16, 2010, from http://www.lyris.com/resources/email-marketing/articles/mobile-email-marketing-tips/

Reisinger, D. (2010). *Mobile game revenue to top $11 billion by 2015*. Retrieved on December 17, 2010, from http://news.cnet.com/8301-13506_3-20024103-17.html

Research, A. B. I. (2010). Online social networking goes mobile: 140 million users by 2013. Retrieved on December 16, 2010, from http://www.abiresearch.com/press/2998-Online+Social+Networking+Goes+Mobile%3A+140+Million+Users+by+2013

Sacco, A. (2007). *A brief history of the mobile phone (1973-2007)*. Retrieved on December 16, 2010, from http://advice.cio.com/al_sacco/a_brief_history_of_the_mobile_phone_1973_2007?page=0%2C0

Sachoff, M. (2010). *Mobile social networking grows 240%*. Retrieved on December 16, 2010, from http://www.webpronews.com/topnews/2010/06/02/mobile-social-networking-grows-240

Teo, L. (2010). *Why Facebook ad click through rates suck – And how to change that*. Retrieved on December 19, 2010, from http://www.ymarketing.com/blog/bid/49898/Why-Facebook-Ad-Click-Through-Rates-Suck-And-How-To-Change-That

TMC. (2007). *The history of SMS messaging*. Retrieved on December 15, 2010 from http://www.tmcsms.com/sms-history.aspx

W3. (2010). *HTML5 differences from HTML 4*. Retrieved on December 18, 2010 from http://www.w3.org/TR/html5-diff/

Wallace, L. (2009). *Blink-182 rocks augmented reality show in Doritos bag*. Retrieved on December 16, 2010 from http://www.wired.com/underwire/2009/07/blink-182-rocks-augmented-reality-show-in-doritos-bag/

Warren, C. (2010). *Mobile social networking usage soars*. Retrieved on December 16, 2010 from http://mashable.com/2010/03/03/comscore-mobile-stats/

Wauters, R. (2009). *There's money in mobile dating*. Retrieved on December 16, 2010, from http://techcrunch.com/2009/01/19/juniper-research-theres-money-in-mobile-dating-services/

Wertime, K. A. (2008). *DigiMarketing: The essential guide to new media and digital marketing*. John Wiley& Sons.

Wireless Internet. (2010). *What's this about 4G?* Retrieved on December 16, 2010 from http://www.wirelessinternet.org/4G-network.php

Chapter 5
Mobile Government and Defense

Jim Jones
Ferris State University, USA

ABSTRACT

The Government, Military, and Intelligence communities of the United States and other countries are adopting mobile technologies almost as quickly as commercial entities, and in some cases are going beyond the applications we see in the commercial space. Government services, such as information access and certain transactions, are rapidly adopting mobile delivery mechanisms. The military is using mobile technology to share static information as well, but is also providing live data feeds and information sharing to support combat operations. Intelligence agencies are using mobile devices as a data collection platform for their own agents, and are also accessing the mobile devices of enemy agents and intelligence targets to collect data surreptitiously. Military operations face unique challenges, given that they are often conducted in regions without existing networks and against an enemy trying to actively disrupt communications. The Government, Defense, and Intelligence communities all face the challenge of securing mobile devices and data in response to regulatory and statutory requirements, as well as a dynamic and evolving threat space of identity thieves, hackers/crackers, hostile military forces, and foreign intelligence services.

INTRODUCTION

This chapter considers applications of mobile technologies in the Government, Defense, and Intelligence communities. Government entities are actively moving services into the mobile domain,

as users demand such access and the technology enables it. The Defense community is delivering similar services, but is also actively expanding mobile device services to provide real-time support for combat and other military operations. The Intelligence community, with its unique mission to collect and analyze data from all available sources, is actively exploiting mobile technology

DOI: 10.4018/978-1-61350-150-4.ch005

as both a deliberate and surreptitious collection platform. Military operations often occur in areas without established network infrastructures and in the vicinity of hostile forces, creating a unique challenge to stand up and defend network infrastructures on demand. All of the communities, Government, Defense, and Intelligence, face the challenge of protecting mobile device data in transit and at rest.

GOVERNMENT SERVICES

Description and Mission

Consider the elements of the United States or another government (federal, state, and local) which provide direct[1] services to citizens and residents. With the exception of the profit motive, these organizations function like many commercial enterprises. Specifically, both commercial and government entities need to conduct bidirectional informational and financial transactions with their "customers". While the two types of entities operate under different sets of laws and regulations, both have requirements to protect information and transaction data, to authenticate users, to provide reliable services, etc. - in short, to protect the confidentiality, integrity, and availability of the information and systems relevant for the services provided. In the examples which follow, we discuss applications of mobile technologies to government services and the relevant assurance criteria, i.e., authentication (none, one, or both parties), transaction security (protected or not), and availability (critical or not). After discussing several current and potential uses of mobile technology for providing government services, the section concludes with a discussion of relevant challenges.

Applications of Mobile Technology

Perhaps the most obvious application of mobile technologies for government services is providing public information such as government office locations, hours of operation, directions, phone numbers, etc. (Harvard, 2003). Such a service does not require strong authentication of either party, nor transmission security, nor high availability. Failure of the service along any of these lines would be inconvenient, but would not violate any legal or regulatory requirements. Most government entities currently provide this information via the web (as well as phone-based 311 services), so the basic service exists if a user simply chooses to use a mobile device browser (or phone) instead of a PC browser (or land line phone). Increasingly, government entities are developing specific mobile applications which provide additional capabilities unique to a mobile device. For example, directions to an office may be directly downloaded to a mobile device navigation app[2], a phone call can be automatically dialed, or a text message sent with information.

In a similar fashion, government entities which provide online (web-based) transaction processing for things like driver's license renewal, vehicle tag renewal, and tax payments are increasingly providing mobile device versions of these services. Since private and possibly financial information is involved, such transactions require user authentication and transmission security. Bidirectional authentication may be optional - while the government entity should be strongly authenticated (to prevent phishing and similar attacks), there is limited value in strongly authenticating each user (are you really going to pay someone else's taxes for malicious purposes?). As with information-only services, high availability is not typically required for these services. A procrastinating user might disagree, but experience with the non-mobile versions of these services has shown that some downtime is acceptable (with the possible exception of the hours leading up to a tax filing deadline). As noted in Nichols (2010), not all services that work on the web are suitable for mobile devices. For example, completing and filing a complicated tax form might be cumbersome on a mobile device, while simple tasks like renewing your vehicle tags might be more suitable. Nichols also reports successful implementation and high usage of mobile apps for financial transactions like inmate trust account deposits and restitution payments.

Law enforcement activities provide a fertile ground for mobile technology. For example, several cities have already implemented crime or tip reporting via mobile apps (Kawano, 2010). Making use of Location Based Services or LBS (GSM, 2003), a city in The Netherlands reportedly considered recruiting city workers to help search for individuals and automobiles of interest to police (Benedict, 2008). The idea was that if an individual or automobile was suspected to be in a particular area (e.g., a bank robbery had just been committed in the area), then all city workers in the area would be notified on their mobile devices to look for the individual or vehicle and presumably to report back using their mobile devices. Location Based Services are critical to this concept in order to limit the message recipients to only the geographic area of interest, and such a notice could be updated in real time based on city worker movement and suspected perpetrator movement. A similar approach could be used for Amber Alerts (Justice, 2010), using government employees or even the population at large.

On a different note, the U.S. Court system currently provides databases of certain offenders and their residence or work locations. For certain offenders (e.g., crimes against children), the information is publicly available. For other offenders, only the victim has access to such information. LBS could be used in two ways: first, a user with a mobile device and LBS could query the system to identify nearby offender addresses (based on the user's location); second, the offender could be required to carry a mobile device, and his or her current location could be integrated into the data provided to the user. Admittedly, requiring an offender to carry a mobile device and verifying that they are co-located might be difficult. However, an automated system might randomly contact the offender and require that he/she take a picture of themselves with the mobile device and send it to the monitoring authority. With associated device GPS data, such offender "check ins" could be incorporated into the public user's query, and the messages might also be archived for later use if necessary (e.g., parole review hearing, crime investigation, etc). A similar approach might be used to implement house arrest, curfews, personal protective orders, and geographic restrictions for offenders. For the concepts discussed here, different levels of security would be required; some information is protected privacy data, and other information would be useful to a criminal.

Bar code apps are available for nearly all mobile devices with a camera (the app interprets the picture as a barcode rather than reading the barcode with a laser) (Waters, 2010)(Mobile Barcodes, 2010). The commercial sector quickly understood the potential for barcode-based user data retrieval, but the government sector has been slower to make use of this mobile device feature. Nevertheless, opportunities do exist for government services, such as CPSC[3] product recalls and FDA[4] alerts. Imagine a parent in a store scanning the bar code of a baby crib to check for recalls, or a grocery shopper scanning the bar code of an egg carton to confirm that the batch on the shelf is not under a current recall. The same actions might be taken on products and food already in a person's home (or day care center, or restaurant food supplies, etc.). Implementation requires (a) building the data-driven association between a barcode and the relevant data, and (b) making the data available on a mobile application. In most cases, a data association already exists between some product identifier and the data, but the product identifier may not be a barcode (and this is where a lot of the work would need to focus).

App development should be straightforward as existing commercial apps can be quickly adapted. The security requirements for barcode and related applications would be minimal, assuming that all data is meant to be publicly available. Following the customer safety theme (but not based on bar codes), a government entity could make restaurant health inspector reports available via a mobile app, possibly using LBS or driven by a user query.

Returning to the 311 theme (information exchange with users), several cities have implemented user-reporting mobile devices apps for items such as potholes, traffic light issues, graffiti, etc. (Klimas, 2010). Similarly, the city of Grand Rapids, Michigan, is employing citizens to help inventory and track the city's trees (Bennett, 2010). Users can look for trees, add trees to the inventory (including pictures), or report on tree health. Services such as these have minimal security requirements and may or may not make use of mobile device LBS capabilities.

Challenges

Providing government services over mobile devices, especially where privacy or financial data is involved, face the same challenges as similar applications in the commercial space. Specifically, the user must be assured that they are connected to the legitimate site and that data is protected in transit. As with any transaction involving privacy or financial data (mobile or not), the data must also be protected at rest. This is not a unique problem on the server (provider) side, but it is a new challenge on the client (mobile device) side. Home computers have at least nominal physical protections, but mobile devices do not. They are easily stolen or lost, and are often used in public spaces where keystrokes and screens are visible to unintended parties.

Malicious applications posing as legitimate apps will continue to be a problem. Consider an attacker who posts an app to ostensibly renew your driver's license using your mobile device. The malicious app might actually forward all information to the legitimate site and actually renew the license, but the attacker has executed a classic man-in-the-middle (MITM) attack, gaining access to any information sent or received in the transaction (credit cards, privacy information, mailing address, etc.). MITM attacks are particularly insidious because the traffic may be encrypted between the user and the attacker, and between the attacker and the legitimate site, giving the user a sense of security. However, this sense of security is misplaced, since the attacker has unencrypted access to all data at the relay (middle) point.

Finally, a user may submit false information to further their own ends. This challenge is common to many information sharing mechanisms but is at least as bad for mobile apps due to anonymity and forgery concerns.. Consider an Amber Alert which goes to the general public. A perpetrator might use a stolen or throwaway phone to report a false sighting, leading law enforcement off track. Similarly, a burglar might make multiple false reports in one part of town to distract law enforcement while committing a crime in another part of town. Previously I mentioned that user authentication was probably not required for a tax payment app since others are unlikely to pay your taxes for you. However, a malicious individual could pay your taxes with a stolen credit card or use a bank account with insufficient funds, creating trouble for you. This is a problem regardless of the means (mobile app or not), but mobile apps make it easier to impersonate someone and to maintain anonymity for the perpetrator. Finally, any information collection operation involving the public suffers from miscreants who simply submit false or misleading data for their own amusement.

DEFENSE OPERATIONS

Description and Mission

The mission of the United States Department of Defense is to "provide the military forces needed to deter war and to protect the security of our country" (Department of Defense, 2010). The U.S. military has a long history of mobile device development and usage, but these devices have historically been designed and used for single purposes (e.g., radios, cameras, sensors, etc.). As evidenced by their research efforts (DARPA, 2010), the military is actively moving in the direction of general-purpose mobile devices and the use of existing as well as custom developed apps for those devices.

Applications of Mobile Technology

A persistent problem in military operations is getting information to and from soldiers in the field. Generally, military bases with high bandwidth and processing capabilities are established within an operational theater, but getting the information available at such bases to the soldiers in the field (i.e., the "last mile") is difficult. These soldiers might be on patrol, reconnaissance, or rescue missions away from the base and without the ability to carry heavy gear. In 2009, Raytheon released the " Raytheon Android Tactical System", or RATS, (Raytheon, 2009) to solve some of these problems. As the name implies, RATS is an Android based mobile device with apps to provide bidirectional data flow for a soldier in the field. Example applications include streaming video, live mapping applications, and communications with other soldiers and facilities. Streaming video might include feeds from Unmanned Aerial Vehicles (UAVs) (Hoover, 2009), and mapping applications can geolocate friendly and hostile forces on a live map and permit annotation and sharing of map data. Geolocation could use the device itself to identify and pinpoint friendly forces carrying the

devices, and integrate data from other sources as well. Communication between soldiers and other facilities using the devices is limited only by the devices themselves and network limitations. Proposed additional applications include reading and uploading license plates (Raytheon, 2009), biometric data collection (including facial recognition), and live querying of intelligence databases. One can imagine a soldier on the streets of Iraq or Afghanistan at an automobile checkpoint. The license plate is scanned and uploaded, pictures or fingerprints of the car occupants are taken and uploaded, and intelligence databases return any potentially useful information. Simultaneously, the location of the car and its occupants at that point in time are recorded for immediate or later use by other military or intelligence agents. If a search is warranted, items in the car or in the possession of occupants (commonly called "pocket litter") can be recorded. Such seemingly minor data points (e.g., a restaurant matchbook or partial phone number) may contribute to understanding a larger operation. Mobile technology facilitates the effective field collection of such data points, as well as access to data points collected by others and access to relevant and more complete intelligence products.

Another Android application aims to provide real-time language translation for soldiers in the field (Gibb, 2010) (Vasdev, 2010). The application is a joint effort between DARPA (Defense Advanced Research Projects Agency) and NIST (National Institutes of Standards and Technology) and is intended to go beyond currently available language translation apps like Google Translate (Google, 2010) and Jibbigo (Martin, 2010). Google Translate and Jibbigo have a text-to-speech capability and can accept voice input, rendering them functionally similar to the Universal Translator of Star Trek fame. The new application was developed under the DARPA Information Innovation Office (I2O) as part of the Spoken Language Communication and Translation System for Tactical Use (TRANSTAC) program, which was launched in 2006 (DARPA, 2006). While the application is not available for public download, it reportedly has voice recognition capability and can translate between English and several Middle Eastern languages.

Less exciting but no less useful, the U.S. Army recently issued mobile devices containing maintenance, repair, and instruction manuals to soldiers at Fort Bliss in Texas (Hodge, 2010). Notably, the system administrators could push updates to the devices in the field, and the Army plans to enable feedback collection from the soldiers (which would also be pushed to other soldiers in the field). For anyone who has ever serviced a car or other complex mechanical device, the benefit of access to other mechanic's notes is well-known, and this device and app provides such access on a grand scale.

Finally, in an effort to tap the internal creative energy of the military, the U.S. Army recently conducted a competition to develop mobile apps (Army, 2010). Over 100 military and civilian Army employees participated, and 35 unique applications were submitted. The top five submissions were apps for individualized physical training, self-monitoring of mental health, map-based disaster relief tracking, route planning, and recruiting.

Challenges

Every regular cell phone user knows the frustration of losing a call in the middle of a conversation. Now imagine that instead of redialing to get the rest of the evening's plans from your friend, you are in the midst of a military operation. You might be having live chat sessions with other military teams, viewing an overhead video feed of your enemy just over the next ridge, or guiding in a helicopter on a rescue mission. In any case, the consequences of an unreliable network and/or lost signal are far greater than anything we experience in our personal lives or in the business world. To complicate matters further, our soldiers need this reliability in geographic locations that typically

do not have existing wireless infrastructures. The military is addressing these issues using portable technology which enables a broadband wireless network to be established in any location. Such networks are commonly called portable mesh networks and are developed and sold by commercial companies like Rajant (2011) for military, law enforcement, first responder, and other uses.

A related problem is the interoperability of different networks and devices. In the commercial space, most of these interoperability issues have been solved. I can transparently send and receive voice and data messages between different networks (e.g., CDMA and GSM, or cell network and land lines) without issues. But a combat environment has a more dynamic network and device landscape, and if the U.S. Army can't communicate with the U.S. Marines in a combat situation using the new mobile apps, then the advantage of having such apps is diminished. In 2009, DARPA hired Raytheon (Finley, 2011) to build a capability called MAINGATE, which stands for Mobile to Ad-Hoc Interoperable Network GATEway. MAINGATE has two purposes: first, it is a mobile ad hoc network, meaning that the network infrastructure is portable and flexible, accepting new nodes (devices) and routing traffic on the fly; second, it provides connections for, and translation between, multiple networks. In an operational environment, everyone brings their own devices (and maybe their own network), and they can all communicate via MAINGATE. As of this writing, video, voice, and data communication at various bandwidths have been successfully tested.

Assuming that a portable, interoperable network has been established in an area of operation, we still have the problem of power. It's difficult enough to find an available power outlet in a busy U.S. airport, but imagine the same challenge while patrolling the mountains of southeastern Afghanistan. Some portable power solutions (e.g., kinetic and solar) do exist, but they can be cumbersome to carry and set up. Worse still, mountainous terrain creates additional obstacles to a reliable connection, and the solution is often to boost the signal of the mobile device, requiring more power. The good news is that the portable power problem is present for commercial mobile technology as well, so significant market energy is being directed at solving this problem.

Finally, the security issues of mobile technology are exacerbated for the military. For example, civilian mobile networks are rarely attacked, yet jamming of an enemy's wireless signals, and even physical destruction of network infrastructure components, is accepted military practice. Further, while losing your credit card to a crafty wireless attacker is inconvenient, allowing your military enemy to read your wireless traffic can cost lives. Consider the revelation in 2009 (Hoffman, 2009) that the U.S. military was aware that enemy combatants could view the video feed from its Unmanned Aerial Vehicles (UAVs). While the control feed and a commander's feed were secured, the video feed typically used by troops on the ground was not, enabling enemy combatants to monitor that same feed. Some steps to secure the feed were taken immediately, but permanent fixes may take longer.

INTELLIGENCE

Description and Mission

Intelligence may be defined as "the skilled analysis of facts and inferences" (Moore, 2003). Further, we'll adopt Robert Clark's (2007, pp. 8-9) ideas that (a) the purpose of intelligence is to reduce uncertainty, and (b) we can differentiate government or military intelligence from commercial or business intelligence based on the additional data collection mechanisms available to governments and military units (e.g., surveillance, wiretaps, controlled satellite imagery, etc.). It will also be useful for our discussion to frame a basic intelligence cycle (Clark, 2007, pp. 10-11). For our purposes, I'll simplify the cycle into collection,

analysis, and dissemination. The intelligence cycle is discussed in depth in multiple sources, but this simplified three-step process will suffice for us. Finally, these definitions and distinctions work whether we're talking about the intelligence activities of the United States government or another nation-state.

Applications of Mobile Technology

First, let's consider mobile technology as a data collection platform for intelligence purposes. Intelligence operations may cast a wide net, collecting as much data as possible, or the operation may target specific information only. In either case, once collected, the data is typically transmitted back to a central location for processing and analysis. Mobile technologies create additional collection and transmission opportunities. Consider a field agent or intelligence asset[5] with a mobile device like a smart phone with camera and wireless Internet access. Imagery data can be easily collected and immediately uploaded to a remote site. This is a significant improvement over the "old days", where the person collecting data had physical possession of collected data (or a copy) for an extended period, often in a hostile environment. Imagine an employee of a foreign government who is collecting data on behalf of a U.S. intelligence agency, perhaps about a nuclear fuel processing facility. Such an employee might have a cell phone with a camera, but may not typically be allowed in sensitive areas of the facility. If access to the sensitive area is gained, the employee can take digital pictures, immediately upload them to a remote site (possibly directly to their handler[6]), and then erase the images from their cell phone. If the employee is discovered in the sensitive area, they will not be carrying incriminating evidence and can plead ignorance. Before the days of mobile devices with wireless access, such an operation would have made use of a small camera which had to be smuggled back out of the sensitive area, increasing the risk of detection for the intelligence asset. We are glossing over counterintelligence capabilities, such as forensic analysis and recovery of deleted images on a digital device, wireless signal monitoring, jamming, hijacking, etc., but it remains true that data collection and transmission are easier with mobile technology. The possibilities are expanded if we consider additional mobile device data collection methods, such as video recording, audio recording, memory card readers, and local image processing applications like bar code interpreters.

A mobile device user doesn't have to willingly participate to still be useful, if a mobile device remote monitoring application is installed on the device. While typically marketed to suspicious spouses and parents, such applications are just as useful to the intelligence community. Once installed, these applications (Spyphone, n.d.) are not visible to the device user. The applications upload call histories, images, video, locations, etc., from the device to a third party site where they can be accessed by the monitoring agent. Stories have been circulating for years that the cell phone network operators (and intelligence agencies) can enable audio and video from a mobile device, remotely, without the user's knowledge (Schneier, 2006). While tools are available for remote audio monitoring using a stealth call (FlexiSpy, 2009), confirmation of capabilities to remotely enable audio and video feeds on a mobile device is not publicly available. Nevertheless, there are no technical reasons why such capabilities would not be possible, given arbitrary software on the device (which may be embedded in the OEM[7] load), and the fact that a mobile device is essentially a radio with video and audio inputs.

Now, let's take a different view and consider mobile technologies as sources of intelligence data themselves. Rather than using a mobile device to collect intelligence data, we might collect data directly from the devices themselves or by monitoring the wireless signals. Devices may be overtly stolen to collect the resident data, but normally we prefer to copy the data from the device without the user's knowledge; the collected data retains more value if the device user is not aware that the information on their device has been accessed by someone else. Such surreptitious collections were historically called *black bag jobs*, as the data collector enters the premises like a burglar, carrying a black bag, and leaves undetected. In the modern version, technologies exist for rapid and undetectable copying of data from a mobile device, at which point the data can be uploaded to a remote site, analyzed, and combined with other data. Our data collector then leaves with an empty black bag (or none at all), since the goods have already been transmitted. The recent emergence of online backup services for mobile devices creates another opportunity for data collection, whereby the intelligence agency accesses the data as an online backup, and direct contact with the device is not required. Similarly, given brief access to a mobile device, an intelligence agent may cause a backup (or selected data) to be uploaded to a remote site, possibly to a site under the agent's control.

The wireless signals which communicate with mobile devices provide a rich source of intelligence data as well[8]. If unencrypted, such signals may be easily decoded and read, in many cases with freely available software (e.g., Wireshark (2011)), and existing hardware (e.g., another mobile device). The eavesdropper would have access to both voice and data communications, to include email, web traffic, text messages, etc. If strongly encrypted (which is increasingly common), the data still has value. For example, a powered mobile device with or without encryption can be tracked to at least the resolution of the network towers servicing the device, and with better resolution if the device's transmission signal can be triangulated. Geolocating a device can be done after the fact as well. The wireless network provider will retain some form of connection history which may be used to establish general locations over time, and the device itself may retain a time-stamped history of towers accessed, although this will be device dependent. GPS-enabled devices provide even higher resolution geolocation, as the device knows its location to within a few feet at any given time. Successfully using any of these techniques to geolocate a device is dependent on data retention (by the network provider or the device), although such retention by the network provider can be required by law enforcement via a subpoena and by the intelligence community (via the FBI) through a mechanism known as a "National Security Letter" (Electronic Frontier Foundation, n.d.). Remember, however, that geolocating a device does not necessarily geolocate the user. Monitoring a mobile device's signal or accessing the provider's records about that device can also establish communication patterns (other devices, usage statistics, organizational structure, etc.), even if the content of the communications is not available.

Returning to the idea of mobile devices as tools (rather than sources) for intelligence agencies, consider the examples from the Defense domain where data is captured locally, uploaded as the basis for a query, and information is returned to the mobile device. In a similar fashion, intelligence agents and analysts may use mobile devices to access classified and unclassified online resources, submit data and run queries, and receive real-time data feeds. A key ongoing initiative in the U.S. intelligence community is information sharing among the 17 member organizations (Office of the Director of National Intelligence, n.d.). Although specific information is not publicly available, we presume that such collaboration efforts are being implemented via mobile devices to the degree that technology and security requirements permit.

Finally, mobile devices provide a rich landscape for deception operations (Heuer, 1999, p. 115)(Tzu, 1963, p. 66). If an intelligence target is using a mobile device, disinformation content may be delivered directly to the target via normal channels (e.g., a text message, Facebook update, etc.). Further, the source of that content may be spoofed or sent from a compromised account

(Segall, 2011). Going in the other direction, an intelligence agency may intercept mobile device communications coming from a target, preventing delivery to the intended recipient, redirecting the communication, or modifying the communication prior to delivery. Looking at the bigger picture, intelligence agencies may use large-audience mobile device communications to spread disinformation, e.g., to deliberately sow unrest in a foreign country, to coordinate mass protests, etc. In 2011, Egyptian authorities disabled Internet and wireless services in Cairo because protesters were using mobile devices and social media sites to coordinate protests (Milian, 2011). It would have been more elegant (and probably more effective) had the authorities co-opted the platform and injected disinformation, sending the protesters in different directions, creating confusion and uncertainty, etc.

Occasionally, deception activities are undertaken to establish bona fides for an intelligence agent, to include the creation of people who don't really exist (Montagu, 1965). As a hypothetical example, consider an intelligence agent purporting to be a government official of a foreign country. At a face-to-face meeting, others at the meeting (the targets of the deception) might use mobile devices to access the Internet and query for the government official's biography or picture to confirm the agent's identity. By intercepting such requests, the intelligence agency can manipulate the returned data to support the deception. Similarly, content on the agent's mobile device may be constructed to support the deception (images, videos, email or text messages, call histories, contact lists, etc.).

Challenges

The primary challenge to the adoption of mobile technology in the intelligence community is the need for security. The United States and other nations have well-intentioned but cumbersome security classification systems to protect critical data, sources, and methods. These systems require assurances for handling data at various classification levels, and such assurances are more difficult to provide in a mobile device environment. Considerations include the transmission as well as the storage of classified data in environments where the communications networks may be under someone else's control, and the security of the device itself cannot be assured. Encryption and self-destruction mechanisms can help but are not guaranteed to be effective.

CONCLUSION

Mobile devices extend the reach of bidirectional communication with users in almost any environment. Services once relegated to stationary computers with local power and high speed wired connections can now be delivered on demand to a user with a device the size of their hand and no local power or communication lines. Government, Defense, and the Intelligence communities are following the lead of commercial entities with regards to mobile devices services. In cases of portable networks and mobile device exploitation, these communities are often surpassing what is being done in the commercial space.

REFERENCES

Army, U. S. (2010). *Top five apps for the army winners recognized at LandWarNet Conference.* Retrieved on January 4, 2011 from http://ciog6.army.mil/AppsfortheArmyChallengeBuilds53Appsin75D/tabid/67/Default.aspx

Association, G. S. M. (2003). *Location based services.* Retrieved on January 4, 2011 from http://www.gsmworld.com/documents/se23.pdf

Benedict, K. (2008). Mobile applications for fighting crime, reporting potholes and birdwatching. Retrieved on January 4, 2011, from http://kevinbenedict.sys-con.com/node/1215961/mobile

Bennett, S. (2010). *GEOTREE: A tree inventory web application for citizen driven urban forestry asset management.* Big Rapids, MI: Master's of Information Systems Management Capstone Project, Ferris State University.

Clark, R. (2007). *Intelligence analysis: A target-centric approach.* Washington, DC: CQ Press.

DARPA. (2006). *Spoken language communication and translation system for tactical use* (TRANSTAC). Retrieved January 17, 2011, from http://www.darpa.mil/i2o/solicit/baa/BAA-06-21_PIP.pdf

DARPA. (2010). *Mobile apps for the military.* Retrieved on January 4, 2011, from https://www.fbo.gov/index?s=opportunity&mode=form&id=6f438e64dcf6bd132987d9554442b851&tab=core&_cview=0

Department of Defense. (2010). *About the Department of Defense: Mission.* Retrieved on January 4, 2011, from http://www.defense.gov/about

Electronic Frontier Foundation. (n.d.). *National security letters.* Retrieved January 27, 2011, from https://ssd.eff.org/foreign/nsl

Finley, K. (2011). *DARPA and Raytheon building new ad-hoc mobile network for the military.* Retrieved January 17, 2011, from http://www.readwriteweb.com/enterprise/2011/01/darpa-and-raytheon-building-ne.php

FlexiSpy. (2009). *How do I use SpyCall / remote listening?* Retrieved January 28, 2011, from http://support.flexispy.com/index.php?_m=knowledgebase&_a=viewarticle&kbarticleid=6

Gibb, K. (2010). *Military testing Nexus Ones for real-time translation in Afghanistan.* Retrieved on January 16, 2011, from http://www.androidcentral.com/military-testing-nexus-ones-real-time-translation-afganistan

Google. (2010). *Google Translate*. Retrieved on January 17, 2011 from http://www.appbrain.com/app/com.google.android.apps.translate

Harvard Kennedy School. (2003). *311 system*. Retrieved on January 4, 2011 from http://www.innovations.harvard.edu/awards.html?id=3670

Heuer, R. (1999). *Psychology of intelligence analysis*. Center for the Study of Intelligence, Central Intelligence Agency.

Hodge, N. (2010). *A combat zone iPhone? Soldiers have an app for that*. Retrieved on January 4, 2011 from http://www.wired.com/dangerroom/2010/03/a-combat-zone-iphone-soldiers-have-an-app-for-that/?utm_source=feedburner&utm_medium=feed&utm_campaign=Feed:+wired/index+(Wired:+Index+3+(Top+Stories+2))

Hoffman, M. (2009). *Fixes on the way for non-secure UAV links*. Retrieved January 17, 2011, from http://www.navytimes.com/news/2009/12/airforce_uav_hack_121809w/

Hoover, N. (2009). An Android app for the military. Retrieved on January 16, 2011 from http://www.informationweek.com/news/government/mobile/showArticle.jhtml?articleID=221200035

Kawano, L. (2010). *DPD launches crime reporting app*. Retrieved on January 4, 2011 from http://www.myfoxdfw.com/dpp/news/100410-dpd-launches-crime-reporting-app.

Klimas, P. (2010). *City of Grand Rapids deploys MyGRCity311 mobile application for customer service*. Retrieved on January 4, 2011 from http://www.grand-rapids.mi.us/download_upload/binary_object_cache/frontpage_311MEDIA%20RELEASE.pdf

Martin, D. (2010). *Jibbigo: Your Star Trek universal translator for iPhone*. Retrieved January 17, 2011, from http://www.cultofmac.com/jibbigo-your-star-trek-universal-translator-for-iphone/44504

Milian, M. (2011). *Reports say Egypt Web shutdown is coordinated, extensive*. Retrieved January 28, 2011, from http://www.cnn.com/2011/TECH/web/01/28/egypt.internet.shutdown/index.html?hpt=T2

Mobile Barcodes. (2010). *QR-code reader and software*. Retrieved on January 4, 2011 from http://www.mobile-barcodes.com/qr-code-software/

Montagu, E. (1965). *The man who never was: World War II's boldest counter-intelligence operation*. Bantam Pathfinder.

Moore, D. (2003). *Species of competencies for intelligence analysis*. Retrieved January 27, 2011, from http://scip.cms-plus.com/files/Resources/Moore-Species-of-Competencies.pdf

Nichols, R. (2010). *Arkansas.gov mobile apps put payment processing on smartphones*. Retrieved on January 4, 2011 from http://www.govtech.com/e-government/Arkansasgov-Mobile-Apps-Put-Payment-Processing.html

Office of the Director of National Intelligence. (n.d.). *Members of the intelligence community*. Retrieved January 27, 2011, from http://www.odni.gov/members_IC.htm

Piike, J. (2010). *ECHELON*. Retrieved January 27, 2011, from http://www.fas.org/irp/program/process/echelon.htm

Rajant. (2011). *Resilient tactical wireless broadband networks for military applications*. Retrieved January 17, 2011, from http://www.rajant.com/solutions/military

Raytheon. (2009). *System runs on Android mobile operating system*. Retrieved on January 14, 2011 from http://www.raytheon.com/newsroom/technology/rtn09_rats/

Schneier, B. (2006). *Remotely eavesdropping on cell phone microphones*. Retrieved January 28, 2011, from http://www.schneier.com/blog/archives/2006/12/remotely_eavesd_1.html

Segall, L. (2011). *Mark Zuckerberg's Facebook page hacked*. Retrieved January 28, 2011, from http://money.cnn.com/2011/01/26/technology/facebook_hacked/index.htm

Spyphone. (n.d.). *Top spyphone reviews*. Retrieved January 28, 2011, from http://www.topspyphonereviews.com/?gclid=CPGJkazV3aYCFY64KgodSB5e1Q

Tzu, S. (1963). *The art of war* (Griffin, S. B., Trans.). Oxford University Press.

U.S. Department of Justice. (2010). *Amber alert website*. Retrieved on January 4, 2011, from http://www.amberalert.gov/

Vasdev, S. (2010). *The 21st-century mobile military*. Retrieved on January 16, 2011, from http://ndn.org/blog/2010/09/21st-century-mobile-military

Waters, A. (2010). *How barcodes and smartphones will rearchitect information*. Retrieved on January 4, 2011 from http://www.bukisa.com/articles/310287_how-barcodes-and-smartphones-will-rearchitect-information

Wireshark (2011). *Wireshark*. Retrieved January 27, 2011, from http://www.wireshark.org/

ENDNOTES

[1] *Direct* services are those in which the government entity interacts directly with the person or group receiving the service (e.g., paying taxes or applying for social security benefits); *indirect* services are those in which the government entity and the service recipient do not directly interact (e.g., national security, treaty negotiation).

[2] Programs or applications written for mobile devices are commonly called "apps".

[3] Consumer Product Safety Commission

[4] Food and Drug Administration

[5] An intelligence asset is a person who provides information to an agent of the intelligence agency. The agency's own employees are not considered intelligence assets for the employing agency (although they may be assets for another intelligence agency).

[6] In the intelligence community, a *handler* is the employee of the intelligence agency who has direct contact with the intelligence asset.

[7] Original Equipment Manufacturer, i.e., software loaded at the factory.

[8] For the U.S. intelligence community in non-US locations, this is simply an extension of the well-reported (and often misreported) ECHELON program (Pike, 2010), which monitors all available electromagnetic signals, including mobile device communications.

Chapter 6
Mobile Healthcare:
Challenges and Opportunities

Ade Bamigboye
Mobile Flow, UK

ABSTRACT

At the beginning of 2010 an estimated 68% of the world's population had access to a mobile device. Successfully placing mobile technology at the centre of any healthcare delivery service could enable innovations in healthcare to be distributed quickly, globally, and equally. For some patients this could mean being able to gain better access to healthcare where previously there was very little or none at all. For others it could mean closer, more convenient healthcare management. For medical professionals, it is about using mobile technology to deliver and manage healthcare services from where ever they happen to be. At a national level mobile healthcare promises to provide solutions to some of the most pressing healthcare service delivery challenges that lie ahead. These include shortage of qualified healthcare professionals, increasing demand for service, and escalating costs. Mobile healthcare is a complex combination of mobile operators, medical device manufacturers, care providers, software developers, funding partners, and regulatory bodies. Each of these stakeholder groups is motivated to participate in the m-health discussion in different ways, but need to work together in order to ensure that this technology can be deployed on a scale that enables benefits to be captured. This chapter presents the challenges that the industry must collectively overcome in order for all stakeholders to be successful.

INTRODUCTION

In conceptual terms mobile healthcare (m-health) is easy to understand. It is about the ability to deliver and manage healthcare services and information through wireless networks. In this context the definition of healthcare services and information is wide. M-health solutions include those that connect medical devices to mobile phones and enable patient data to be transmitted to remote clinics or healthcare systems for further processing. Technology that enables field based medical professionals to record patient data either for local use or for transmission to a

DOI: 10.4018/978-1-61350-150-4.ch006

remote processing facility can also be classified as m-health solutions. Some of the mobile apps that are downloaded to consumer handsets and that provide answers to medical questions without necessarily connecting to a remote data centre can be classified as m-health applications. In practice, m-health is complex, multifaceted and confusing. The absence of large-scale deployments leads many healthcare professionals to question the validity of claims that m-health will deliver widespread benefits to health systems. At the same time, other healthcare professionals see great opportunities. Ganapathy (2010), speaking at MedHealthWorld in 2010 spoke of the tremendous potential for the expansion of m-health technology in India, a country that has 1/6 of the world's population. The ability to deliver healthcare services to regions and communities that normally only have limited access to services is one area of great potential. Another is in being able to reduce the costs associated with delivering healthcare to a wider community.

All of the participants in the m-health value chain contribute to the pace and direction of m-health development in different, sometimes conflicting ways. Investment and entrepreneurial communities are drawn to m-health because of the potential for good returns on investment as the world's population continues to age and as care providers out of necessity adopt more technologically driven solutions. Viewpoints suggesting that the return on investment potential of m-health should be the main driver are counterbalanced by those that see access to healthcare as a fundamental right of every person and feel that m-health opportunities should be developed on the basis of social enterprise rather than a free market. (Bellina, 2010) reports how a patent for a method that uses Multimedia Messaging Services (MMS) to provide tele-diagnostic services was filed in 2008 so that it could not be commercialised but distributed freely. SANA, an MIT-led team of students, volunteers and sponsors are building a

vision in which open source technology components are made freely available to any organisation wishing to launch their own m-health platforms (Denison, 2009).

The opportunity presented by m-health is driving changes in the structure of the healthcare industry. Organisations from a number of industry sectors are interested in participating more closely in the development of m-health. Some, such as mobile operators see this as a natural extension of the services that they already provide. Others, such as telephone call centres and utility companies who are already connected to millions of consumers see new revenue generating opportunities from their ability to provide support services to users of m-health services.

The interaction between these new entrants and established healthcare providers could accelerate the exploitation of emerging research and understanding in which case the market for m-health solutions would benefit greatly. However, increased competition along with a growing need for clarity, accountability and regulation poses a threat to the development of m-health beyond the niche that it currently occupies. Critical challenges such as implementing appropriate regulatory frameworks, defining policy that sets m-health as a core part of healthcare delivery systems and creating business models that are acceptable to all participants must be overcome if m-health is ever to become an established as a mainstream component of healthcare systems.

This chapter presents m-health as the complex combination of mobile operators, medical device manufacturers, care providers, software developers, funding partners and regulatory bodies that it is. It combines the challenges faced by each of the stakeholder groups to provide a more holistic view of issues that need to be resolved if m-health is ever to be established as a core component of healthcare delivery. These issues include the lack of regulatory direction on matters relating to m-health adoption which are causing confusion for technologists and care providers. They also include the way in which the political motivations of the stakeholder groups differ and the effect that this has on the development of m-health. The manner in which innovation and trends in technology are shaping m-health is considered along with the key social and demographic trends that are driving the need for better, more technology driven healthcare. From a financial perspective, the need for radical reform in healthcare economics is discussed since large scale commercial viability of m-health depends upon private sector investment rather than continued support of the government and research grants that fund most of m-health development.

A number of companies and products are mentioned in this chapter. These serve to provide the reader with concrete examples of the concepts that are being discussed and do not in any way represent endorsements of these companies or products.

THE STATE OF MOBILE HEALTHCARE

M-health applications cover a broad spectrum. At one end solutions can be as simple as sending a text message to remind a patient to attend a clinic for a check-up. At the other end solutions can involve automatically collecting a number of different physiological, physical and environmental parameters that are then transmitted to a remote data centre for analysis and possibly instant feedback.

Whilst the discipline of m-health is new, the concepts driving it are not. M-health has evolved from other disciplines that include Telecare, Telemedicine and Telehealth. In 1967 for example, medical staff stationed at Massachusetts General Hospital provided medical care to patients attending an unmanned medical station situated in Boston International Airport (Allan, 2006). In this application, medical personnel provided consultation remotely, 24 hours a day, 7 days a week

through an audio-visual link. In 1993, American TeleCare was established as a commercial concern and started providing video-based Telecare solutions. Their solutions have evolved over the years and they now provide a range of remote care solutions targeting a wide variety of population types and disease conditions.

Since the 90s, billions of dollars and much medical, clinical and technological expertise have been devoted to the development of remote patient care solutions. Investment has been provided through government grants, private investment and charitable donations. Now, there is a great deal of understanding as to how remote monitoring technology can be applied to alleviate problems across all geographies. Yet Telecare, Telemedicine and Telehealth, which have all been promoted as being technologies that would make a significant difference to the availability and affordability of a number of healthcare services, have failed to gain widespread adoption and general use. Reasons for this are many and well documented. Medical professionals are much more conservative than technologists and require a higher level of proof of the efficacy of the technology. In reviewing the UK's Department of Health's Whole System Demonstrator, (Clark, Goodwin, 2010) suggest that the reason for a limited uptake in Telehealth is the lack of robust evidence on its' cost-effectiveness. Another reason often cited is that the equipment required is expensive, difficult to deploy and not necessarily patient friendly. Turning to the question of the market for m-health, a critical question is, "what, if anything makes the success of m-health more probable?"

m-health has attracted the attention of businesses that are well established in the healthcare industry as well as new entrants. It is a global phenomenon with 75% of countries able to report at least one m-health program (International Telecommunications Union, 2010). The pace and scope of development continues to increase as healthcare systems everywhere have become overwhelmed with demand for services and as costs continue to escalate. The adoption of new technology is seen by many as the only way for care providers to be able to continue to provide quality services when faced with so many challenges.

After Telecare and Telemedicine, m-health has become the latest candidate technology to provide solutions for some of the problems faced by healthcare services. It has been a technology-led quest, with developments in communications and chip technology being the key drivers. This has led some observers to critique that m-health is a technology at the top of a hype-cycle looking for a problem to solve, (Linkous, 2010). Whilst developments in technology have been racing ahead of the industry's ability to absorb it, much less effort has been invested in developing the regulations required to support the technology's general adoption by healthcare services. This has not stopped the many m-health stakeholders from maintaining high levels of interest. Investors and innovators remain interested in m-health, attracted by the opportunities for generating large investment returns as IT and mobile communications start to transform healthcare services. Social entrepreneurs on the other hand believe that m-health is an opportunity to provide equitable access to healthcare throughout the world. Never the less, an estimated USD 233M was invested in m-health technology companies during 2010 (Dolan, 2010) which indicates that interest is now being converted to commitment.

Industry views on the potential of m-health are largely positive and policy makers who believe in the ability of m-health to reduce costs and deliver personalised healthcare to everyone that needs it can be found across most regions. These views are balanced by those that urge the need for a more cautious approach. Some of this caution focuses on the potential for information overload that arises from so much data being generated and wonder whether this can be processed in any meaningful way. At a mobile healthcare industry summit in London, a senior lecturer at London's UCL stated, "The reality is that extra data is not what doctors want", (Potts, 2010). Here, the context was the way in which devices worn or carried by patients are constantly generating data that needs to be processed and analysed. Questions over the liabilities faced by medical professionals who engage in m-health are also unanswered. Fundamentally, the patient-medical professional relationship potentially changes when care is provide remotely or in an automated fashion. (Buckner, 2004) asked whether a Telemedicine physician assumes a duty of care for a remote patient, an important question since it has a direct impact on liability.

A growing number of care providers are developing clear views as to how m-health could help them provide better services through on-going monitoring of patients and faster automated responses to urgent situations. Patients that would normally be tethered to a doctor's surgery, yet want to remain mobile have begun to understand how m-health can benefit them and their families.

At a macro level, there is little difference between Telemedince and m-health. In both cases technology is used to connect a patient with a remote care facility suggesting that m-health should not be treated as a separate discipline. At a micro level there are huge differences. One of these is that m-health is better placed to provide access to healthcare in rural environments with little fixed infrastructure. Other differences include the fact that m-health solutions are better able to collect data from a wider range of patients and that medical advice and treatment can be dispensed directly to patients in real-time wherever they happen to be. With 68% of the world's population now in possession of a mobile device, there is a real, scalable channel through which m-health can be delivered. In this regard, few technologies offer the same ability to provide healthcare services that cover a wide range of disease conditions and address populations across all geographies.

In high and middle income countries much of the technology and ideas driving m-health remain within the boundaries of research-led environments or are used only in trials. In low income countries there are many examples of operational projects and these tend to focus on single disease conditions or enable healthcare educational programs.

Across low, middle and high income countries the large number of projects that are actually taking place has led to calls for global coordination of projects and trials to reduce duplication of efforts and consolidate learning. Whilst this is good in theory, it is necessary to understand how m-health can be made to operate locally and to this extent duplication of efforts will be an ongoing requirement.

The type of companies getting involved is changing and this is further driving innovation and competition. Companies involved in the early days of m-health were often University-led spinouts or research-led start-ups. Now, with the promise of large, stable, global markets, which some analysts predict will grow to be in the region of $50 billion (McKinsey & Company, 2010), bigger multinationals such as Intel and IBM who can provide global manufacturing and operational capacity are getting involved. This could enable m-health to be deployed on a much broader scale.

The long-term viability of m-health in terms of cost effectiveness is still not proven. While there are many projects that can clearly demonstrate how it can be beneficial, the real issues of scalable deployment, manageability and affordability remain largely unanswered. Lessons to be derived from the various Telemedicine initiatives are still being collated and evidence to show that these technologies can be made to work on scale that will have a notable, measurable impact across the spectrum of healthcare services is still being gathered. Without the benefit of this learning, models and case studies for m-health need to be developed bottom-up.

MOBILE HEALTHCARE TECHNOLOGIES

The popular view of m-health is of one or medical devices connected to a Bluetooth enabled phone. The phone, usually a smart phone, transmits data to a remote medical centre. Once the data is processed diagnosis and recommendations can be sent back to the phone where appropriate actions can be taken by the patient or local medical staff. In another scenario, patients or medical professionals enter data manually into a smart phone application. The application either processes the data on the phone or transmits it to a remote centre for processing. Both scenarios are valid but given the range of technologies that can be used to enable m-health the actual opportunities extend far beyond these scenarios. Technology being used in m-health applications includes digital pens that can be used to digitise and transmit written information gathered in the field. Cameras, a standard feature of most smartphones can be used to capture and transmit images of a patient or an affected area. Accelerometers, gyroscopes and GPS chips also standard features of smartphones can be used without any attached medical devices to detect when a patient is falling or has wandered way from a specified zone.

A complete assessment of m-health technology needs to take into account the mobile networks over which data is transmitted, the remote data centre and applications that process the data along with medical personnel who are involved in using the data. The way in which the data is shared across other clinical applications and between different medical disciplines must also be considered and increases in importance as the adoption on m-health solutions continue to grow.

Technological innovation is occurring in all areas of m-health. Industries that have never been involved in healthcare, but each with their own area of expertise, vision and ambition are now getting involved. This is leading to new types of collaboration. Some collaborative efforts will change the way in which certain healthcare services are delivered. One example is 24eight, a US based company has combined technology and footwear to develop products that combine sensor technology for measuring walking speed, gait and balance with footwear in order to track a patient's likelihood of falling.

Technology used to enable m-health solutions includes mobile networks, mobile messaging services, software applications installed in remote data centres, mobile apps, devices and sensors as well as all of the technological "glue" that binds these components together. Each of these components is discussed briefly below.

Mobile Networks

Mobile operators have been involved in delivering m-health for a number of years. In 2002 Qualcomm, a US based mobile communications technology company provided mobile network connectivity to CardioNet a specialist cardiac healthcare provider. Qualcomm extended its involvement in m-health in 2007 through the creation of LifeComm LLC, a mobile Virtual Network Operator (MVNO). LifeComm provided a complete package of medical devices and a monitoring platform. By establishing LifeComm, Qualcomm provided early insight as to how far operators could get involved in delivering m-health services. The mindset of many mobile operators has now changed from that of being a communications only participant to rolling out more complete commercial services. Orange Austria for example launched a full commercial m-health service during 2010. The service has been developed for patients with chronic illnesses that require regular monitoring and includes a package of devices, mobile data plan and access to a web based monitoring platform. Consumers can access the service through the network of retail stores. Spain's Telefonica launched a global eHealth initiative during the summer of 2010 and is currently participating in over 80 projects.

For operators that do get involved in m-health, the extent of the services that they will be able to provide is dependent on the capacity and quality of their networks. The capability of mobile networks can generally be described in terms of generations. Second generation networks (2G), are capable of supporting text messaging and low-speed internet only. 2G networks started being replaced in 2001 by third generation networks (3G), capable of much faster data transmission and more advanced network services that included MMS and Mobile Video. Some mobile operators are starting to roll out fourth generation (4G) networks which will provide much greater bandwidth and faster data transfer enabling a critical, real-time applications to be supported. When Sprint, a US mobile operator launched its 4G network late in 2010, one of the m-health applications demonstrated was American TeleCare's LifeView patient self-help station, (Reed, 2010) which enabled real-time video consultation to be carried out over a mobile network. 4G networks are unlikely to be widely available before 2012. Table 1, provides examples of the type of m-health solutions that can be delivered across each type of network.

Network quality of service and capacity that an operator can guarantee are important issues. Most mobile phone users will be familiar with the inconsistent quality that can occur when using their mobile phones. Whether or not data can be transmitted securely and reliably over mobile networks in a timely manner is an important question for care providers. Many mobile operators remain on 2G or 3G mobile networks. Where networks remain on 2G the range of m-health services that can be delivered will be limited to the simpler text message based services.

Software Applications and Platforms

A wide range of data centre software applications and services are being developed to process the enormous amounts of additional data that will be generated through the uptake of m-health. Simple

Table 1. Services supported by each type of mobile network

Mobile Channel	Network Type		
	2G	3G	4G
SMS	Text alerts for appointment reminders, prompts to take medication, health information updates		
MMS	-	Send and receive images and short video clips enabling images to be used in diagnosis or to deliver medical instruction	
Mobile Video	-	Enables more personalised remote consultation between care provider and patient; Provides a platform for delivering more detailed healthcare information or education	
Mobile Internet	Simple text based web pages providing medical checklists and basic healthcare instructions	Enables access to more complex information such as patient records in the form of text, charts, images, video clips and document attachments	
Mobile TV	-	-	Multi-channelled services accessible by an unlimited number of viewers that provide support for a range of targeted demographics or diseases conditions

applications include those that send mobile messages to notify patients to take medication or to attend a clinic for a check-up. More complex applications combine data from multiple sensors and attempt to determine the most appropriate individualised care plans for patients. Table 2 provides examples of some common m-health applications.

Mobile Applications

The first wave of m-health applications developed were typically proprietary, vendor specific applications that could be used by medical staff to track and review patient data through the use of web-based applications. As patients and medical professionals increased their levels of confidence in the use of the internet to access patient data, vendors were encouraged to provide patient friendly interfaces to their applications. Apple's introduction of the iPhone and the success of app store encouraged many software developers to extend this paradigm and create healthcare and wellness mobile apps that provide similar levels of functionality. Most of these do not connect to sensors or medical devices and users are required to enter data manually. These types of mobile app

provide functionality such as managing weight loss programs, provide medication reminders or very general diagnosis. There are also an increasing number of more sophisticated healthcare apps that give more detailed medical advice from the data provided. Apps are now being replicated for the Android handsets and further development for Blackberry and Windows app stores is almost certain.

As the number of mobile apps increases so too will the number of issues arising. Standards and regulations governing the distribution of healthcare apps are still in the early stages. A critical question here is, "Who has the authority to highlight and remove apps that could cause harm to patients?" In the US, as early as 2008 the FDA made a request for an app to be removed from Apple's appstore (Dolan, 2010) and during 2010 made further requests for the removal of apps were made after the FDA received reports of their inaccuracy.

Medical Devices and Sensors

Where medical devices are used in m-health solutions they are typically connected to a mobile phone via Bluetooth and the phone provides a

Table 2. Example m-health solution categories

Application Type	Summary
Remote monitoring and diagnosis	Data generated automatically from medical devices or through manual data input is processed locally on the device or delivered to a remote data processing centre; Comparing data with pre-existing patient records allows diagnosis and more personalised recommendations to be made
Medication adherence	Dispatches scheduled messages to remind patients of their medication regime;
Disease outbreak surveillance	Receives and collates information from field based medical staff; Produces real time reporting of disease conditions across geographical regions
Dissemination of public health information	Delivers routine or emergency health information
Personalised health records	Enables consumers to record health and wellness information; Recommendation engines can provide suggestions relating to diet, exercise or suggest need to visit relevant specialist; Data could be shared with qualified medical personnel.

gateway over which data for the transmission of data between the device and a remote data processing facility. Blood pressure monitors, precision weight scales and spirometers are amongst the types of devices that are used in this way and solutions can be developed to monitor patients who have complex conditions such as congestive heart failure or chronic obstructive pulmonary disease. Patients with multiple or severe conditions are likely to require many devices which increases the learning curve, the cost of providing a solution and creates a greater support requirement. The situation is beginning to change however as innovation and competition drive an increasing level of functionality. Device and sensor manufacturers are delivering innovations in embedded wireless technology, miniaturisation, data processing capacity and durability that enable the development of more compact solutions. This new technology coupled with a better understanding of how patients would actually use monitoring devices in practice is enabling new formats to be designed. Examples of the most recent innovations include an FDA approved ingestible sensor launched by US based Proteus Biomedical during the latter part of 2010. The sensor is capable or measuring the body's response to medication and transmitting information from within the body to a receiver embedded just below the surface of the

skin. At the time of the launch, the receivers were capable of monitoring patient's internal organs for up to 48 hours. This is a much better scenario than keeping the patient under close surveillance in an expensive medical facility.

Wearable devices are another device class emerging from research and development. Typically used for long-term data collection, wearable devices incorporate a number of sensors including accelerometers to detect motion and falls, GPS to track location and distance travelled as well as thermometers and heart monitors. Depending on the configuration, data can be transmitted continuously to a remote monitoring station whilst a patient goes about their normal business.

Whilst the devices discussed above are patient-centric, devices used in m-health also include those support busy medical professionals such as digital pens that allow medical notes collected in the field to be digitised and transmitted wirelessly to remote centres. Mobile phone cameras can be used to take pictures or video clips of wounds for remote analysis. Table 3 classifies devices and sensors on the basis of their physical form and mode of use.

Integrating m-Health with Mainstream Healthcare Systems

Ensuring that m-health solutions work with existing healthcare systems is becoming increasingly essential to ensuring their successful uptake. As more medical professionals start to adopt m-health solutions, the lack of ability to easily move patient data across systems is viewed as significant barrier. Patients and medical professionals using m-health solutions do not exist as a separate group. Care providers need to be able to access relevant patient data whether it is derived from visits to the surgery or remotely through m-health. M-health solutions need to be based on standardised patient health records and consistent clinical language. Health Level 7 (HL7) established in 1987 provides a strong basis for interoperability of clinical systems. Electronic Patient Records (EPR) also referred to as Electronic Health Records (EHR) also provide an opportunity to ensure that standardised patient data is used across m-health and existing healthcare solutions.

It is not clear whether or not current m-health solutions have adopted these standards but future solutions should do so by default.

Information and Data Security

The requirement to integrate m-health and mainstream healthcare systems gives rise to the need to ensure that relevant rules for data and information security are maintained within each system and across system boundaries. These rules can be quite complex and differ from country to country. In the United States, rules regarding information and data security are covered by Health Insurance Portability and Accountability Act of 1996 (HIPAA). Applied to m-health, these laws require solution suppliers and care providers to ensure that features such as data and message encryption, audit control and formalised methods for retrieving patient data into applications

Table 3. Classification of devices and sensors

Form	Description	Examples
Traditional	Medical devices connected to a mobile phone via Bluetooth and which use the phone as a data gateway.	Blood pressure monitors, Precision weight scales, Spirometers
Implantable	Implantable medical devices embedded just below the surface of the skin and which use a small transmitter to send data to a gateway attached to the patient or located in their vicinity. Implantable devices are typically used to monitor disease conditions such as chronic diabetes where 24 hour a day constant monitoring is required.	GlySens - wireless enabled glucometer
Wearable	Body sensors and wearable devices contain one or more small, embedded sensors capable of monitoring different physiological and physical parameters.	BodyMedia Armband and monitoring system
Ingestible	Small micro-transmitters in chemical agents, which react with the contents of the stomach to generate small electrical signals which are then transmitted via a data hub to a medical station.	Proteus ingestible sensor and monitoring system
Mobile Peripherals	Devices that can be plugged directly into a mobile phone and which turns the mobile phone into a new medical instrument	AgaMatrix device manufactured by Sanofi Aventis and which is used to monitor diabetes
Mobile Apps	Applications that are developed to replicate functionality that is normally provided by a dedicated medical device	iPhone Stethoscope app that records a heart rhythm and plays back a cardiogram.

are implemented and maintained. In the UK the rules governing the security of patient data are covered by a number of laws including the Data Protection Act of 1998 and Common law duty of confidence. For care providers and technology suppliers with global ambitions, the challenge is in ensuring that m-health solutions are compliant with all of the rules that are required in countries that they intend to supply.

Technology Standards

Technology standards in any industry serve to ensure that different products and services conform to a consistent pattern of functionality. This allows the market to understand how a particular vendor's products and services work and how they fit in with all the other products and services in the market-place. Historically there have been very few universally accepted technology standards across the healthcare products. Medical device manufacturers have typically used their own proprietary software and communications protocols to transfer data to a connected network or other device. There is general acknowledgement that lack of standards is having a negative effect on the ability to create end to end m-health solutions from the many building blocks that are available. This situation is now being addressed and in recent years a number of new industry bodies have been created to work alongside existing authorities in order to create and manage relevant standards.

Device Standards

Medical device standards address the safety and usability of devices that are made available in the market but there is no explicit extension of these standards to cover devices in the context of m-health. Now, with the continued interest in m-health, work is in progress to start to extend some of the existing standards or create new standards that will provide patients and medical profes-

sionals with adequate protection. The authorities that have a role to play in the development and maintenance of medical device standards include:

- European Medical Devices Directive (MDD) - sets out the obligations required for medical device manufacturers to place products in the European market
- Food and Drug Administration (FDA) - provides validation and certification of devices that will be marketed in the US
- Conformité Européenne (CE) - provides certification to show that products meet European safety requirements.
- Federal Communications Commission (FCC) –collaborating with other agencies to develop guidelines concerning the use of converged communications in healthcare devices

Wireless Standards

Wireless technology is at the centre of most device driven m-health solutions. Those that are applicable to m-health and for which industry standards and certifications exist include:

- Bluetooth (Medical Device Working Group) – create, publish and promote standards that enable compatibility and interoperability across devices using Bluetooth for communications
- ZigBee Alliance (ZigBee Health Care) - create, publish and promote standards that enable compatibility and interoperability across devices using ZigBee for communications

Data Standards

Data standards relevant to healthcare information systems and which should be factored into any m-health solution include:

- Health Level 7 (HL7) – provides a standard for consistent handling of clinical data
- Electronic Patient Records (EPR) – intended to provide a standard for the consistent representation of medical data; whilst there is no single EPR standard, openEHR is widely used

M-Health System Standards

Whole systems standards focus on ensuring a robust architecture and fail-safe processes for deployed m-health solutions. To address the lack of standards in this area, The Continua Alliance, was established in 2006 with the aim of creating device interoperability standards between connected devices used in personalised healthcare solutions. The Continua Alliance provides a certification process that covers wireless device connectivity and personal health records. Although the Alliance is technically driven and managed by the members rather than any independent body, products that carry the Continua logo are guaranteed to work with other certified products.

More recently, June 2010, the European MHealth Alliance was established with one of its stated aims being to assist the industry in providing cost effective, interoperable solutions.

Practical Use of Technology Standards

Taken as a whole, this collection of standards is challenging to understand and for most medical professionals a more concise framework will be required in order to make a confident assessment of the technology available. One approach is that taken by the UK's National Health Service (NHS). Here, there is a centralised list of Telecare, Telehealth and m-health products and suppliers that meet required standards. Care providers can use this list to select appropriate products without having to individually assess them.

MARKET DYNAMICS

For some commentators, the concept of healthcare delivery being part of a market in which access, availability and affordability are determined by the natural forces of supply and demand is wrong. Access to healthcare should be a right and m-health provides an opportunity to ensure that those healthcare services that can be delivered in this way are made available to every citizen that needs it. Others see the emerging m-health market as a substantial business opportunity. At the beginning of 2010, 68% of the world's population, (5.5 billion people), had an active mobile phone subscription, (United Nations, 2010). Mobile phone ownership is high even in developing nations and it is for this reason that m-health has been labelled as a technology that could make a significant contribution to solving healthcare problems across all economies, geographies and demographic groups.

The challenges faced by the world's healthcare systems are different across the various economies, geographic regions and demographic groups. M-health solutions need to be regionalised in order that they fit the requirements of the local community. In advanced economies, ageing and lifestyle are the key challenges. Ageing populations place huge demands on the budgets of healthcare systems through the need for more long term care and high rates of chronic illness. Increasingly, insufficient or expensive residential care space is driving demand for care in the community. In the euro zone, 24.1 million citizens, (5.2%) of the population are expected to be over the age of 80 by 2014 (Leonardi, 2008). These trends contribute to an expected substantial increase in national healthcare budgets in the years to come. M-health could be used alongside other technologies such as Telecare and Telemedince to monitor patients remotely and this will be of great value in planning and managing scarce resources. Lifestyle choices made by people across all demographic groups have contributed to the huge growth in the number of people with long term chronic conditions such as diabetes and obesity. M-health could be used to monitor patients' adherence to prescribed medication regimes or to send out personalised alerts at timely intervals in order to help patients adhere to those regimes. These applications of m-health can be valuable in terms of preventing lapses in lifestyle choices that result in expensive care episodes.

There is a global shortage of qualified medical personnel. The situation is particularly acute in developing countries such as those in Africa and Asia where the estimated deficit is 4.25 million health workers, (Word Health Organisation, 2009). The often underdeveloped healthcare delivery infrastructure and lack of training programmes for medical staff exacerbates the problem. These countries are challenged by diseases that could easily be treated if resources were readily available. Applied to these situations m-health can help make better use of scarce resources by enabling trained members of the local community to collect data from patients and transmit this via text messaging, multimedia messaging or video clips to specialists for further analysis.

Whilst a 68% mobile phone penetration rate is a strong driver it is not the most important. At the beginning of 2010, only 14% of mobile phone subscribers had access to the faster 3G networks and in developing countries only 3% of the population had access to either fixed or mobile broadband (The International Telecommunications Union, 2010). If mobile networks are central to the delivery of m-health services, then the range of solutions that can be delivered in a given region is limited by the capabilities of the mobile operators serving that region. Message based services are all that can be supported

in 2G networks whereas 4G networks will be able to support real-time video conferencing and intensive multi-sensor monitoring applications. This is not to say that the industry should withhold from establishing m-health services in geographic regions where they do not have access to fast networks. In actual fact, the opposite is true and a disproportionately high number of m-health projects have been established in regions that are amongst the poorest in the world. This is clearly good for those regions. It establishes precedents and allows them to learn how m-health should be deployed and managed. It also provides an excellent showcase for the rest of the world, demonstrating what can be achieved and what can be measured. More importantly, it provides real, much-needed healthcare services to regions that typically have very little or even no access to on-going healthcare.

The influence of NGOs has been critical in driving adoption in developing countries. In 2002, the United Nations commissioned a set of Millennium projects. A number of those projects, many of which are still on-going, have m-health embedded as a core principal. These projects are well targeted at specific demographic groups. In Bangladesh, for example, m-health technology has been used to monitor and guide pregnant women during the early days of pregnancy. This has helped to cut down the risk of infant mortality. In 2010, the World Economic Forum added m-health to their agenda. Experts from around the world discussed the state of the technology and what measures should be taken to ensure its global distribution and deployment.

In developed countries, recent adoption of m-health has been driven by more commercial factors as value chain participants wishing to capitalise on this emerging opportunity start to establish business units to focus on creating products and services. In most cases mobile operators are taking the lead as they already have established channels and distribution networks in place. During 2010, Telenor Group, a Norwegian mobile operator established partnerships with Serbia's Health Ministry and Telefónica, a Spain-based operator launched a global e-health unit.

Regional healthcare policy is a significant factor in determining where m-health solutions are likely to be deployed. Typically, low and middle income countries, where much need for healthcare resources is required, a large number of m-health solutions are being deployed. In these regions the m-health projects are driven by NGOs and charities. These projects tend to focus on specific demographic groups and disease conditions which are low-risk but deliver high value. Disease tracking, medication alerts and alerts delivered to change behaviour are typical of the projects that are carried out.

Estimates of the market size of m-health take into account the cost of medical devices provided to patients, fees for monitoring services and data communication charges. As with any fledging market estimates vary widely. One estimate suggests that 400 million sensors will be used in medical monitoring applications by 2015 (Fayad, 2010). McKinsey & Company estimate that the value of the global market in 2010 was USD 50 billion (McKinsey & Company, 2010).

The size of the market has not gone un-noticed by consumer electronics manufacturers some of whom have already created products for the healthcare market. Nintendo, for example, have already developed a number of games for their Nintendo Wii machine that are focused on wellness and healthcare. Further involvement from consumer electronics companies is to be welcomed. Their manufacturing scale and expertise along with consumer product design and distribution capability will provide opportunities for sensor manufacturers wishing to licence their technology for use in bigger markets and could enable the creation of more appealing products.

Although m-health solutions have been shown to be effective across many areas of healthcare, they have not been deployed on a scale that can show serious impact across the broader spectrum of diseases, populations or geographies. For all the trials that have taken place, the investment that has been made, and the evidence that has been gathered, it is still difficult to make generalisations regarding the ability of m-health to have a major impact on healthcare delivery. To overcome this, projects need to scale across a wider variety of health conditions and a greater number of patients. Since market forces alone have not been sufficient to drive adoption of m-health to a much larger scale policy intervention at the highest levels may be required.

THE POLITICS OF MOBILE HEALTHCARE

Healthcare issues across the world are high on political agendas. Some regions focus on their ageing societies and the high proportion of citizens with long-term chronic conditions. Other regions are more concerned with their ability to provide basic healthcare to their citizens.

Amongst national healthcare agencies there has been an absence of clearly defined policy on the extended use of m-health and as a result innovation has remained confined to research, small-scale trials or the deployment of low-risk inexpensive m-health solutions such as those involved in basic monitoring or medication alerts. Systems that are more complex, such as those that are required to monitor and manage long-term chronic illnesses are not widely deployed. Non-governmental agencies on the other hand have been very clear in their policy statements. The World Health Organisation (WHO), whilst working on alleviating the world's most pressing humanitarian problems have implemented over 50 m-health projects in low and middle-income countries (Garten, 2010). Collectively these projects provide extensive, first-hand testimony as to how much benefit m-health can deliver in terms of saving and improving lives.

Within this uncertain political environment stakeholders who have remained interested in m-health have continued to work as best they can within the framework of existing healthcare regulations. All stakeholders recognise that significant policy decisions are required to accelerate the uptake of m-health but their different viewpoints and motivations raise a wide range of sometimes conflicting issues.

Investors remain interested but have been unwilling to commit without the regulatory clarity that allows them to assess which products will be acceptable to healthcare authorities. Clearly stated policy guidelines from healthcare authorities and regulatory bodies are required in order to improve investor sentiment.

Lobbyists that support the more extensive use of m-health within the context of healthcare reform are arguing for tax rebates and Government reimbursement to help drive uptake. On the other hand, there are those who are less enthusiastic and focus on demanding further proof of efficacy, of cost savings and of better patient outcomes.

Within the care provider community even the most supportive have been slow to adopt and roll out m-health solutions. In the absence of any clear guidelines they run the risk of taking an approach that may prove to be incorrect when policy and regulatory frameworks finally fall into place. While they watch and wait for official regulatory guidelines to be made available the investment in training, systems development and process re-engineering that needs to take place in order to be ready to support m-health solutions is not happening. Care providers for the most part are still not ready to adopt m-health solutions.

Mobile healthcare is rising in prominence at a time when the problems facing healthcare systems are becoming critical. The situation as of mid-2010 is that policy makers at all levels and across all regions continue to receive demands for clarification and guidance from care providers and technology innovators. Several initiatives involving regulatory authorities from across the world are now under way to redress the policy void. Ultimately, these will manifest themselves in terms of new and updated rules, regulations and policy guidelines. A key element to that clarity relates to what should be regulated, how it should be regulated, and by whom.

REGULATING MOBILE HEALTHCARE

There is no doubt that the m-health industry needs to be regulated. Patients and care providers need assurances that the products and services that they are using are safe. Suppliers need to understand the standards to which their products must conform, their responsibilities and their liabilities. There are currently few regulations that apply to either to the components of m-health or whole systems. However, the absence of a regulatory framework has not prevented the deployment of trial and operational m-health solutions. Of the operational solutions that are in place, most are being carried out in developing countries and are driven by NGOs. Usually, they are not dealing with complex chronic conditions so the lack of regulations has not been an issue. In developed countries, there are few examples of large scale operational solutions that deal with chronic conditions and one reason for this is that the lack of regulatory clarity does not encourage wider deployment. The situation over the last few years can best be described as being self-regulatory and has really developed from the pioneering days of remote care. Healthcare professionals who could be persuaded to participate in trials or who had a specific interest in delivering remote care did so under carefully controlled conditions of their own design. Without a significant body of experience or evidence to rely on, it was necessary for them to design their own studies, perform their own risk assessments, determine which conditions to monitor and, more importantly, which patients would be suitable for inclusion in any trials or pilots that they carried out. External audits or oversight were provided by peer review whenever work was published and brought into the public domain. The medical devices used in these projects were typically already commercially available and therefore already approved by the relevant authorities. There was never any need to regulate the other components of remote patient monitoring solutions such as the networks over which data was carried, or the software that was used to receive, process and interpret the data. This situation has stayed with the industry throughout its development and is still prevalent today.

Care providers who are motivated to explore commercial opportunities for m-health have no single authority refer to. When attempting to navigate the m-health landscape they are faced with a wide variety of devices, applications, services and business models. Determining which solutions meet acceptable standards is challenging and removes the motivation for care providers to adopt new technologies especially when they are still in the process of being developed and for which they cannot fully understand the scope of their liabilities and obligations. A clear set of regulations will help care providers, suppliers and patient representatives to understand how responsibility and accountability is distributed across the m-health value chain. Figure 1 summarises the landscape for potential regulation in m-health.

Regulating Devices and Sensors

There are a number of regulations that apply to medical devices and sensors used in a clinical setting. Whilst extending functionality of devices by embedding wireless technology that enables remote connections does not alter their clinical functionality, it has generated debate as to whether or not wireless enabled devices should be regulated in a different way. The critical issues in this debate are

- which combination of devices, applications and services should be regulated
- at what level should they be regulated, assuming that the FDA classification system of class I, II and III devices is used

When looking at how sensors or devices are used in m-health solutions it is difficult to determine in advance all of the possible use case scenarios and how solutions may develop over a period of time. Connecting one device to mobile phone that enables the user to monitor their own well-being for example can be regarded as a wellness application and therefore either does not requires regulation or requires to be regulated at the lowest level, Class I. If an automatic data processing function was added to the same device that triggered additional medical interventions when certain conditions were reached, a different, more critical level of functionality is introduced which may require regulation to ensure that the correct analysis was being generated.

Whether or not the regulatory authorities that will be responsible for m-health publish a full set of guidelines in the near term is fast becoming irrelevant when viewed from a commercial perspective. An increasing number of technology companies are pre-empting the situation and seeking FDA approvals for their products. Those that have gained FDA approvals in recent months are much better placed to promote their products to care providers who are likely to be more receptive to adopting solutions.

Regulating Mobile Apps

There are currently no regulations that apply to mobile apps that are used as part of an m-health solution yet thousands of apps are already available. Some of these have been developed for use by patients, others specifically for use by medical professionals. An article in Technology for Doctors Online, (Lombardi, 2010) identifies in excess of 1,500 medical smartphone applications. The large

Figure 1. Landscape for m-health regulation

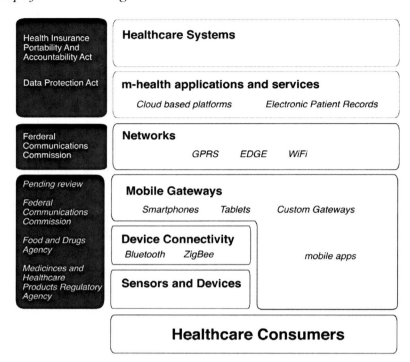

number of apps that are being downloaded and used has started to cause concern amongst medical professionals and regulators. In April 2010, Scientific American (Wapner, 2010), referring to the use of smartphone m-health apps stated, "Currently, there are no clear federal regulations in place, however, to guarantee their quality and accuracy."

In reality, most of the wellness apps currently available are unlikely to required regulation but the FDA and other authorities in the process of assessing and defining the regulatory process.

Regulating Mobile Networks

The most appropriate standards that can be applied to mobile network operators in the context of m-health relate to quality of service, specifically, the ability of the networks to ensure data security, guarantee delivery and manage the large amounts of data that will flow across their networks. Whilst

operator licenses incorporate statements covering the expected quality of service levels that should be provided, there are no obligations that relate specifically to healthcare. As mobile operators continue to take more prominent roles in m-health, issues relating to liability when data is not transmitted in a timely fashion or when networks are not available need to be addressed. Specifically the question that needs to be answered is, where does the liability lie if the failure to send or receive data causes a patient harm? Whilst regulation is not likely to answer this question, it should contribute to a clear understanding of the limitations of the data communications element of m-health.

Establishing Regulatory Frameworks

One of the challenges involved in creating appropriate regulatory frameworks is that industry is attracting a range of new entrants. Currently, mobile operators and mobile app developers fall outside the remit of existing healthcare regulation. Key questions remains as to whether or not they should be regulated and if so how.

Regulatory authorities and newly formed industry groups started responding to calls for clarification during 2009. A European consortium, European MHealth Alliance (EuMHA) was established during 2010 to develop guidelines for software development related to wireless medical devices. In the UK, the Medicines and Healthcare Products Regulatory Agency (MHRA) set up its own medical device technology forum in July 2010 with the aim of establishing best practice for new technologies including wireless enabled medical devices used in m-health solutions. The approach that they are taking is based on reviewing each submission on case by case basis. The stated reason for this is that each m-health solution is complex in its own right and as the regulators are not entirely sure what needs to be regulated reaching a settlement on a case by case basis is the most practical approach. In the US, the Food and Drugs Administration and the Federal Communications Commission met in July 2010 to explore how far wireless medical devices should be regulated. In the latter part of 2010, a m-health regulatory coalition was established by a number of commercial companies with vested interests in developing m-health as a viable business. The companies included handset manufacturers, mobile operators, pharmaceutical companies and medical device manufacturers. Their stated aim is to prepare views on what should be regulated and what should not be regulated by the FDA and the FCC. Their report, expected during the latter part of 2011 will be presented to the FDA and the FCC. It is hoped that this will form a core part of a legal and regulatory framework for the m-health industry.

It is early days and although there is still much to accomplish across all geographies some regulatory decisions specifically relating to m-health have already been taken. During the summer of 2010 the WellDoc DiabetesManager System and Proteus Biomedical's Personal Health Monitor were both given FDA approvals. In February 2011, Mobisante gained FDA approval for it's Smartphone based ultrasound system.

Once regulatory frameworks are in place, the challenge will be in maintaining these in-line with technology advancements. The FDA last updated its guidance on the use of technology in healthcare in 1955. Since then use of information technology across the healthcare industry has changed. Anyone working to the FDA 1955 standards is working with an outdated information technology dictionary. It does not for example mention any of the modern wireless technologies such as Bluetooth or ZigBee that enable many connected medical devices. Future updates will need to be issued far more frequently in order to support the latest set of proven technologies that are being used in commercial applications.

MOBILE HEALTHCARE ECONOMICS

In general the business case for m-health has two elements. The first is the huge cost savings that m-health can bring about as a result of being able to communicate with and monitor patients more closely wherever they happen to be. Savings can be derived in a number of ways. These include an improved ability to pre-empt adverse conditions that might be experienced by patients with chronic conditions and which would otherwise be expensive to manage. Sending medication and appointment reminders reduce the amount of time that is wasted by medical professionals and therefore contribute to overall cost savings. Estimates as to how much can be saved vary. One estimate suggests that the global amount that could be saved from monitoring alone is in the region of USD 1.9 billion to USD 5.8 billion (Cox, 2010). The GSMA in its European Mobile Manifesto (2009), estimated total cost savings across Europe to be in the range €50 billion to €78 billion over a five year period from 2010 to 2015 (Conway,

2009). The second element of the business case is the more efficient use of scare resources. This is of particular benefit to populations where there is little or no access to healthcare services and where efficiencies can be derived from being able to use the technology to process patient data remotely.

To achieve these benefits, the deployment of m-health has to be fully funded but the question of who pays for m-health remains unanswered just as it was with Telemedicine and Telecare. In current models healthcare costs are generally met in the following ways:

- Healthcare insurers covering the cost from the proceeds of insurance policies
- Governments, where the policy is to provide public funding
- Patients themselves where they do not have any insurance or who prefer to purchase care privately

For healthcare insurers and Governments existing models for funding healthcare are already stretched to breaking point. Extending these to include funding for m-health is not a realistic option but alternative funding models have been difficult to achieve because the stakeholders all have different business models and different motivations.

Economic Impact of m-Health on the Stakeholders

m-health stakeholders will be impacted in different ways. Regulatory bodies for example will suddenly find themselves with much more work to do and the costs associated with providing the expertise required to evaluate products and systems so that approvals can be granted or complaints be reviewed will be high.

Mobile operators will introduce new costs to healthcare services through data charges for moni-

toring services as all of the data that is generated in deployed m-health solutions will be delivered over the mobile operators' networks. Whilst m-health represents a great revenue opportunity for operators it creates costs for healthcare consumers.

Sensor and device manufacturers could be major beneficiaries of a move to m-health through increased sales especially in developed countries where personal medical devices are more likely to be used as key components in m-health solutions. These additional costs will be new to healthcare systems and although in theory they will be offset by lower operating costs in the future, the lack of business models that are acceptable to all participants in the value chain has stalled the wider deployment of m-health and has reduced the short-term business opportunity for manufacturers. The medium to longer term, should m-health be established as a core part of the healthcare delivery system, will provide device manufacturers with a global market that is estimated to be USD 30 billion, more than half of the estimated overall market for m-health technology and services.

Care providers have difficult decisions to make with regard to how and when to roll out m-health services. They are, and will remain, the primary point of contact for the patients that they serve, irrespective of how much technology is actually used. It is the provider who needs to determine the value of the technology that they have been asked to use for the benefit of their patients. It makes sense to do this in the context of the complete range of services that they provide and the economies and regions within which those services are provided. This means that a complete understanding of the cost of deploying and operating m-health solutions has to be considered in the context of the benefits that will accrue to patients and the impact on all of the other care services that they provide.

If the test for value is passed, deployment becomes an investment decision. Who funds the resources that are required to enable the service?

For the least complicated healthcare scenarios, such as delivering medication alerts to change behaviour, and where the only requirement is for a mobile phone with a data service, the investment decision is relatively easy to make. For more complex scenarios that require the use of multiple sensors or expensive medical devices the investment required is much greater and decisions can be harder to justify. The cost of providing expensive equipment to all patients who might benefit is not built into the healthcare cost models that exist today. Care providers will also need to consider how they provide the relevant education and training to healthcare professionals involved in deploying and managing m-health. Taking all of these elements together, the adoption of m-health services represents a serious cost burden to care providers. An additional complication is that there is not much research to indicate what proportion of a budget should be allocated to m-health.

The reimbursement policies that are likely to convince care providers to deploy m-health on a much wider scale are being discussed across the world but have not been fully worked out.

Centres of excellence will emerge in the same way that clinical centres of excellence have emerged over the years as a result of acquiring deep knowledge and understanding of the practical applications of m-health and this is likely to attract significant investment from venture groups and corporate investors.

Since m-health will ultimately be able to provide a direct replacement for some existing services skills and procedures that are easily replaceable by m-health alternatives will start to become redundant and decommissioning them will become another source of conflict. Additionally, in pursuit of the cost savings promised by m-health, patients who currently have access to medical professionals on a regular basis might find that they will have to place more reliance on technology. This may not be acceptable to some medical professionals, patients or their families.

Alternative Models

There is little evidence to suggest that even stakeholders that see m-health as a substantial business opportunity are willing to provide the levels of investment required to enable m-health to go mainstream before they can see guaranteed revenue streams. Major changes to the current models for healthcare economics are likely to be required before much further progress can be made.

In a study released by PwC's Health Research Institute (PwC, 2010), 40% of the patients that where surveyed said that they would pay for remote monitoring devices and also subscribe to regular monitoring service. In reality, this sentiment only applies to countries in which the per capita income can support the payments. If self-payment is the preferred option then m-health will not be available to many of the people that would benefit most. Medical devices represent the biggest capital cost for all but the most basic m-health services. Estimates as to how much monitoring equipment can cost varies from $1,000 to $10,000 per patient. The additional cost of on-going monitoring and support will be added on top of this. The PwC study indicated that potential beneficiaries were not keen on spending more than USD 10/month so it is not difficult to see how the cost of rolling out m-health on a wider basis can be prohibitive especially where the focus is on chronic conditions and complex care scenarios. Whilst this can be a lucrative opportunity for medical device manufacturers, it is not necessarily best for the m-health industry. If the cost of uptake remains high, the net effect is a rationing of m-health services that require these devices in order to provide benefits.

Medical device manufacturers could change their business models so that they operate on the basis of service revenues rather than product revenues. In this scenario rather than sell their products for price plus profit, they could lease their products for a fixed term, take payments by instalments or provide the equipment free and take a share in revenues that would be generated.

CONCLUSION

A review of relevant research literature and of the information available on the many companies that have been established in recent years to specialise in m-health indicates that m-health is ready for the mainstream. This research, together with the many thousands of mobile apps that have been developed and that are in use by both patients and members of the medical profession further indicate the presence of a well-developed, growing market. In reality, the market for m-health is just beginning. A relative measure of the market's youth is that it was only during 2010 that the first FDA approved products were launched and that only during 2009 did serious effort start to be directed towards the formalisation of regulatory bodies with a remit to address the lack of regulation covering m-health technology.

There is still much to be done before m-health can truly claim a place in mainstream healthcare. The USD 233M investment in start-ups and established businesses that was made during 2010 shows a growing believe in m-health as a solution that will deliver benefits and generate good returns. The FDA approvals made during 2010 provide an insight as to how companies and products might be developed in a way that will create useful products and deliver real-benefits even where care requirements are complex.

Of all of the issues to be addressed, technology is probably the least important. Finding appropriate business models is probably the most important. Across all economies niche markets for m-health services based on expensive devices and associated monitoring services can already be supported but this not where m-health is targeted. Business models need to be found that can support a wider range of demographics and for that all stake-holders need to be prepared to embrace wholesale change.

REFERENCES

Allan, R. (2006). A brief history of telemedicine. *Electronic Design*. (July 2006).

Ascari, A., & Bakshi, A. (2010). *McKinsey & Company - Global mobile healthcare opportunity*. Retrieved from August 23, 2010, from http://www.mckinsey.it/storage/first/uploadfile/attach/141765/file/global_mobile_healthcare_opportunity.pdf

Bellina, L. (2010). *Mobile diagnosis*. Telecare Soapbox. Retrieved October 15, 2010, from http://www.telecareaware.com/index.php/telecare-soapbox-mobile-diagnosis.html

Clark, M., & Goodwin, N. (2010). *Sustaining innovation in telehealth and telecare*. Whole Systems Demonstrator Action Network Briefing Paper. The Kings Fund.

Conway, R. (2009). *The European mobile manifesto: How mobile will help achieve key European Union objectives. November 2009*. GSMA.

Cox, A. (2010). *Mobile healthcare opportunities: Monitoring, applications and mHealth strategies 2010 – 2014*. Juniper Research.

Denison, D. (2009, October 14). Using cellphones to change the world. MIT project leads to programs that help health workers, farmers in developing countries. *Boston Globe*. Retrieved November 12, 2010, from http://www.boston.com/business/technology/articles/2009/10/14/mit_program_looks_at_ways_to_change_the_world_using_cellphones/

Dolan, B. (2010). Interview: The iPhone medical app denied 510(k). *mobihealthnews*. Retrieved August 11, 2010, from http://mobihealthnews.com/6932/interview-the-iphone-medical-app-denied-510k/2/

Dolan, B. (2010). *Investors pumped $233M into mobile health in 2010*. Retrieved February 2, 2011, from http://mobihealthnews.com/10087/investors-pumped-233-million-into-mobile-health-in-2010

Fayad, G. (2010). mHealth – Just what the doctor ordered. *Comm*. Retrieved from September 15, 2010 from http://comm.ae/2010/10/10/mhealth-just-what-the-doctor-ordered/

Ganapathy, K. (2010, July). *An mHealth perspective from India*. Paper presented at The World Congress 2nd Annual Leadership Summit on Mobile Health, Boston, MA.

Garten, M. (2010). *United Nations compendium of mHealth projects*.

King, R. C., & Buckner, F. (2004). Telemedicine. In ACLM (Ed.), *Legal medicine* - 6th edition, (pp. 424–431). Philadelphia, PA: Mosby, Inc.

Leonardi, M., Catterji, S., et al. (2008). Functioning and disability in ageing population in Europe: What policy for which interventions? *European Papers on the New Welfare, 9*.

Linkous. (2010). *It's mHealth but will it be a revolution?* Retrieved December 12, 2010, from http://americantelemed.blogspot.com/2010/06/its-mhealth-but-will-it-be-revolution.html

Lombardi. (2010). Mobile technology – Should medical smartphone applications be regulated? *Technology for Doctors*. Retrieved August 20, 2010 from http://www.canhealth.com/tfd-news0065.html

Potts, H. (2010). *All bets are on m-health, but where to start to cash in on opportunities*. London: Mobile Healthcare Industry Summit.

Price Waterhouse, Healthcare Research Institute.

PwC. (2010). Healthcare unwired: New business models delivering care anywhere.

Reed. (2010). Sprint shows off 4G video apps. *IDG News*. Retrieved October 12, 2010 from http://www.oswmag.com/article/sprint-shows-4g-video-apps

Toure, H. (2010). *Monitoring the WSIS targets: A mid-term review*. World Telecommunication/ICT Development Report 2010.

Vital Wave Consulting. (2008). *Sizing the business potential of mHealth in the global South: A practical approach*. Presented at The eHealth Connection. Bellagio, Italy.

Wapner, J. (2010). iRegulate: Should medical apps face government oversight? *Scientific American*, (April): 2010.

ADDITIONAL READING

Berg Insight AB (2009). M-health and home monitoring.

Broens, T., Halteren, AV et al. (2007). Towards an application framework for context-aware m-health applications. International Journal of Internet Protocol Technology. Volume 2 Issue 2, February 2007

Deloitte, 2010 Survey of Healthcare concerns. World Economic Forum, 2010, Advancing m-health Solutions, Proceedings of the m-health Summit at the World Economic Forum, 2010

Free, C., & Phillips, G. (2010). The Effectiveness of M-health technologies for improving health and health services: a systematic review protocol. *BMC Research Notes, 3*, 250. doi:10.1186/1756-0500-3-250

Istepanian, R. (Ed.). (2005). M-health: Emerging Mobile Health Systems (Topics in Biomedical Engineering. International Book Series). Springer.

Mechael, P. (2009, Winter). The Case For m-health In Developing Countries. *Innovations: Technology, Governance, Globalization*, *4*(1), 103–118. doi:10.1162/itgg.2009.4.1.103

Mechael. P., Batavia, H. et al (2010). Barriers and Gaps Affecting mHealth in Low and Middle Income Countries: Policy White Paper. Center for Global Health and Economic Development. Paper. Earth Institute, Columbia University

mHealth Regulatory Coalition (2010). A Call for Clarity: Open Questions on the Scope of FDA Regulation of mHealth. December 2010.

Olla, P. Tan. J (2005). The m-health reference model: an organizing framework for conceptualizing mobile health. *International Journal Of Healthcare Information Systems and Informatics (IJHISI)*. Volume 1, Issue 2. pp. 1 – 19.

Pharow, P., & Blobel, B. (2008). Mobile Health Requires Mobile Security: Challenges, Solutions and Standardization. S.K. Andersen et al. (Eds.), *In eHealth Beyond the Horizon – Get IT There* (pp. 697 – 702).

Stockdale, R., et al. (2008). Mobile Health and Chronic Disease Management: Moving Towards a Holistic Approach. Presented at Australasian Conference on Information Systems M-Health and Chronic Disease Management. 3-5 Dec 2008, Christchurch

Vital Wave Consulting. (2009). mHealth for Development: The opportunity of mobile technology for healthcare in the developing world, D.C. and Berkshire, UK: United Nations Foundation - Vodafone Foundation Partnership, 2009.

Chapter 7
mHealth:
Mobile Healthcare

Barbara L. Ciaramitaro
Ferris State University, USA

Marilyn Skrocki
Ferris State University, USA

ABSTRACT

Mobile Healthcare, or mHealth, involves the use of mobile devices in healthcare. It is considered a revolutionary approach to delivering health care services such as diagnosis and treatment, research, and patient monitoring. Much of its revolutionary reach is due to the widespread adoption of mobile devices such as mobile smart phones and tablets such as the Apple Ipad. It is estimated that there are over five billion mobile devices in use throughout the world. In terms of demographics, in the United States, it is estimated that five out of seven Medicaid patients carry a mobile smart phone. One result of this mobile reach is the ability to provide healthcare services to people nonambulatory and isolated in their homes, and in underdeveloped and emerging countries, in ways that were previously cost prohibitive. mHealth is also seen as a way to emphasize prevention through mobile monitoring devices and thereby reduce the overall cost of healthcare. mHealth is viewed as changing the healthcare landscape by changing the relationship between the patient, healthcare provider, and between healthcare providers. "A new generation of eHealth products and services, based on wireless and mobile technology, is putting diagnosis and treatment management into the hands of the patient" (The Mobile Health Crowd, 2010). There is clearly a growing interest in, and emphasis on, mobile healthcare applications in the world today by vendors, physicians and patients. It is predicted that the mobile health application market alone will be worth over $84 million, and that by the year 2015, more than 500 million people will be actively using mobile health care applications (Merrill, 2011; Merrill, 2011b).

DOI: 10.4018/978-1-61350-150-4.ch007

INTRODUCTION

At the time of writing, the annual Consumer Electronics Show (CES) is being held in Las Vegas. This conference showcases the newest technology gadgets and tools that will soon be available to the consumer. Historically, it is filled with amazing innovations such as those providing 3D and other immersive applications, new smartphones, or motion-sensing games. What is unusual this year however is the number of devices on display that focus on mobile applications for health care from personal health records (PHR) systems, patient monitoring systems, to eHealth games. In fact, it is expected that "healthcare devices and applications will increasingly move closer to the center stage of CES." (Beaudoin, 2011, p. 1). This prediction reflects the growing interest in, and emphasis on, mobile healthcare applications in the world today by vendors, physicians, hospitals, and patients. It is predicted that the mobile health application market alone will be worth over $84 million, and that by the year 2015, more than 500 million people will be actively using mobile health care applications (Merrill, 2011; Merrill, 2011b).

Mobile Healthcare or *mHealth* involves the use of mobile devices in healthcare. It is considered a revolutionary approach to delivering health care services such as diagnosis and treatment, research, and patient monitoring. Over the last decade, "Mobile computing devices are becoming as commonplace in the practice of medicine as stethoscopes" ((Hau, 2001, p. 1). For example, OptumHealth, a division of United Health, a medical insurance provider, advertises the following; "With NowClinic[SM] online care, you can talk to a doctor like you would in an exam room. Share your symptoms, receive a diagnosis, and even get a prescription, if clinically appropriate. Available anytime, anywhere you have Internet access." They preface the advertisement indicating the diagnosis can be made if clinically appropriate and that no controlled substances may be prescribed and the availability of other prescriptions may by restricted by law. However, it indicates how telemedicine is desired by many and warrants a product by the insurance company to meet their needs. (OptumHealth, 2011) Much of its revolutionary reach is due to the widespread adoption of mobile devices such as mobile smartphones and tablets such as the Apple Ipad. It is estimated that there are over five billion mobile devices in use throughout the world (Griffith, 2010). In terms of demographics, in the United States, it is estimated that five out of seven Medicaid patients carry a mobile smartphone (Griffith, 2010). One result of this mobile reach is the ability to provide healthcare services to people nonambulatory and isolated in their homes, and in underdeveloped and emerging countries, in ways that were previously cost prohibitive. mHealth is also seen as a way to emphasize prevention through mobile monitoring devices and thereby reduce the overall cost of healthcare.

Of course, one key question with the introduction of any new technology is the adoption rate by its targeted users. Will health care providers and patients be willing to deliver or receive health care services through mobile devices? From a patient perspective, a recent study found that 41% of consumers would prefer to have more of their health care delivered to them through a mobile device; 31% stated that they would willingly use a mobile application to track and monitor their health; and 40% would be willing to pay monthly subscription fee to have their medical data such as heart rate, blood pressure and blood sugar, automatically transmitted to their doctor through their mobile device (Lewis, 2010). From a physician perspective, 57% reported that they would use remote devices to monitor their patients' health and vital signs; 56% report that mobile devices assist in expediting decision-making; and 40% claim that remote monitoring and messaging could reduce office visits by up to 30%. However, physicians in the study did report a concern about being provided with too much patient data through these remote transmissions and would prefer to see filtered information or exception reports detailing abnormal results (Ibid).

mHealth is viewed as changing the healthcare landscape by changing the relationship between the patient and the healthcare provider. "A new generation of ehealth products and services, based on wireless and mobile technology, is putting diagnosis and treatment management into the hands of the patient." (The Mobile Health Crowd, 2010, p. 1). It is interesting to note that a key barrier to adoption of these mobile health applications is not the patient, but rather the health worker who fear that this new use of mobile technology will make their jobs less secure (Ibid). Let us examine some mHealth applications and their current and potential impact on patients and healthcare providers.

MHEALTH APPLICATIONS AND DEVICES

In general, mHealth applications and devices are categorized as administrative, physician-centric or patient-centric. Administrative applications are those that assist in increasing efficiency in health care operations such as patient registration and admitting. Physician-centric applications refer to those applications that support the physician in research, diagnosis and treatment. Patient-centric applications are those that use mobile technologies to deliver healthcare services to patients such as remote monitoring of vital signs, medication and office visit reminders. Administrative and physician-centric applications appear to have captured the mHealth market early on. As the emphasis on electronic health records (EHR) has grown, so has the availability of mobile application extensions to these EHR systems to administrative staff, nurses, and physicians. Physicians, nurses and staff were soon provided with mobile applications such as drug interaction checks, correct dosages, Medicare formularies and billing code references. Patient-

centric applications have more recently entered the market and focus more health on information rather than interactivity such as medication and office visit reminders. We are now seeing growth in more interactive patient applications such as the monitoring of vital signs and monitoring of cardiac arrythmias.

One important caution in the development and use of mHealth devices and applications is that they are often regulated under the eyes of the FCC (Federal Communication Commission) and the FDA (Federal Drug Administration). "Every medical device that uses wireless communications technology, whether implanted, worn on the body, or ingested, falls within the FCC's authority to manage the electromagnetic spectrum." (Barritt, 2010, p. 1). Similarly, many mHealth devices and applications must pass the 510K standard of the Federal Drug Administration before it can be released. Among other points, this standard requires that medical device manufacturers prove the safety of their devices. As it relates to mHealth devices, "Class III devices are the highest risk. These include many implantable devices, things that are life-supporting, and diagnostic and treatment devices that pose substantial risk." (Deyo, 2004, p. 1). Therefore, it is important to recognize that the constraints on mHealth applications and technologies far surpass those that are placed on mobile devices and applications unrelated to healthcare.

Administrative Applications

There are several mHealth mobile applications that can reduce errors and increase efficiency in healthcare organizations. One area of mHealth that is reaping significant benefits is the use of SMS text messages to send reminders to patients to take their medicine or remind them of their upcoming appointments. Another area that has great promise is the monitoring of medicine and other medical supplies through the use of wireless tags and readers. This mHealth application has been found to reduce the loss of supplies and creates an accurate inventory which results in more timely replacement of necessary medicines and supplies as well as saving for the medical provider.

Physician-Centric Applications

One area of mHealth applications of particular interest to physicians is the ability to transmit and view medical images to their mobile device. Although some ability to remotely view medical images on mobile devices has been available over the last 10 years, in many cases the limitations of the mobile devices, applications and network infrastructure limited their use (The Mobile Health Crowd, 2010c). Currently, Infinitt (www.infinitt. com) is offering a web-based mobile viewing application that includes a broad range of viewing applications, including radiology information system (RIS), cardiology picture archiving and communications system (PACS), radiology PACS and reporting applications. Importantly all of the images are transmitted over a secure encrypted network in order to be HIPAA-compliant. (Imaging Technology News, 2011). Mobile medical devices are also replacing the use of traditional hardware such as handheld mobile ultrasound units or echo cardiograms (Topol, 2010).

Other examples of mHealth devices and applications are the use of small wireless devices that are used inside the body to examine and diagnose possible health conditions. One recent development has been the use of a small capsule that is swallowed and transmits images as it travels through the stomach and small bowel. Similarly, implantable pacemakers can now transmit ongoing information for monitoring the patient's heart condition. Other implantable devices are able to assist in controlling bodily functions such as heart rhythms, provide or subdue nerve stimulation, and monitor cranial pressure. (Barritt, 2010).

Patient-Centric Applications

An intriguing area of patient-centric mHealth applications involves wearing wireless devices that can transmit vital signs to a central monitoring station. For example, BodyKom sold by Teliasonera (www.teliasonera.com), a Swedish company, is a mobile heart monitoring service that transmits the ECG of heart patients to a monitoring center via the mobile network. The monitoring center also receives GPS (Global Positioning System) data in order to identify the location of the patient.

In a talk at the TEDMED conference in 2009 (Topol, 2010), Eric Topol describes the emerging uses of wireless mobile healthcare applications that are transforming patient and personal care, and supporting the emerging emphasis on consumer driven healthcare. In addition to the ability of healthcare providers to monitor the vital signs of patients, individuals will have the ability to be proactive and monitor their own vital signs, glucose levels, caloric intake, and sleep quality on a daily basis. One could also monitor vital signs of an unborn child. Amazingly, these mobile sensors are the size the band aids.

Another growing area in patient mHealth is targeted toward the growing elderly population. Rather than be confined to healthcare institutes or other long term care, there is a current effort focused on the development of sensors, applications and supporting technologies that create "smart homes". The homes become smart in the sense that the movement and vitals of its inhabitants can be tracked. If the sensors identify a fall or other medical emergency, the appropriate healthcare provider is notified. (The Mobile Health Crowd, 2010d).

Health insurance companies support disease management programs. Disease management is very dependent upon mobile devices and the monitoring of patients by telephone. Beginning in 2011, health insurance companies will be required to spend 80-85 percent of the premiums they collect on medical claims and other activities that improve member's health. (Aizenman, 2010) If mHealth devices are proven to improve member's health, a large purchaser of medical devices will be insurance companies.

Much of this wireless mobile health technology is encompassed under the infrastructure of a Body Area Network (BAN) which consists of wireless communications between small Body Sensor Units (BSU) and a Body Central Unit (BCU) worn on the human body. The sensors send a signal to a gateway such as a smart phone or a dedicated gateway where it is transmitted to a healthcare provider. (Topol, 2010). A BAN uses very small sensors that are either placed on the skin or worn on the human body. In most cases these sensors are comfortable and do not inhibit and normal activity. The sensors collect physiologic data which is transmitted wirelessly to an external monitoring unit which then transmits the data to healthcare providers. The BAN is a two way communication network as a healthcare provider could respond to the data received by informing the patient of appropriate directions. The BAN is just one part of the mHealth mobile network as we will discuss in the next section.

MHEALTH NETWORK INFRASTRUCTURE

Mobile networks to support mHealth are relatively straightforward to implement. They rely most often on existing wireless networks within healthcare institutions and those provided by telecommunication vendors to the public. Most mHealth applications actually transmit only small amounts of data and can be operated on most existing networks in the world (The Mobile Health Crowd, 2010b). It is not necessary to wait for the availability of higher bandwidth 3G or 4G networks in order to deploy most mHealth applications. Of course, there are some applications, such as remote viewing of MRI and other test results that may require higher bandwidth, throughput and transmission speed.

The organization and structure of data and patient management in a mHealth environment usually involves an infrastructure that includes a call center, centralized patient records, healthcare personnel, monitoring and response units. Two important components to a mHealth network are the inclusion of mobile devices that send and transmit data to and from the patient and the healthcare provider. The second addition is the use of intelligent agents (IA) that are able to evaluate the data received and respond appropriately such as contacting a physician or response unit. One example of a mHealth network infrastructure is provided below.

As can be seen in Figure 1, the patient in a mHealth environment retains the ability to contact the healthcare provider through the nursing staff or call center when desired. However, the patient also communicates health information on a regular basis through their mHealth mobile devices. Both of these flows of information allow the healthcare provider to retain a history of the patient's health status and assess the urgency of the call or data transmitted through the use of nurses and intelligent agents.

CHALLENGES

Although mHealth application can contribute to improved healthcare, there are several challenges in its implementation. These challenges generally fall within the categories of usability, mobile networks, application management and support, privacy and security.

Usability

Usability remains a concern primarily due to the lack of screen space on smart phones which can limit the types and amount of information available to review. For many healthcare providers, the adoption of the Apple Ipad, with significantly more viewable screen space, has provided a work-

Figure 1. Patient in a mHealth environment (Reprinted with permission from The Mobile Health Crowd [2010])

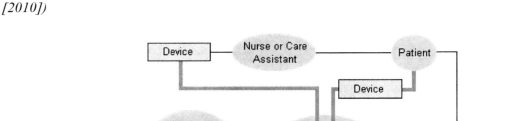

able alternative to the use of smartphones and is being adopted at a high rate by hospital and medical staffs (Moore, 2010; Mobility Management, 2010). Another aspect of usability particularly with the use of smartphones is the small keyboards or keypads that make accurate data entry difficult. Once again the Ipad can assist but in many cases the environment may not be suitable for data entry on any mobile device. One proposed solution is the development of an accurate mHealth voice recognition system that can mitigate the challenges of data entry. The life of a mobile device battery can also provide challenges to users of mHealth applications and devices. Depending on the device, its use and applications, and environmental conditions, and the availability of a recharging station, the battery on many mobile devices may run out too quickly for healthcare providers to find it useful. Lastly, although there is growth in Internet web sites designed specifically for mobile devices, the far majority of healthcare related web

sites remain optimized for personal computers. This may result in an inability to view or navigate through material provided. Additionally, unless a specific mHealth application is used, many lab, pathology, x-ray and other reports are difficult, if not impossible, to view on most handheld devices. (Bhutkar, 2009)

Mobile Networks

The use of mobile devices and mHealth applications are increasing during emergency patient care situations. As a result, the speed, coverage, and throughput of the mobile network are critical. Although the coverage of mobile networks is increasing around the world, not all are developing at the same rate of throughput and speed. The reach and functionality of mobile devices depends on their underlying network infrastructure and the capabilities of the mobile device or handset. Limited bandwidth results in limited functionality. Limited reach results in limited access and communication capability. The ability to effectively market to mobile device users requires an understanding of the strengths and limitation of their network infrastructure and the design of a marketing plan that accommodates those strengths and weaknesses. Over the past 15 years, the mobile network infrastructure has evolved. However, it is important to recognize that on a global basis, not all regions and countries have evolved similarly and many underdeveloped areas continue to rely on less current network technologies.

It was with the release of the third generation of mobile networks, referred to as 3G that much of the capability used for mHealth came into existence. Although the first 3G network was introduced in Japan in 2001, it was not until 2005 that the use of **3G** networks became widely adopted. The demand and use for mobile devices skyrocketed during this era, so that by 2006, the number of mobile subscribers exceeded 2.5 billion (Membridge, 2008). 3G actually refers to a set of standards that meet ITU (International Telecommunication Union) requirements (Union, 2005). In addition to voice, 3G is able to provide a wide variety of application services including mobile internet access, multimedia, location-based services, video calls, and mobile television. Its data transmission rates are rather robust at 384 Kbps (Krum, 2010). In addition to the voice and text messaging capabilities of the 3G era, a significant added functionality to mobile handsets was that of the mobile web browser. The introduction of the web browser expanded the use of the mobile device so that Internet information could be searched for and accessed on demand by the user.

The fourth generation of mobile networks, referred to as 4G, provides high speed data transmission and high levels of data throughput at 100 Mbit/s for high mobility communication (such as from trains and cars) and 1 Gbit/s for low mobility communication (such as pedestrians and stationary users). The 4G system provides a 100% IP based broadband network and allows for

the delivery of high demand applications such as Voice over IP and streamed multimedia (Krum, 2010). 4G networks are developed to meet QoS (quality of service) requirements in that they are able to appropriately prioritize the various types of network traffic for the most optimum results.

Ideally, to reach their maximum effectiveness, mHealth applications should run on a 4G network. Unfortunately, for most of the world that does not yet exist. Most mobile networks still run on 3G infrastructures and in some parts of the world, there are mobile networks that have not yet reached 3G capabilities. As a result, many mHealth applications are limited in their use based on the local network availability.

Management and Support of Applications

"Two thirds of physicians say they are using personal devices for mobile health solutions that aren't connected to their practice or hospital IT systems, but nearly a third said their hospital or practice leaders will not support the use of mobile health devices." (Commins, 2010, p. 1).

One of the key challenges of mHealth is the management and support of mobile devices and applications. Healthcare regulatory requirements place the responsibility of ensuring the privacy and security of patient data on the shoulders of the healthcare organizations. Therefore it is essential that a healthcare organization begin their mHealth efforts with policies and procedures aimed at securing patient data while providing the benefits that mobile devices and mHealth applications can provide. The healthcare organization needs to be able to support that diverse collection of mobile devices used by healthcare professionals, doctors and staff. If this widespread support is not possible, then an approved set of mobile devices must be mandated for use with patient care. Similarly, although one of the benefits of mobile devices is the ability to download applications as desired, this can present a tremendous challenge for support both in guidance in their use and protection against malicious software. In many cases, it may be necessary to only allow approved applications to be downloaded onto mHealth devices.

However, most hospitals appear to dedicate more resources on reactive than proactive measures. Reacting to a breach, or leak of patient data, appears to get health care administrators attention more than preventing one. This is understandable as penalties for HITECH violations can reach as high as $1.5 million dollars. (Himss Analytics, 2010) When healthcare providers were asked "which of the following initiatives are likely to be your organizations top IT security priorities over the next 12 months", regulatory compliance was number one, with data security a close second. (Forrester, 2010)

Another significant challenge of support mHealth is the integration of mobile applications with existing legacy systems within the healthcare organization. For mHealth applications to be effective, they must by linked to the patient data that are stored in the various data repositories including laboratory and radiology test reports, patient progress notes, medication, diagnosis, and current patient status. The application support must provide access to tools that can aid in both diagnosis and treatment without requiring the healthcare provider to re-enter patient information or to continuously logon on to different applications to complete a patient visit. In order to be effective and usable, mHealth applications must support both existing legacy system repositories of information and new functionalities provided by mHealth mobile devices and applications. With ease of access, is the topic of awareness of the security issues that mobile devices bring. When hospitals were asked to "rank which of the following are among the top three issues your organization faces in supporting a mobile workforce", they ranked the number one challenge as "understand the risks and vulnerabilities" (Forrester Consulting, 2010).

Security and Privacy

The legal requirements that are mandated by HIPAA (Health Insurance Portability and Accountability Act of 1996) and the HITECH (Health Information Technology for Economic and Clinical Health) place tremendous security and privacy constraints on the storage, transmission, and sharing of patient data. HIPAA set national standards and requirements for the protection of individually identifiable health information. Among other requirements, HIPAA mandates that that patient information is unavailable to unauthorized users during storage and transmission. Therefore, mobile devices and mHealth applications must have the ability to use encryption software to protect patient data. Many healthcare users are unfamiliar with using encryption software and therefore the responsibility falls back on the shoulders of the organization. Healthcare organizations "must seek out a data protection solution that will automatically encrypt data and enforce security policies across all devices" (Credant, 2010, p. 1).

HITECH provisions within the American Recovery and Reinvestment Act (ARRA) signed into law by President Obama in February 2009 provides a number of objectives related to the secure and wide-spread us of electronic health data between health care organizations. Among many financial incentives to health care providers, it also addresses many issues that have arisen since HIPAA was implemented in 1996. HITECH creates a new, tiered penalty structure of violations to HIPAA or HITECH requirements. A willfully committed violation that is not corrected within 30 days may be penalized at least $50,000 PER violation, not to exceed $1.5 million per calendar year. (Berry, 2010)

The most common security threat associated with mobile devices is their propensity to become lost, stolen or misplaced. In fact, it is estimated that over 8 million phones are lost with 32% of them never recovered (Jansen, 2008). Unless a user specifically establishes security precautions on their mobile device, once it is in the hands of unauthorized users they have the ability to access and use any of the data and functionality provided by the mobile device. The optimum solution to this threat is to install software on the mobile device that is able to erase the contents of mobile devices and deactivate their status when they have been reported as lost or stolen. Lookout (www.mylookout.com) is a software product that provides a set of services for lost or stolen mobile devices. If your mobile device is equipped with GPS, then Lookout can locate your phone. It can also issue a loud audible alarm. Most importantly it can either remotely lock your mobile device or remotely eliminate all of its contents before it falls into the hands of malicious users.

Closely related to the loss of a phone is the security threat of unauthorized access. If a mobile device comes into the hands of an unauthorized user, their first goal will be to access the information and resources stored on the mobile device. In order to protect their device against unauthorized access users should establish a password that is prompted when they activate their mobile device. Additionally, a mobile device should become inactive after 2 to 3 minutes of idle time, once again requiring a password when it is reactivated each time for use.

It is also crucial that healthcare users of mobile devices and mHealth application receive training on mobile security threats and the policies and practices that have been implemented to prevent or minimize them. Compliance with enterprise polices require monitoring and enforcement to be effective and the enterprise must be willing to put these elements into practice in order to protect their enterprise from mobile breaches and malware.

CONCLUSION

We have come a long way since the health care industry's only reliance on wireless technology was the ability of doctors to wear beepers on the golf course." ((Barritt, 2010, p. 1). We are certainly in the midst of a mHealth revolution. As we discussed, much of its revolutionary reach is due to the widespread adoption of mobile devices. We are seeing the widespread adoption of mHealth applications by administration, physicians and nurses, and most importantly, by patients. These mHealth devices and applications are certainly transforming the world of medicine but they are not without their risks and challenges. We must first ensure the safety of these devices that are being used for patient care, diagnosis and sometimes treatment. We must also be aware of the need to protect the security and privacy of patient information on these mobile devices and during transmission of patient data. The enthusiasm for mHealth is extensive and vendors and hospital administrators see a pot of gold at the end of the rainbow. However, as exciting as this technology is, as efficient and productive it may make us, we must make sure that we take the steps necessary to protect our patients from harm.

REFERENCES

Aizenman, N. C. (2010, July 20). Insurers tout disease management programs, but critics are wary. *The Washington Post.*

Analytics, H. I. M. S. S. (2010). *Healthcare industry continues to overlook critical gaps in data security, according to new bi-annual report.* Retrieved on January 23, 2011, from http://www.himssanalytics.org/general/pr_20100421.asp

Barritt, K. (2010). *Wireless medical devices: Navigating government regulation in the new digital age.* BNA's Medical Devices, Law, & Industry Report.

Beaudoin, J. (2011). *Healthcare's increasing presence at CES harbinger of things to come.* Retrieved on January 9, 2011, from http://www.healthcareitnews.com/blog/healthcares-increasing-presence-ces-harbinger-things-come

Berry, D. McNeil & Parker. (2010). *Analysis of HITECH provisions in the American Recovery & Reinvestment Act*. Retrieved on January 23, 2011 from www.bdmp.com

Bhutkar, G. K. (2009). *Major challenges with mobile healthcare applications*. Retrieved on December 28, 2010, from http://www.bjhcim. co.uk/features/2009/909004.htm

Commins, J. (2010). *Physicians: Mobile devices expedite decision making*. Retrieved on December 29, 2010, from http://www.healthleadersmedia. com/content/TEC-256203/Physicians-Mobile-Devices-Expedite-Decision-Making##

Consulting, F. (2009). *Managing and securing mobile healthcare data and devices*. A custom Tech Adoption Profile Commissioned by MaaS360 by Fiberlink. Retrieved on January 21, 2011 from http://www.informationweek. com/whitepaper/Mobility/Mobile-Business/ managing-and-securing-mobile-healthcare-data-and--wp1269371115712

Credant. (2010). *Tips for securing healthcare data on mobile devices*. Retrieved on January 11, 2011, from http://www.executivehm.com/article/Tips-for-securing-healthcare-data-on-mobile-devices/

Deyo, R. (2004). *Gaps, tensions and conflicts in the FDA approval process: Medical devices*. Retrieved on January 17, 2011, from http://www. medscape.com/viewarticle/474285_4

Griffith, A. (2010). *Revolutionizing healthcare delivery with mobile health (mHealth)*. Retrieved on December 29, 2010, from http://www.suite101. com/content/revolutionizing-healthcare-delivery-with-mobile-health-mhealth-a243935

Hau, S. (2001). *Moving to mobile - Five key strategies for managing handheld devices in healthcare settings - Technology information*. Retrieved on December 20, 2011, from http:// findarticles.com/p/articles/mi_m0DUD/is_7_22/ ai_76548959/.

Imaging Technology News. (2011). *Mobile PACS application added for smart phone access*. Retrieved on January 17, 2011, from http://www. itnonline.net/node/38592/3/

International Telecommunications Union. (2005). *Cellular standards for the third generation: The ITU's IMT-200 family*. Retrieved on December 15, 2010 from http://www.itu.int/osg/spu/imt-2000/ technology.html#Cellular%20Standards%20 for%20the%20Third%20Generation

Krum, C. (2010). *Mobile marketing: Finding your customers no matter where they are*. Indianapolis, IN: Que.

Lewis, N. (2010). *Mobile devices to transform healthcare*. Retrieved on December 29, 2010, from http://www.informationweek.com/ news/healthcare/mobile-wireless/showArticle. jhtml?articleID=227400122

Management, M. (2010). *iPad is overwhelming tablet of choice in healthcare*. Retrieved on December 29, 2010, from http://www.visagemobile. com/news/news/managing-mobile-devices-news/5964/ipad-is-overwhelming-tablet-of-choice-in-healthcare/

Membridge. (2008). *History of cell phones*. Retrieved on December 16, 2010, from http://www. historyofcellphones.net/

Merril, M. (2011). *Kalorama: Medical mobile app market worth $84.1M*. Retrieved on January 9, 2011, from http://www.healthcareitnews.com/ news/kalorama-medical-mobile-app-market-worth-841m

Merrill, M. (2011b). *Report: 500M to use mHealth apps by 2015*. Retrieved on January 9, 2011, from http://www.healthcareitnews.com/news/report-500m-use-mhealth-apps-2015

Moore, J. (2010). *mHealth in the enterprise set to explode*. Retrieved on January 9, 2011, from http://www.healthcareitnews.com/blog/mhealth-enterprise-set-explode

OptumHealth. (2011). *NowClinic*. Retrieved January 25, 2011, from http://www.optumhealth.com/solutions-services/care-solutions/wellness/nowclinic/

The Mobile Health Crowd. (2010). *Introduction - The tipping point*. Retrieved on December 28, 2010, from http://www.themobilehealthcrowd.com/?q=node/7

The Mobile Health Crowd. (2010b). *101 applications - Engaging the mobile telecoms industry*. Retrieved on December 28, 2010, from http://www.themobilehealthcrowd.com/?q=node/9

The Mobile Health Crowd. (2010c). *PACS - Getting the picture at last*. Retrieved on December 28, 2011, from http://www.themobilehealthcrowd.com/q=node/13

The Mobile Health Crowd. (2010d). *Smart homes for the elderly*. Retreived on December 28, 2010, from http://www.themobilehealthcrown.com/?q=node/79

Topol, E. (2010). *Eric Topol: The wireless future of medicine*. TEDMED Conference 2009. Retrieved on January 17, 2011, from http://www.ted.com/talks/eric_topol_the_wireless_future_of_medicine.html

Chapter 8
Potentials and Challenges of Mobile Augmented Reality

Joerg H. Kloss
Consultant, Germany

ABSTRACT

In order to deliver a foundation for orientation and decisions in the growing field of Mobile Augmented Reality (MAR), this chapter intends to deliver a realistic picture about MAR, where it comes from, where it currently stands, and where it is going. It distinguishes between the variants of MAR, explains their similarities, differences, and different development stages and outlooks. While describing potential application fields and their current limitations, the opportunities and challenges of MAR become obvious, as well as how much work is still to do. The chapter intends to categorize the challenges for MAR, covering also related issues in the mobile hardware, software, and operator industry, and their efforts in standardization and open interfacing. Besides the technology-driven discussion, a strong emphasis will be taken also to the essential aspect of user experience. With ubiquitous MAR, the technology will become more and more secondary, and the user and her individual context moves into the center of attention.

INTRODUCTION

One of the most promising technologies for the mobile future seems to be the Mobile Augmented Reality (MAR). Leaving the laboratories, and based on the work of more than two decades of research, the technology and interface approach

DOI: 10.4018/978-1-61350-150-4.ch008

of Augmented Reality (AR) tapped the pulse of the current age only a few years ago. MAR started in early 2008 with printed markers held in PC webcams, and overlaying the recorded video stream with virtual 3D objects that follow the user's movements in real-time: the first public wow-effect was created. Almost simultaneously, and enabled by the emergence and distribution of powerful as well as affordable handhelds with integrated sensors (e.g. GPS, compass, gyroscope), some of the AR approaches have been transferred successfully also to the new generation of smartphones. In late 2008 one of the commercial pioneers, Mobilizy, released its MAR browser "Wikitude" for the Android platform, and for the first time MAR was available to the public masses, almost anywhere and anytime.

Since then, many more MAR apps have entered the mobile marketplaces for different hardware platforms and operating systems. Some marketing agencies are overturning in predicting the rosy future of MAR-based apps and business, and Gartner names AR as one of the ten "disruptive technologies" for 2008 to 2012 besides social networks, cloud computing, ubiquitous and contextual computing (News Release, 2008). Also the estimations by other market researchers for MAR in 2014 are promising: while ABI research predicts a market volume of $350 million (ABI, 2009), Jupiter Research anticipates even $732 million (Bergman, 2010), partly leveraging on hyper-local advertising. Not surprisingly that more and more startups, fancy apps, and manifold participants are entering the promising MAR ecosystem. However, with the short-term success and the public attention of MAR also the expectations increased, with the risk of disappointment in case of delay or failure. Gartner pushed AR in its Hype Cycle for emerging technologies from "more than 10 years" in 2008 (Martin, 2008) to "5 to 10 years" in 2009 (Gartner, 2009) before mainstream adoption, and from a "technology trigger" to the "peak of inflated expectations" in 2010 (Martin, 2010). Accordingly, current venture capitalists, startups, as well as consumers are expecting even faster and more disruptive results.

To make the right decisions about an engagement in the young field of MAR, it is essential to have a complete picture about its current situation. There are various variants of MAR already available, each with its more or less own stage of development and time bar, but always somehow interrelated. The spectrum of MAR and therefore its potential is broad, it ranges from simple geo-tagged Web links up to a mobile multi-user ubiquitous mixed reality, or maybe beyond, that could even change our current comprehension of information technology all in all. MAR also depends on and interrelates with further technical (computer vision, image recognition, RFID, NFC etc.), infrastructural (3G, WiFi, LTE networks etc.), social (Web 2.0 interfacing, friend tagging etc.), security (user tracking, profiling etc.) and many other areas and their developments. However, the grains of MAR are in the earth, and we see the first spears, but there are forests to come.

BACKGROUND

At this early stage of MAR, there are already different approaches, suppliers, technologies, and systems on the market that are forming or at least delivering their part to the growing MAR community. 'Older' players like the German metaio left the research labs and entered the commercial market in 2003 with their PC and camera based AR system "Unifeye" before launching their MAR app "junaio" in 2009. Other players like the Austrian company Mobilizy entered the market right away with their MAR service "Wikitude" in 2008. Some of the other providers and their MAR browsers are listed in Table 1, following Perey (2010).

Table 1. MAR providers and their browsers

Provider	Country	MAR browser
Mobilizy	Austria	Wikitude
Layar	Netherlands	Reality Browser
metaio	Germany	junaio
Tonchidot	Japan	Sekai Camera
Nokia	Norway	Point and Find
Iryss	USA	Tagwhat

For most of the existing MAR browsers, the number of supported Mobile Operating Systems (MOS) is permanently extended according to the tendencies on the highly dynamic MOS market. Some MOS make it easier for developers to access the hardware, sensors and also the features of the mobile network infrastructure by offering open programming interfaces. So it was not a surprise that the first MAR browser from Mobilizy was released on Google's open MOS called "Android", that was originally developed under the auspices of the Open Handset Alliance (OHA), an industry group with currently 76 prominent members from the mobile operators, semiconductor, software and commercialization industry (OHA, 2010). Along with the current trend that Android becomes more and more widespread on many different devices from miscellaneous manufacturers with favorable prices; the availability of MAR enabled devices increases tremendously.

This is also true for some of the other preconditions of a successful MAR market like the continuous performance progress of the mobile devices (CPU, memory, battery etc.), and their technical facilities like enhanced sensors (GPS, EGPS, AGPS etc.), as well as the mobile infrastructure like mobile networks (3G, LTE, WIMAX etc.) or short distance environments (RFID, NFC etc.). Also camera resolution is relevant especially for marker-based MAR, where simple 2D markers like bar or QR codes are recognized and overlaid with 2D or sometimes 3D graphics. However, for the current mainstream of marker-less, location-based MAR apps only three core technologies are required in order to display the relevant information at the right position on the mobile's camera display: a GPS system to determine the current user position, a digital compass to recognize the user's orientation, and a gyroscope to measure the pitch and yaw of the mobile device held in the user's hand.

Basic Approaches

Most of the currently available MAR apps are based on these three hardware sensors, and on the central concept of the so called "Airtags" (Tonchidot, 2010). In order to augment the reality the user points the camera of her smart phone to real objects in the surrounding environment. The recorded video stream is displayed on the mobile device and is overlaid in nearly real-time with small graphical 2D icons or buttons at almost exactly the position where virtual information is available for a real object in the physical world. While the device calculates what the user is looking at through the mobile's camera lens based on its position, orientation and viewing angle, the device is also aware about the geographical distance to the targeted object, and can adapt the size of the Airtag accordingly (the closer the object the bigger the displayed Airtag).

The definition of Airtags is somehow similar to the definition of hyperlinks in HTML. The main difference is that a classical hyperlink has its position on a dedicated text line on a webpage, while an Airtag has its position on a geographical coordination point in the real world. If an Airtag is tapped on the mobile's display a hyperlink to a webpage with the according location-based information can be called via the mobile's data link to the Internet. This concept is usually also called *marker-less* or *location-based* MAR (Lamb, 2010), since no visual information from the surrounding is required nor evaluated by the MAR app in order to display the Airtags on the camera's picture. It is used in one or the other way also in the most popular MAR browsers like Sekai Camera, Wikitude, junaio or Reality Browser.

On the other hand, and still more rarely in the field of mobile AR, are the so called *marker-based* MAR apps. Originally coming from the stationary PC-based desktop AR this kind of MAR approach is not based on the sensor information about the user's current position and orientation, but on markers that are recognized within the current camera's picture. Based on elementary methods of pattern matching and image recognition predefined graphical 2D patterns (e.g. QR codes) are recognized, and their position and alignment are captured in real-time according to the recorded camera stream. Based on these calculations 2D and also 3D graphics can be overlaid and adjusted according to the markers, so that a user can move his camera around the marker to examine a bigger virtual 3D object from all sides, or to move a smaller object with the marker in his hands in front of the camera lens.

The marker-based image recognition and overlay of aligned 3D graphics to the camera stream in real-time is still a challenge for today's mobile hardware performance of usual smartphones, but it is already possible on an elementary level, as metaio's "junaio" extension "glue" demonstrates (metaio, 2010). Moreover, the combination of adjusted 3D graphics and marker-less MAR currently seems to be even more far from mass utilization due to the limitations of mobile CPU performance, camera and localization exactness, as well as fast network access to rich 3D geographical databases. But potential solution approaches are already underway also in this field with its associated peripheral and core MAR technologies, as demonstrated by Layar's arcade game trial that augments a real soccer field with 3D pac-man figures (Layar, 2010b).

With the approach of marker-less MAR (also in combination with marker-based MAR) many different apps in the context of tourist information, user navigation, hyper-local marketing, museum tours, real-estates, games, or social activities can be easily imagined, and are being released since 2008. So metaio's "junaio" allows users to see and find the nearest ATM machine from Addison Avenue, or the nearest public viewing place for the next soccer match, or even to shoot virtual zombies in real environments with their MAR game "Zombie ShootAR" for Nokia (metaio, 2010). Mobilizy's "Wikitude" augments holiday trips worldwide with references from Lonely Planet Compass

Guides, overlays the city of London with tourist informations, or shows the names and heights of the Alp's crests to a user looking at the surrounding panorama (Mobilizy, 2010). Iryss' MAR app "Tagwhat" allows the user to follow her friends from social platforms in the real world (Iryss, 2010), and with the app "Recognizr" from the Swedish company The Astonishing Tribe (TAT) even faces of arbitrary passerby's can be recognized and tagged with links to their actual social network profiles (Gaerdenfors, 2010).

One essential step ahead in the development of MAR apps was the introduction of the concept of "layers". A MAR layer typically merges Airtags of one topic, group, company, region, or whatever information provided from an institution of interest. The Airtags of interest could be all ATM machines worldwide, all historic buildings of a country, or all free flats of a realtor in a city. Layers can be chosen and loaded into a MAR browser when they are needed, alone or in combination with others. So the user only sees the information he is interested in at a current point of time and situation. Furthermore, layers can be built without programming a MAR app from scratch. The operators of layer-enabled MAR browsers and servers, like Layar or Mobilizy, are offering open interfaces to content providers to develop and upload their own layers and MAR points of interests (POI) to public available servers. However, the final selection of the best fitting layer(s) is still a manual process triggered and carried by the user, responsible for filtering the appropriate type and amount of information by herself, and depending on the exactness and correctness of the information provider.

After the Hype

As of writing this chapter the still small MAR community already experienced their first hype phase. During the enthusiasm of 2009 and the further growth in 2010 most of the plenty articles, blogs, or conferences in the context of the mobile future mentioned at least also the topic of MAR, or even made it to its central subject matter. Also the increasing number of MAR apps for fun, marketing as well as first commercial services nourishes the impression that MAR has definitely left the research labors and is finally on its irresistible triumphal procession. But has today's MAR really developed that far?

Some of the most successful providers of MAR apps and browsers are in doubt themselves. Although they ride on the current wave of public interest and mobile enthusiasm they are aware that a hype phase is usually followed by a valley of disillusion and prove that decides about the long-term success and establishment of a new technology, service approach, and finally market. MAR is currently leaving its hype phase, having raised big hopes, but is forced to provide proof of concept now to not disappoint its believers and investors (Lens-Fitzgerald, 2010). Also metaio misses some fundamental progress in MAR, advises against stagnation in development, and reminds that the first mobile AR system based on GPS and compass was already developed more than ten years ago. Additionally today's MAR systems are still suffering from the same well-known problems, like restricted usefulness of MAR content, insufficient indoor tracking and precision, restricted 3D capabilities, and other issues mainly due to technical limitations (Meier, 2010). While some of the players suspect the end of the AR-hype bubble in 2010 (Lamb, 2010), others proclaim quite contrary that the current offspring of MAR are just some early forms of pseudo-AR, with 'true' MAR still to come (Théreaux, 2010).

The idea that the current system approaches are just some elementary ancestors of future MAR systems is quite common in the MAR community. Some of the predictions for advanced MAR systems seem to be more available, conceivable and maybe realistic than others with a broader vision of future usage of information technology in general. Robert Rice from the MAR provider Neogence Enterprises bridges in one of his articles from today's available technologies via conceivable research approaches towards the visionary future of advanced MAR in a form of augmented vision in a so called decade of ubiquity. He distinguishes between four development levels within different time frames considering also the origin of MAR in the general field of AR (Rice, 2009).

- **The Past:** After Ivan Sutherland invented and built the first data helmet in 1968 the civil AR systems as well as its military ancestors experienced its first public hype in the early 1990's, with still clunky hardware like semitransparent head mounted displays (HMD), and its destiny closely related to the sister technology Virtual Reality (VR). The dedicated high-end systems of this era are still in use in professional environments of the automotive, architecture or medicine industry.

- **The Present:** Heralding today's hype the current AR approaches of the second generation respectively MAR clearly have their roots in the Past, but are running on today's high-capacity and affordable hardware systems available at relatively low costs for the average user. Their basic properties have been described in this chapter before. The first two development Levels are dominating the AR market as known today.

  ○ **Level 1:** Based on the modern desktop hardware the success of AR on stationary PC's with markers held into the desktop camera was re-invented. Markers are used crucially to position and / or to select an appropriate graphical item on the monitor.

  ○ **Level 2:** With the appearance of today's smart phones the stationary AR became mobile and the success story of current MAR apps began. Hardware became small, mobile and convergent, i.e. integrates former separated systems like computer, camera, GPS etc. into one single handy device. At the same time wireless networks like 3G or WiFi were rolled out, and the necessary broadband data access to the Internet became available nearly anywhere and anytime.

- **The Future:** In its Level 3 MAR would transform itself into something totally different as we know it from today. It would change our way how we use and perceive technology as it moves closer to the user, experienced somehow as part of the environment. Although radical in its overall perception the future MAR is still mostly based on the successful convergence of already known and to some extend existing single disciplines and technology approaches. By wearing lightweight eyeglasses with semitransparent displays and other pervasive equipment in ubiquitous environments the human vision becomes almost naturally augmented by virtual information, and former AR transforms itself to a more immersive "Augmented Vision" (AV).

But AV is not just an advanced generation of sophisticated hardware. It is much more related to new software and network approaches in retrieving, processing, connecting and providing of relevant information to the user. The relevance of information strongly depends on the individual context, preferences and other backgrounds of the user at the current time, location and situation. This contextual and personalized information has to be selected from the enormous amount of general data available on the Internet about the space-time continuum in the user's mixed reality of physical and virtual social life. Rice expects that the necessary convergent technologies of pervasive wireless broadband, semantic search, image recognition, intelligent agents, persistent multi-user data clouds etc. will be commercially

realized in about 2-3 years, while their domination of the mass market will at least took another 5-7 years from now.

According to Rice, the MAR of Level 3 will have a huge impact to our life equal or greater than the effect of the Internet and the Web combined. It could have the potential to raise entirely new industries, and change the existing ones fundamentally. In the so called "Decade of Ubiquity" between 2010 and 2020 the real world would develop to a seamless interface to the information space, with any and everything tagged, labeled, connected, tracked, interpreted, stored, and filtered, dynamically and interactively in real-time.

- **The Distant Future:** Although for the Future it already seems to be difficult to distinguish between conceivable research targets and speculative science fiction, Rice affiliates an even more long-term outlook for his Level 4. At this far point of time the last barriers between man and machine will finally overcome. Contact lens displays would be just the transition to "direct interfaces to the optic nerve and the brain". However, at the University of Washington work on a contact lens for bionic eyesight is already underway (Parviz, 2009). This will be the point of time where "multiple realities collide, merge and we end up with the Matrix" (Rice, 2009).

To evolve MAR successfully to the future is a mission-critical task for the young MAR community. They must overcome the upcoming valley of disillusion after the hype with as less damage as possible in reputation and expectation to this promising new technology and market. But the risks are significant. In one of her articles Christine Perey from Perey Research & Consulting, defines several categories of risks that could have an impact to the growth of MAR. She also delivers estimations about the timeframes when the different risks could probably become relevant (present, in 6-12 months, beyond 12 months) and suggests institutions and industries who could best take care on the open issues (Perey, 2010b).

- **Technology:** As in the many years of basic AR research, the current discussions about the challenges of MAR are still dominated by technological aspects and challenges. As stated before, these are actually general issues like restricted CPU power, short battery capacity, security, insufficient network coverage and data capacity, or also more specific aspects like inaccurate or delayed GPS localization limited to outdoor usage that could be addressed by manufactures of mobile equipment as well as mobile network operators. One of the main issues is the increasing diversity and fragmentation of mobile devices and their MOS that tightens the problem of interoperability also for MAR apps even further. This is also true for the required location-based content and geo-map databases that are available from competing providers in not always compatible formats (e.g. Google Maps, Apple Maps, Ovi Maps). Same problems emerge also for medium and long-term technology trends like image recognizers and 3D databases as basement for advanced MAR apps that are even more far from consolidation and standardization.

- **Usability:** At first glance the handling of today's MAR apps seems to be quite intuitive for the average user pointing her smartphone camera to the object of interest and receiving the associated information right away. Though, this usage of information technology is new and quite unusual for most of the users who are used to their keyboard and mouse sitting in front of her PC or Laptop instead of walking around while holding up and looking through their mobile phone. Additionally, today's limited technological quality in exactness of

localization, slow data access, delays in displaying, and potentially too many items in view due to immature mechanisms of content filtering and label selection could even more disrupt the user experience, and compromise the acceptance of MAR apps for everyday usage as well as their added value. As one of the central risks, first-time users of MAR could be attracted by the current hype, become disappointed due to the current limitations, and never try out MAR again in the future.

- **Profitability:** The potential market success of MAR is closely interrelated with the challenges in technology and usability. To reach the critical mass of users in order to justify investments necessary to advance the still rudimentary MAR infrastructure, as much as possible potential customers have to be attracted precociously. If the average users try out MAR too early they could get disappointed, and otherwise, if they hesitate too long, not enough investors will be interested to finance the necessary improvements for the MAR user experience. Although this situation seems to be a dilemma, it is a quite usual situation for almost all new technology approaches in their start phase - for the winners as well as for the losers. On the other hand the business cases are also depending on the agreements between the many involved participants in the MAR ecosystem. Will the mobile operators provide sufficient data capacity required for high quality MAR services? Will this infrastructure be co-financed from MAR system operators who benefit from high-performance mobile networks? Will both parties participate from the revenues of the MAR content providers? Will the user pay an extra fee for MAR services, or are they cross-financed totally or partly by advertising or other third parties? Resolving these types of questions is necessary to develop a stable and long-term market, to reduce the risk of bad investments, and to obtain the necessary funding for new startups and the entrepreneurs pushing the further MAR development.

- **Socio-cultural and legal:** The MAR approach by itself compasses some of the most emotional aspects of today's technology usage. Based on the inherent mechanism of tracking the user's position (GPS) plus the actual focus to points of interest (compass and gyroscope), and maybe their selection. (Airtags) delivers some very personal and therefore critical information about the user and her private affairs. This personal data tracked and recorded can be used to help the MAR app provider to optimize the service by deriving individual preferences and to select the most valuable information for the user's particular situation. The same information can also be used to update automatically the user's social web profile, so all friends stay up-to-date about the user's current location and what she is doing there. By the connection of the personal tracking data with other platforms and services many different applications are conceivable. While some of them will 'just' change the transparency and self-conception of the individual in her personal environment with her compliance to some greater or lesser extent, some others could actually misuse this highly private data. This could range from harmless location-based advertising up to illegal forms of observation and other criminal offenses. Until now the total number of MAR users still seems to be too small to give a cause for a broader discussion by public organizations, government or legal institutions, but this will follow for sure with a growing market penetration.

Challenges and Solution Approaches

As a consequence of today's technical limitations and the different risks for the still young MAR market, the new players of the MAR ecosystem are confronted with essential challenges to develop their common future. As with the grouping of the risks also the different resulting challenges can be categorized in order to deliver an orientation about some of the main foci in the current discussions and working groups. Although the following discussion of today's challenges and solution approaches makes no claim to be complete, it presents an overview of the diversity of questions arising in the context of MAR.

Collaboration

One complex challenges of MAR is the structures of cooperation between the different participants. While mobile infrastructure providers are in one of the toughest phases of consolidation, convergence, and cutthroat competition, the investments into the new mobile data infrastructure are immense at the same time. Having spent billions of Euros for the UMTS licenses and the 3G network rollout over the past years, the mobile data capacity already reaches its limit again. Therefore high investments into the next mobile network generation like LTE (Long Term Evolution) are required. Thanks to this sophisticated mobile data infrastructure the hardware manufacturer industry experiences a new disposal peak for smartphones and touch pads after a phase of depression during the global economic recession. Besides, a whole new market for apps emerged rapidly from a niche product to an already weightily, area. According to the consultancy Booz & Company the App Store from Apple with more than 140.000 apps expects revenues of about 2.4 billion Euros for 2010, with one third of the revenues flowing directly to Apple. For 2013 Booz estimates a yearly growth in sales of about 73%, with more than one billion of Internet-enabled smartphones worldwide then, and a pure app download revenue volume of about 17 billion Euros (Booz, 2010).

On this multi-billion Euro market the traditional and the new players will have to organize themselves in order to establish a stable and long-term mobile economy. The discussion about cost sharing for the usage of the mobile infrastructure between network operators like Deutsche Telekom and service providers like Google with its data-intensive video platform YouTube have just begun (Heise, 2010). Also the mobile data flat rates are already questioned in their function as market enablers on the one hand and regarding the cumulative discrepancy of revenues versus network costs resulting from an increasing demand for capacity on the other hand. The global network operators, service providers, handset manufacturers, and research institutions are forming strategic alliances to bundle their strengths in order to enable the common growth market together. One prominent example is the Open Handset Alliance (OHA) where the world's biggest telecommunications companies are working together on the mobile future e.g. by developing open mobile operating systems like Android to push the application development (OHA, 2010).

However, today's MAR community is still forming a very small niche within that huge evolving mobile ecosystem. They have the potential to generate a strong impact to the many different kinds of mobile apps and information services as a new kind of human computer interface in general, besides the inherent potential as dedicated MAR service itself. To gain the necessary attention in the overall discussions during the upcoming segmentation of the mobile market the relatively small MAR startups would be well advised to bundle their activities, efforts, and votes. Certainly individual preferences and products are highly important in order to diversify the MAR portfolio further and to attract the necessary venture capital for the short run. But to achieve success,

the MAR entrepreneurs will have to compete against the big players who will also try to get their piece of the action with all the financial and marketing power they already have available. As an example, IBM already experimented with their MAR app "Seer Android" in the context of Wimbledon 2009 (IBM, 2010), while Nokia and Intel just established a Joint Innovation Center at the University of Oulu in Finland for the development of rich, immersive and augmented services using 3D on mobile devices (Intel, 2010). Although today's MAR community already has its forums at conferences and online blogs, it could make sense to form other kinds of official MAR alliance in order to represent the common interests, and to provide a strong counterpart as well as partner to the other mobile players.

Customer-Orientation

As discussed in the previous sections about usability, socio-cultural and legal aspects the future of MAR is also highly affected from the acceptance of the new technology approach by the users in their specific situations, environments, and societies. Although there definitely 'was' a first hype about MAR and more people than ever know about this kind of new application, the big market breakthrough is still missing. The current discussions at conferences and blogs are still dominated by technical issues not least due to the enormous challenges in this area that still lie ahead of the MAR community. But after demonstrating the pure rudimentary technological approach to the public the MAR entrepreneurs are now pressured to develop true added values for their future MAR customers, both in the private as well as in the business consumer area. This does not mean a reduction in the efforts into technological development, but to build up and strengthen at least a second pillar aside to build a stable MAR market on. Dazzled by the first enthusiasm in building and demonstrating working MAR apps now is the time to build up also the customer base willing to pay an extra amount of money for pure or extended MAR apps. And finally this is what the venture capitalists are expecting, too.

For the early MAR pioneers and entrepreneurs, this additional requires a change in perspective from a technology-centric innovation to a supplementary customer-centric marketing approach. Both approaches are important and are influencing themselves reciprocally, but need to be balanced. According to Kozick, a well-balanced interaction of marketing and innovation should be the basis to generate true added value of new technological services for customers in a three-step approach: a) marketing researches human needs b) innovation serves the needs, and c) communicates the innovation to the users. However, understanding human needs should not only mean to detect more and more ideas for even more technological innovations on the one side, but to identify and filter those innovations that really fulfill the true needs of the customers on the other hand. Furthermore, innovation should not end with the development of a new technology, but additionally with its creative commercialization. In this context Kozick proclaims an "art of innovation" and "science of marketing", and calls on the MAR community for a reality check to prove if the current MAR apps and approaches really meet the customer needs (Kozick, 2010).

While strengthening the customer's added value of MAR products and services also the drawbacks and risks should be taken into consideration, and maybe should be confronted and communicated pro-actively and transparently. Concerned by the current discussions about privacy of online data, most users are sensitive about providing the data that location-based MAR apps inherent. It could be a good measure for the players of the MAR ecosystem to actively provide confidence about their services, and the secure handling of personal user data. MAR has the potential to change the users' personal life. Together with other new technologies like social networks, Internet,

over-layered environments, and advanced forms of ubiquitous MAR apps with face recognition and other applications, the situational setting of the user could change dramatically. It could result in a perception that it is getting more complicated and irritating by infiltrating and penetrating the user's daily life. Schroll names this phenomenon the "social nakedness" of users in an "augmented world", and expects political and legal influences to the so called "augmented citizen" (Schroll, 2010). It could be worth to consider these potential social developments already today, and to be open for discussions also with federal institutions and related organizations.

Innovation

As already described in the context of the four Levels of MAR, according to Robert Rice (Rice, 2009) the present development phase of MAR is somewhere near to Level 2. Since late 2008 the first AR applications have been successfully transformed from stationary PC-based systems to mobile handhelds and therefore heralding the hour of birth for MAR. Although on a very early development stage in relation to the two additional Levels to come, most of the current elementary technologies for MAR apps are still far from their accomplishment. Poor CPU and battery power of mobile devices, low mobile data coverage and bandwidth, inaccurate and outdoor-only localization and orientation sensors, increasing fragmentation of hardware devices and MOS, and additionally still expensive mobile data rates, as well as more or less quite exotic MAR apps with low usability. However, on most of these issues, huge efforts are being undertaken to resolve these issues. As a side effect of the general improvements on the multi-billion Euro mobile market, MAR will profit from the enormous investments of the big telecommunication and supplier companies who will enhance the capabilities of their mobile networks, hardware components and devices anyway in order to ensure or even increase their market shares, revenues and margins. While some MAR relevant technologies like CPU, battery, coverage, and bandwidth will profit more, some others like exactness of GPS, gyroscope, and compass will profit less from the general developments. The better the MAR community manages to push 'their' technologies to a general focus and relevance, the bigger the chance to ride on the crest of the wave.

To improve the basic technologies for today's MAR apps does not seems to be sufficient enough to proceed on the next development stage towards Level 3 MAR. Within the next level of MAR fundamental technologies will have to join the foundation of MAR apps with additional sensors, audio and tactile input, and most notably techniques for computer vision and object recognition, like patterns, images, and faces. But instead of replacing the current technologies and approaches, the new modules will be added to complete the tool box for building future MAR apps. At this stage questions like, "what is the best approach for MAR, marker-based or marker-less?", will become more and more dispensable in favor of questions like "what is the best combination of MAR technologies for a specific application and situation?". One of the success factors will be to integrate MAR within the already connected world of the user's digital environment and life, and to generate additional, rich and valuable interrelations. Devices will become more active in processing information from the user's real and virtual surroundings as the so called "eyes of MAR" (Ryu, 2010).

As an example ubiquitous MAR equipment will perceive objects and people by visual recognition and RFID sensors while retrieving and sending intelligently selected information permanently back and forth between the user and the different services on the Web. Not a single sophisticated MAR technology approach will probably support all of these requirements, but the combination of

existing and new technologies that are already used to some extend today by tomorrow's MAR customers.

Also the work and research on these new adding technologies for MAR seems to be already on its way. In many different contexts the technologies of computer vision, information retrieval, semantic web, artificial intelligence, human computer interaction, ubiquitous computing, embedded systems, image and face recognition, ultra-precise tracking and positioning, virtual worlds and social networking, and other relevant fields, are proceeding both as single disciplines and common efforts within research institutes, small startups as well as at the R&D departments of the big mobile players (Marimon, 2010). MAR will also profit from those advancements, and once again the MAR players have the chance to bundle these efforts under the framework of the multidisciplinary approach of MAR, Augmented Vision, or whatever it will be named in the future. Additionally, there are other framework disciplines that are already bundling some of the fundamental technologies in interdisciplinary efforts and teams.

One prominent candidate is the so called discipline of Contextual Services (CS), where mainly mobile services are "designed around your situation", like the mobile service architects from SPRXmobile define it. They generally see the mobile user within different contexts, like her current environment, activity, social, spatio-temporal, mental, physiological, real as well as virtual auto-perception, that all changes dynamically over time. In a fulfillment process a CS app conciliates and anticipates between the context specific need of the user in the physical world and the relevant information in the virtual world (Boonstra, 2010).

While the CS approach could be handled independently from some of the core MAR technologies, the latter would be definitely the ideal human computer interface for this kind of context-aware information and communication technology. That both approaches are highly interconnected becomes also obvious within Nokia's vision for future mobile services, where MAR merges with location-based and context-aware services (Nokia, 2009). Context plays the central criteria for individual information selection, while the substance of information has a stronger focus on commercial content (media, music, gaming, navigation, messaging) due to the business and customers of the company. This brings another important participant into the MAR ecosystem that the community is well advised to take into closer consideration: the traditional content providers. Without the 'right' content also MAR will be just another access technology, or even worse just an empty pipe. So it is essential to also talk to the big content owners to make both sides more sensitive for the specific features, requirements, and success factors of geo-based and visually recognized content for MAR apps as a new distribution channel. Again, the better the business, research and technological interrelationship between the core players and flanking industries are, the better for the success of MAR itself.

Standardization

The more heterogeneous the components, participants and interests within a highly interdisciplinary ecosystem are, the more important is the role of common technological standards. This basic principle applies for almost the whole value chain in the anticipated MAR ecosystem. The consumer wants to use different MAR apps with just one single mobile device, the app developer wants to build an app only once usable on many platforms, the hardware supplier wants to support as much as possible apps, and content providers want to re-use their information for multiple channels. Furthermore, the lack of standards bears the risk of bad investments by betting on the wrong horse and leading to a dead-end, resulting in frustration and financial disaster. Therefore it is of essential

interest for all of the participants to setup and agree on international standards to guarantee the highest compatibility and interoperability between the hardware and software systems as possible. This pressure but also the common hope for additional future business opportunities makes strong competitors working together in industry alliances, like the Open Handset Alliance (OHA, 2010).

So far the standardization approaches within the MAR community are still on a very rudimentary level. Following the active Web concept of participation instead of just consumption MAR browsers like Layar's Reality Browser became customizable by adapting templates for user-defined layers of Airtags (Layar, 2010). Currently some of the MAR providers are granting restricted access to their platforms via defined API's, like metaio to their junaio server platform mainly for third-party content developers in the area of location-based games (Meyer, 2010). Although this still has nothing or less to do with a standard itself it is a first step into the right direction to open proprietary systems for other MAR participants. Of course the flexibility and functionality in creating MAR apps in this way is totally limited to the feature set of the provider's service framework.

Mobilizy goes one significant step further by defining a generic Augmented Reality Modeling Language (ARML) to configure MAR apps for their Wikitude platform by external developers. Lean towards the Web language HTML its AR pendant ARML builds on the Keyhole Markup Language (KML) used for describing geo-data for Google Earth and Google Maps. Since KML is dedicated to the description of what is where (latitude, longitude, altitude) the tag set was exempt from non-AR-relevant tags and extended by AR-specific tags for interaction with the geographical points of interests (POI). Within two data areas the content provider is determined followed by the description of the POI's based on three gradual namespaces *kml* (basis), *ar* (generic AR tags), and *wikitude* (manufacturer-specific AR tags). The data format of ARML files is based on the Web standard XML and is therefore highly compatible to other Web-based applications. ARML was submitted to the World Wide Web Consortium (W3C) for evaluation as a possible Web standard for generic AR content (Lechner, 2010). Also the open source industry collaboration BONDI with its Open Mobile Terminal Platform (OMTP) intends to build upon Web standards. Instead of realizing MAR browsers as proprietary systems on dedicated MOS platforms the intended setup as generic Web applications respectively widgets could help to reduce or even to avert the app fragmentation. Running within ordinary Web browsers the AR widgets could make generalized API calls to the hardware components (e.g. GPS, camera, gyroscope) of a specific mobile device, and therefore enabling highly compatible and platform-independent MAR apps running on almost any mobile device (Rogers, 2010).

All of the efforts for standardization are especially crucial for the content providers. To generate and provide geo-based and augmented content, enhanced or dedicated for MAR apps, is a complex and expensive task that needs to be as cost-effective as possible. This applies even more for advanced MAR apps using 3D overlays. If content has to be adapted at worst manually to each of the different mobile hardware platforms, operating systems and single MAR apps the big content owners (e.g. producers of news, films, games) will hesitate to invest for entering the market. With a generic description language like ARML there could be a basis also for them to overcome the current obstacles and lower their risks.

Most of today's online data is already based on the W3C standard XML. And XML files can be easily transformed from one XML language to another by using stylesheet translators (XSLT), or even mixed with each other by using namespaces, like e.g. with the 3D Web standard X3D (Kloss, 2010). That makes the already existing data easily reusable also for MAR apps. But still the specific geo-coded data for MAR apps has to be produced extra, and for most of today's data on the Web

there is neither geo-data nor 3D data available, yet. So it is a reasonable approach for today's MAR apps to migrate and interlink what is available already today (e.g. transportation timetables, estate addresses, social network information, user localization data), and to foster the development and establishment of an independent and comprehensive MAR standard, as well as to support and motivate a gradual and participative assembling and extension of Web data with MAR specific components together with the other participants of the MAR ecosystem.

FUTURE RESEARCH DIRECTIONS

While discussing the many aspects that are involved the development of the MAR, it became obvious that MAR is not a separated technology from others, but is in fact a melting pot that has the potential to bring together some of today's most exciting technologies such as Web services, and communication trends together, and to advance them fundamentally. With MAR not only the reality of the users can be augmented, but also the added-value of today's Web trendsetters can be enhanced with a new kind of human computer interface. And MAR is not just about bringing formerly separate technologies together, but furthermore to harmonize the many different people, industries, and interests involved into the current and future ecosystem of MAR. That implies that today's still quite homogeneous MAR community will probably develop itself to a much bigger and more heterogeneous group or maybe alliance - quasi 'augmenting' itself - in order to get the weight that is needed to become finally successful on a wide base and in long-term (see current mainstream adaptation of reinvented stereoscopic 3D by the big entertainment industry for comparison). Instead of discouraging the young MAR startups and entrepreneurs these circumstances should be taken as a chance to position themselves as a competence center and as managers of collaboration. To avert the destiny of MAR as just another app, huge and collective efforts are just simply necessary.

However, there is still much specific work to do also in the single disciplines. As discussed there are not just challenges in the hardware and software features for MAR services, but also according to their standardization, ergonomics, security, business case, socio-cultural, and also legal aspects. Researchers and developers should consider also the bigger framework in their specific work areas, and more comprehensive disciplines like Contextual Services or Augmented Vision should monitor the single achievements and assemble them to an overall picture. In a three-step approach specific results from single R&D disciplines could be bundled to new versions of MAR services that could be commercialized to the customers. New releases would therefore reflect the current overall status in research, technical development, collaboration, and business models of the whole MAR ecosystem. Just a small part of these releases will be controlled or managed by a central instance, and most of it will happen due to unmanageable factors. But it is a good advice to all of the participants to keep the big picture in mind while working on their own single targets for future MAR.

CONCLUSION

While writing this chapter and investigating the different aspects relevant to MAR it became clear that we currently see just a very small tip of the iceberg. This seems to be surprisingly at first since (M)AR is actually an 'old' discipline being researched for more than two decades. But during this time the world changed dramatically, and it happened that MAR fits to most of the current technological achievements especially in mobile information and communication technology. The chances for a successful rebirth of that promising approach are better than ever before. Still quarrel-

ing with some of the teething troubles, MAR now grows up in a whole new environment of powerful mobile networks and miniaturized devices, with objects and users already multi-connected with and within the virtual world, online anytime anywhere. Although there is no guarantee that MAR will succeed this time, there is also none that it will not.

REFERENCES

ABI. (2009). *Dramatic growth for AR via smartphones.* Retrieved September 1, 2010, from http://www.abiresearch.com/press/1516-ABI+Research+Anticipates+%93Dramatic+Growth%94+for+Augmented+Reality+via+Smartphones

Bergman, C. (2010). *AR will be big in 5 years, says study.* Retrieved September 1, 2010, from http://www.lostremote.com/2010/01/06/augmented-reality-will-be-big-in-5-years-says-study/

Boonstra, C., van der Klein, R., & Lens-Fitzgerald, M. (2010). *Contextual services in mobile.* Retrieved September 23, 2010, from http://www.slideshare.net/Thinkmobile/contextual-services-in-mobile-presentation

Booz & Company. (2010). *App-Downloads generieren 2013 Umsatzvolumen von 17 Mill. Euro weltweit.* Retrieved September 7, 2010, from http://www.booz.com/de/home/Presse/Pressemitteilungen/pressemitteilung-detail/48203889

Gaerdenfors, D., Haliburton, J., & Stark, P. (2010). *AR for the masses.* Retrieved September 1, 2010, from http://www.perey.com/MobileARSummit/TAT-Augmented-reality-for-the-masses.pdf

Gartner. (2008). *Gartner identifies top ten disruptive technologies for 2008 to 2012.* Retrieved September 1, 2010, from http://www.ehomeupgrade.com/2008/05/28/gartner-identifies-top-ten-disruptive-technologies-for-2008-to-2012/

Gartner. (2009). *Gartner hype cycle 2009.* Retrieved September 1, 2010, from http://www.gartner.com/it/page.jsp?id=1124212

Heise News. (2010). *Telekom will Diensteanbieter zur Kasse bitten.* Retrieved September 7, 2010, from http://www.heise.de/newsticker/meldung/Telekom-will-Diensteanbieter-zur-Kasse-bitten-1042726.html

IBM. (2010). *Augmented reality projects at IBM.* Retrieved September 7, 2010, from http://www.perey.com/MobileARSummit/IBM-Recent-AR-Project-Experiences.pdf

Intel. (2010). *Intel and Nokia: A new lab for mobile 3D.* Retrieved September 7, 2010, from http://blogs.intel.com/research/2010/08/today_intel_along_with_nokia.php

Kloss, J. (2010). The role of standards for e-commerce in virtual worlds. In Ciaramitaro, B. (Ed.), *Virtual worlds and e-commerce.* Hershey, PA: IGI Global. doi:10.4018/978-1-61692-808-7.ch015

Kozick, Z., & Gettliffe, C. (2010). *Why AR needs a reality check.* Retrieved September 21, 2010, from http://www.perey.com/MobileARSummit/Omniar-Augmented-Reality-Reality-Check.pdf

Lamb, P. (2010). *Ever wider and deeper: Augmented reality in 2010.* Retrieved August 18, 2010, from http://www.perey.com/MobileARSummit/ARToolworks-Ever-Wider-and-Deeper-AR-in-2010.pdf

Layar. (2010). *Website.* Retrieved August 14, 2010, from http://www.layar.com/

Layar. (2010b). *Layar reality browser adds 3D to its platform.* Retrieved August 18, 2010, from http://site.layar.com/company/blog/layar-reality-browser-adds-3d-to-its-platform/

Lechner, M., & Tripp, M. (2010). *ARML - An AR standard.* Retrieved September 23, 2010, from http://www.perey.com/MobileARSummit/Mobilizy-ARML.pdf

Lens-Fitzgerald, M. (2010). *Was there movement on the AR hype cycle curve?* Retrieved September 1, 2010, from http://www.perey.com/MobileAR-Summit/Layar-Was-there-movement-on-the-AR-Hype-Cycle.pdf

Marimon, D., et al. (2010). *Mobile visual recognition, the future of MAR.* Retrieved September 23, 2010, from http://www.perey.com/MobileARSummit/TelefonicaR&D-Mobile-Visual-Recognition.pdf

Martin, H. E. (2008). *Gartner's emerging technologies hype cycle 2008.* Retrieved September 1, 2010, from http://hemartin.blogspot.com/2008/08/gartners-emerging-technologies-hype.html

Martin, H. E. (2010). *Gartner's hype cycle emerging technologies 2010.* Retrieved September 1, 2010, from http://hemartin.blogspot.com/2010/08/gartners-hype-cycle-emerging.html

Meier, P. (2010). *Mobile augmented reality 2010.* Retrieved September 1, 2010, from http://www.perey.com/MobileARSummit/metaio-Mobile-AR-in-2010.pdf

Metaio. (2010). *Press release.* Retrieved August 18, 2010, from http://www.metaio.com/media-press/press-release/

Mobilizy. (2010). *Press.* Retrieved August 18, 2010, from http://www.wikitude.org/category/07_press/press-press

Nokia. (2009). *Location, context, and mobile services.* Retrieved September 23, 2010, from http://research.nokia.com/files/insight/NTI_Location_&_Context_Jan_2009.pdf

Open Handset Alliance. (2010). *Home.* Retrieved August 14, 2010, from http://www.openhandsetalliance.com/index.html

Parviz, B. (2009). *Augmented reality in a contact lens*. Retrieved September 3, 2010, from http://spectrum.ieee.org/biomedical/bionics/augmented-reality-in-a-contact-lens/0

Perey, C. (2010a). *Cross-platform AR: The market need and potential solutions*. Retrieved August 18, 2010, from http://www.perey.com/MobileAR-Summit/PEREY-CrossPlatform-AR.pdf

Perey, C. (2010b). *Clouds on our horizon? MAR risks and obstacles*. Retrieved September 3, 2010, from http://www.perey.com/MobileARSummit/PEREY-Clouds-on-the-Horizon.pdf

Rice, R. (2009). *Augmented vision and the decade of ubiquity*. Retrieved August 18, 2010, from http://curiousraven.squarespace.com/future-vision/2009/3/20/augmented-vision-and-the-decade-of-ubiquity.html

Rogers, D. (2010). *BONDI augmented reality*. Retrieved September 23, 2010, from http://www.perey.com/MobileARSummit/OMTP-BONDI-Augmented-Reality-David%20Rogers.pdf

Ryu, J., & Lee, K. (2010). *How can we make the eyes for the MAR services*. Retrieved September 23, 2010, from http://www.perey.com/MobileARSummit/Olaworks-Eyes-for-MobileAR.pdf

Schroll, W., & Romescu, D. (2010). *Augmented citizen*. Retrieved September 21, 2010, from http://www.perey.com/MobileARSummit/Romescu-The-Augmented-Citizen.pdf

Tagwhat. (2010). *Iryss*. Retrieved August 18, 2010, from http://tagwhat.com/

Théreaux, O. (2010). *Beyond the glorified tour guide and the dystopian future*. Retrieved September 1, 2010, from http://lab.pheromone.ca/2010/02/09/mobile-augmented-reality/

Tonchidot. (2010). *Home page*. Retrieved August 18, 2010, from http://www.tonchidot.com/

Chapter 9
The Use of Embedded Mobile, RFID, and Augmented Reality in Mobile Devices

Greg Gogolin
Ferris State University, USA

ABSTRACT

The proliferation of mobile devices such as smart phones and other handheld devices has stimulated the development of a broad range of functionality, including medical, retail, and personal applications. Technology that has been leveraged to enable many of these uses includes radio frequency identification (RFID), embedded mobile, and augmented reality. RFID involves communication between a tag and a reader. Mobile RFID extends the technology by tagging the mobile device with an RFID tag to perform tasks on the device. Embedded mobile refers to preprogrammed tasks that are performed on a mobile device. Personal care and monitoring is one of the most common uses of embedded mobile. Augmented reality involves the use of computer generated or enhanced sensory input such as audio and visual components to enhance the perception of reality. This is commonly used in situations such as video games where there is feedback in the game controllers.

INTRODUCTION

The focus of this chapter is to examine the impact of the rapid growth of mobile technologies such as embedded mobile, radio frequency identification (RFID), and augmented reality, and explore how this growth is extending the use of Internet technologies to create a more convenient and accessible experience for its users. Mobile technologies present many challenges and opportunities in a number of domains including education, healthcare, banking, emergency response, and commerce.

DOI: 10.4018/978-1-61350-150-4.ch009

This chapter will investigate the use and capabilities of current and emerging technologies such as embedded mobile, RFID, and augmented reality in mobile devices in the context of the opportunities and challenges presented. These technologies will be presented according to the following emerging framework:

1. Historical and Current State
2. Trends and Opportunities
3. Risks, Challenges, and Limitations
4. Global Implications
5. Cultural Implications

The chapter will conclude with a summary and a look at emerging trends and future direction in embedded mobile, rfid, and augmented reality in mobile devices.

BACKGROUND

An embedded system is a computer system that is integrated within the hardware and software of a computing device. Embedded systems are pre-programmed for a specific task within the device it operates. Embedded systems have been in use for many years in a variety of industries including automotive, health care, consumer electronics, video games, and a variety of other applications. Embedded mobile (EM) refers to the use of one or more embedded systems in a mobile device. The rise of EM has occurred largely as a result of the advancement in cellular communication and the proliferation of the Internet.

Radio frequency identification (RFID) is an electromagnetic wave communication technology that uses readers and tags. The reader incorporates an antenna that can read a tag wirelessly, and then transmit the information from the tag for processing. RFID technology is often used to track inventory by placing a tag on the product to be tracked. A reader reads the information on

the tag for interpretation. An example would be tagging cases of products in a warehouse with RFID tags. A reader interprets the tag which allows a worker to quickly determine information related to the case.

Mobile rfid (M-RFID) are services provided by mobile devices equipped with an rfid tag over a telecommunications network. M-RFID can be use to execute a service or mobile phone functions such as messaging or calling when an RFID tag is sensed. This can include making phone calls for those unfamiliar or uncomfortable with mobile phones. Emergency situations and assisting physically disabled individuals are common uses of M-RFID.

Augmented reality involves the use of computer generated or enhanced sensory input such as audio and visual components to enhance the perception of reality. This is commonly used in situations such as video games where there is feedback in the game controllers. Mobile Augmented Reality Systems (MARS) has become popular in devices equipped with a global positioning system (GPS), compass, accelerometer, camera, and ability to share or manipulate information. Applications can include pointing the devices camera at an object and then receiving feedback. For example, pointing the camera at a restaurant and then receiving reviews or using the device to take a tour.

HISTORICAL AND CURRENT STATE

RFID

RFID evolved from radar, which was discovered by Sir Robert Alexander Watson-Watt in 1935 (Rau, 2005). Radar was used extensively in World War II to detect aircraft, but one of the problems was that it was difficult to detect if the aircraft were friend or foe. The Allies began using radar in combination with a tag on their aircraft to help distinguish them from other aircraft (RFID Journal, 2005). This is widely viewed as the beginning of RFID technology as we know it today.

RFID tags evolved to the point where they could be rewritten and used for multiple purposes. As the costs associated with the technology decreased, use became common in many applications and industries including supply chain, retail, and medical. It became possible for workers to locate products that they could not see, and security and loss prevention initiatives had the ability to become much more stringent. The cost of RFID tags has reached a point where items costing only a few dollars can be labeled with one time use tags for inventory and loss prevention purposes.

Security uses of RFID include using badges to unlock doors and provide other forms of secured access. A shortcoming is that tags can be read or copied by devices other than those intended, which can lead to security breaches. This is one reason why two factor authentication should be considered when stringent security is required. Two factor authentication refers to the use of more than one means of verifying the authorized user of the device. It usually entails the use of a password along with either a thumbprint or a physical card or token.

A more obscure and not yet mature potential use of RFID is to track items after purchase. RFID tags are typically deactivated upon checkout at a retail establishment. Similar to website tracking for marketing purposes, and transaction tracking by financial companies, what is done with RFID enabled information is unclear. Financial institutions such as banks and credit card companies have privacy policies. Although customers are given the opportunity to either opt in or opt out, more often than not, the default for tracking and sharing information is to opt in. Marketing companies and other corporations purchase this information from the financial institutions and sell it to almost anyone that is willing to pay for it. If a customer chooses to opt out of information sharing, they often need to express this on a recurring basis or

else they will be changed to opt in status. In other instances, a customer's choice to opt out results in an inability to use the service or application. A Google search of your own name will likely lead to an incredible amount of information about yourself that you didn't realize was freely available, much of it gathered through these types of collections and sales of personal data.

RFID tags can be very small, and have the ability to be hidden on items. The technology exists to enable the tracking of RFID tags after an item has been purchased. If RFID tags were improved and RFID readers were more prevalent and powerful, people could be tracked real time based on the RFID tags embedded in the items they have purchased. Anyone that has purchased an item and then had an alarm go off when leaving a store, while entering another store with the purchase item, or returning an item to the purchase location is fully aware that RFID tags are not always permanently deactivated upon checkout. Since a 64 bit RFID tag can be assigned a signature from a potential pool of signatures in the many billions, it would not be difficult to assign each RFID tag – and therefore each tagged item – a unique signature, thereby enabling the capture of who purchased each item. A database could be designed to track RFID tags just like each item purchased with a credit card is tracked. The difference is that now an individual could be tracked and identified real time based on the tags they are wearing or carrying. Clearly, a discussion on ethics needs to occur before the problem becomes as problematic as web tracking and credit card purchase tracking.

Embedded Systems

Embedded systems are preprogrammed for a specific task and have existed for many years in areas such as health care, consumer electronics, appliances, automobiles, and traffic control signals. Devices such as the Apple iPhone and Blackberry have greatly increased the functionality and implementation of embedded mobile systems (EM). One of the key features that the iPhone and Blackberry have is the ability to house multiple EM applications in one device, thereby condensing and integrating the technical footprint to something that is easily manageable by most individuals. In many ways this has also helped remove cost as a barrier to adoption.

The actual definition of an embedded system has somewhat of a grey area because embedded systems can often be altered or modified in some fashion, or a number of embedded systems may be integrated in one device. A popular embedded system would be an MP3 player. Most MP3 players can be programmed in some fashion such as to randomize playback, creation of favorite songs lists, and sort files in a variety of ways.

There are a variety of embedded systems in the average household. Household broadband connections are often augmented by a router to provide wireless connectivity to multiple devices. The software that controls embedded systems is often referred to as firmware. Upgrading the firmware is a common way to update the functionality of an embedded system.

Augmented Reality

Augmented Reality (AR) differs from Virtual Reality in that part of the scene is real and part is computer generated in an AR environment whereas in a Virtual Reality environment the entire scene is frequently generated. Initially, Augmented Reality (AR) systems commonly took the form of a headset to view the physical world with virtual enhancements. Movie theaters began providing glasses for three dimensional movie effects in the 1950's. Other common AR formats include spatially augmented and handheld installations. The spatial augmented format uses digital projectors to display information on physical objects, thereby detaching the display from the user and integrating it within the environment. Size, weight, power consumption, scalability, higher resolution

and larger field of view are much more favorable with spatial display than handheld or headset AR (Bimber, 2005). Handheld AR adaptations include game devices and camera equipped cell phones.

AR applications for mobile apps combine virtual and real world data, often using some combination of the mobile device's compass, GPS, and camera. AR applications can do things like tell you where you are when you point your devices phone at a building, act as tour guides by identifying landmarks and providing information about them, identify where Twitter messages originate, and provide enriched Facebook services. In order for AR to recognize objects and locations the information has to be properly tagged. In other words, objects have to be prepared and classified by a proper process in order for the AR application to recognize them.

AR on smart phones is a relatively new phenomenon. There were virtually no users in 2008, but the number of users grew to approximately 600,000 in 2009. By 2012 the number of smartphone AR users is projected to grow to between 150 – 200 million (King, 2009). One of the main reasons AR has only recently gained a foothold in smart devices is that the devices really were not capable of supporting the technology. The Apple iPhone 3GS debuted in June 2009, signaling the arrival of devices that could deliver AR functionality.

An AR facial recognition application created by The Amazing Tribe (TAT), a Swedish software company that became part of Research In Motion in 2010, uses the Polar Rose facial recognition software. The application, known as *TAT Recognizr*, can recognize a person's face if the person is registered with the company. Additionally, the application can recognize and list the social networks the person is connected to. This type of technology has been of great interest to law enforcement and security personnel because it a means to identify someone of interest that may not be known to the investigator.

For AR to function properly, information has to be correctly categorized, stored, and geo-tagged with location specific data. A building may look different at night than it does during the day, and it may take on a different appearance as the seasons change when landscaping changes. A person may not be recognized if they grew facial hair and the source compare picture is the person without facial hair. So challenges lie not only with capturing the information, but storing it in such a way that an application knows to look for the proper picture and attributes. Database and content management skills are critical behind the scenes support technologies that can separate a marginal AR application from a strong one.

TRENDS AND OPPORTUNITIES

Embedded systems are one of the fasting growing segments in the computing industry, and mobile devices are one of the largest areas fueling that growth. Smart phone applications and medical devices are two key areas that support this growth. The integration of internet technology with cell phone technology provides a strong machine to machine communication platform that enables the development of applications where continuous connectivity is required. Pacemakers and other medical devices that monitor patient health can now provide information remotely, thereby reducing or eliminating the need for face to face meetings. In addition to this, vital information can be tracked perpetually, providing valuable longitudinal data that can better inform health care decisions.

Smart phones such as the iPhone, Blackberry, and Android enjoy a rapidly increasing assortment of applications for seemingly any purpose. The Apple's Application Store and the Android website provide thousands of applications, many at no or low cost. Some of the capabilities provided, such as Global Positioning Systems, can embed the coordinates of where pictures were taken within the digital picture file. Similarly, positioning co-ordinates can track a person or device in a medical or emergency situation, or even track kids on a school bus. Coupled with AR techniques, this can be coupled with mapping techniques for visual status and location capabilities. Smart phones can also be equipped with secured key capabilities that can be used as a key for an automobile, house, or similar application.

Social media such as Facebook and Twitter are also gaining a synergistic benefit from applications that provide bidirectional communication and updates from handheld devices. Many of the games and other features of sites such as Facebook have AR enhancements that extend to mobile devices. Mobile Twitter feeds and handheld integration are further examples of ubiquitous proliferation.

Smart phones and similar devices are able to use *near field communications*, which is a technology that allows the transfer of information between devices when they are within 10 centimeters of each other. Consumers can pay for purchases by passing their smart phone near a reader to debit their account, thereby completing the transaction. In many ways this advancement can be more secure than using a credit card because it will allow a credit card company to geographically track where a phone is used to make a purchase. This can not only reduce fraud, it can provide the location of the individual of the fraud perpetrator.

Near field communication in handheld devices is frequently secured in tandem with the *Subscriber Identity Module (SIM)* card. SIM cards are found on *Global System for Mobile (GSM)* devices, but not on *Code Division Multiple Access (CDMA)* devices. Near field communication relies on high frequency (HF) magnetic coupling, which limits its transmission range to a few centimeters. RFID commonly operates using inductive coupling on ultra high frequency (UHF), which has a range of about 10 meters. The antenna for UHF is currently too large and expensive to be implemented in a handheld device, which in part drives the reason for the limitations and capabilities of near field communication.

Other similar communication technologies are Bluetooth and infrared. Bluetooth is a wireless technology for transmitting data over short distances. Bluetooth uses short wavelength radio transmissions that are generally limited to about 10 meters or less, although there are ways to increase this distance. Infrared is also used to communicate between devices that are in close proximity, although infrared requires a clear line of sight in order to communicate and can be limited to communicating with one device at a time.

Wal-Mart became a heavy adapter of RFID technology to track inventory in the years leading up to 2005. Bill Hardgrave headed a University of Arkansas study on Wal-Mart's RFID initiatives and found that it reduced out of stock inventory by 30%. Fast selling items (6 to 15 units per day) were the items that benefited most from RFID inventory tracking, whereas items that sold slow (less that 0.1 units sold per day) received virtually no benefit (Collins, 2006). Reducing out of stocks can not only improve sales at the retail establishment, it can have a ripple effect of increasing a supplier's sales as well.

RFID technology can be utilized for security in a variety of ways. Parents can track their kids in a crowded environment such as a park or beach. Computer hard drive manufacturers use an RFID tag to serve as a decryption key for information stored on the drive. RFID is also used for traffic management to prevent collisions in crowded locations such as airports and ports, often in combination with embedded systems.

Healthcare is an area that realizes considerable benefit from EM, RFID, and AR. Personal care and monitoring is one of the most common uses of EM. Diabetics and cardiac patients can have their vital signs measured and tracked, with the information being relayed to a health care provider. This not only provides longitudinal data to provide

better healthcare decisions, it also streamlines the monitoring process and can provide efficiencies that allow a provider to treat more patients.

The use of RFID in healthcare is widespread in areas such as patient care and safety, education, and monitoring of standards. Patient care assets such as equipment, drugs, health care providers, and patients themselves can be tracked with RFID systems, thereby assisting in their identification and location. Hospitals can be large, complex environments, and maintaining inventory and location information of things ranging from beds to monitors to medications can be an arduous process that can create significant overhead and increase costs. An RFID system can be used to provide immediate location and utilization information.

Healthcare treatment standards exist for several things including patient care levels, insurance, and accreditation. Certain patients may require that a nurse or other provider visit a patient on a regimented schedule. A provider that is equipped with an RFID tag can be tracked throughout a hospital and each patient contact can be tracked through the system. This can reduce manual patient charting requirements and also remind the provider when a visit is scheduled.

Augmented Reality is used to help surgeons visualize conditions and parts of the body, assist radiologists in reading CAT scans and MRI's, and provide a simulated training environment. Particularly in microsurgery, where success or failure of a procedure can be determined by a very small distance, AR provides enhanced 3-D visualization and navigation assistance (Lapeer, 2004). Interpretation of images from CAT scans and MRI procedures is often reliant on subtle changes in shading or a pattern. These changes can be enhanced by AR, thereby improving recognition and diagnosis decisions.

Augmented Reality is also a technology that has shown great promise in education. Early adoptions were for specific training purposes, including training in the automotive industry. The core competency areas of math, science, social studies, and language arts all could benefit from AR. In a world where video games are common, the use of AR in education may be the one of the few ways to engage a significant portion of the population.

Mobile technology is beginning to have a measured impact on printed media, and can be particularly evident in the advent of eReaders. An Amazon Kindle 3 can hold approximately 3,500 books and a Barnes & Noble Nook WiFi e-reader allows for micro SD card storage. Within a few short years, the equivalent of the Library of Congress will be able to be carried around on an eReader. The eReader market is still maturing, and many of the features that make an eReader attractive are features that are not attractive in a smart phone. The screen size of a Blackberry, iPhone, or Android are too small for serious reading, whereas an eReader, or a device such as an Apple iPad, have the screen real estate to be effective for long duration reading, but the device is too cumbersome for many of the things attractive for a smart phone design. Finding one device that can serve many diverse applications effectively is one of the most difficult design challenges facing electronics.

RISKS, CHALLENGES, AND LIMITATIONS

As with many technologies, EM, RFID, and AR are not without their challenges. There are many risks including over reliance on a device, malfunctions, and security concerns. Over reliance on a device is when someone assumes the device is correct without critical analysis. This can be a particular concern if the device is used in a critical situation such as a medical device where misinformation can be life threatening. A similar situation can arise if a device malfunctions and the necessary intervention does not occur. This can be illustrated in the case of a pacemaker, which helps ensure a safe heart beat, doesn't work correctly.

Security concerns include the loss of personal information, unauthorized access to a device or facility, mistaken identity, and breach in information security. RFID tags in most retail applications have no encryption or easily breakable encryption. This means that someone could walk through a location with a RFID reader and determine products carried and inventory levels. Similarly, RFID credit cards have been reported to be hacked with an $8 reader (Patel, 2008). Near field communication has a similar eavesdropping security hole. Because of the integration of RFID with United States Passports in 2006, information on the passport is also at risk. Travelers should keep their passport in an RFID shielded case or wrapped in aluminum foil to prevent lost of information.

The nation's automated teller machines (ATMs) are a frequent target of robbery or other violence. The 2010 *Report On Emergency Technology For Use With ATMs* by the Bureau of Economics was mandated as part of the Credit Card Accountability Responsibility and Disclosure Act of 2009. It required the Federal Trade Commission (FTC) to analyze current or emerging technology that would allow a distressed ATM user to send an electronic alert to a law enforcement agency. While the cost effectiveness, contribution to safety, and deterrence to commission of a crime are not clear, the fact remains that customers at ATM machines are at risk. While not in the report, it is worth noting that the use of near field devices for financial transactions rather than debit cards may provide relief to this risk as near field devices do not require a trip to an ATM to obtain money. Additionally, if the near field device used in financial transactions is a smart phone or similar device, a variety of customer security features already exist including the ability to record audio, take pictures and/or video, obtain global positioning coordinates, and make 911 emergency calls.

ATM cards are susceptible to a number of threats including double swiping, cloning, and skimming. Near field technology payment systems have a similar risk, but skimming may be even more at risk. For ATM skimming, the perpetrator places a card reader over the face of an ATM, which reads the information on the magnetic strip and transfers it using Bluetooth or cellular technology. Since a near field device only needs to be near a reader as opposed to being inserted into a reader like an ATM card, skimming can take place by simply placing a reader in close proximity.

Embedded systems are often linked together over a network. One potentially serious issue is a software defect masquerade fault, which is a condition where an embedded system on one node on the network sends information that is interpreted as coming from another embedded system on at a different location or node (Morris and Koopman, 2003). The implications of this are many, including reacting incorrectly because the information is wrong. An example would be to send an emergency response unit to the north end of a complex when in fact the issue requiring response is at the south end of the complex.

Embedded systems are often on a closed network, so it is not uncommon for the applications to have a lower degree of security. In other words, incorrect assumptions are often made that the closed nature isolates the systems from malware and other exploits from outside influences. Power plants are just such an environment, and the 2010 Stuxnet computer worm that targeted industrial infrastructure is an example of a serious and far-reaching compromise. Stuxnet caused the Microsoft Windows based Siemens control software to encounter many widespread problems. The complexity and sophistication of Stuxnet was such that it is often referred to as the first instance of cyber warfare where the computer was the primary vector and target.

There are other recent examples of embedded systems being targeted as part of a war effort. The 2008 invasion of Georgia by Russia was preceded by pointed computer system attacks, the 2007 cyber attacks on Estonia, and the Titan Rain attacks starting in 2003 and aimed at American computer systems illustrate the serious implications of computer system compromise.

Augmented Reality is not without its security risks. By its very nature, AR is meant to augment what is perceived. Misinformation can be a key security exploit of AR. If the augmented information is in reality incorrect information, the response elicited may have negative repercussions. For example, if the AR installation was part of a radar system used in air traffic control, aircraft could appear to be in different locations and incorrect pilot instructions could be given that result in a crash. There are many other scenarios that could be drawn from malfunctioning AR systems that could have even more tragic results.

Risks can present themselves in many ways, but perhaps the greatest risk of having a single smart device that has so many capabilities and so much information is simply losing control of the device. This could manifest itself as physically misplacing the device, having it suffer a functionality outage, or something nefarious such as a virus or malware. Organized cyber crime is already far beyond the capability and capacity of law enforcement (Gogolin, 2010), so smart devices would be easy targets for cyber criminals. In the United States, businesses are also at risk as the link between business and cyber crime law enforcement is also precariously limited (Gogolin & Jones, 2010).

A risk that is often overlooked but often felt is embracing a technology that has a limited life span. History is rife with examples of promising technology that missed the mark or hit the mark but failed to adapt. From the Sony Betamax to the Palm Pilot, from OS2 to Firewire, from floppy disks to Iomega disks to perhaps even hard disk drives, great technology has a limited lifespan. The painful risk is when the lifespan of the technology is short and your investment is high. Many emerging technologies try to lock you in to a particular solution that has limited or no support outside its framework. Apple has been slow to open up its application interface (API) for cameras and other features that make widespread AR application development by non-Apple vendors difficult. Amazon's Kindle does not support e-pub format, which is a barrier in creating eReader standardization. These risks have all been seen before, but with the efforts to create a true multifunction smart device, barriers of this type don't make the ability to deliver one an easy thing to accomplish.

GLOBAL IMPLICATIONS

As the use of embedded mobile, RFID, and augmented reality in mobile devices increases, the impact on people's lives will be widespread. One smart device will have the potential to contain all of the keys that a person needs for their home, business, automobile, as well as all of the pertinent personal information people need for everyday life. Driver's licenses, passport, birth certificates, education records, medical records, insurance policies, financial information, and entertainment are already making the transition to smart devices. Consolidating all of this information and more into one device is a logical progression.

New industries will be spawned that enable these types of applications, and those that are architected for interaction on a global scale will change the way the world interacts. To truly enable global interaction many things have to be considered. Character sets will need to support Unicode, thus allowing for use of most of the world's alphabets. Displays need to be able to accurate display and interpret these character sets. Input devices need to embrace these requirements as well.

In the late 1990's the general population of most countries of the world began to know the Internet and the World Wide Web. Though most still don't know the difference between the two, they do know how the exchange and ability to

access information has been profoundly impacted. Small devices are leveraging these technologies and others such as cellular communication are utilization capabilities that seemed like fairy tales just a few short years ago.

CULTURAL IMPLICATIONS

A challenge of relying on a few devices and applications is that cultural manners may be diminished. A simple handshake or other greeting may be replaced by an augmented reality enhanced exchange of information between smart devices. This is already seen in many circles where cell phone calls and text messages interrupt face to face conversation. Many are already offended when a conversation is interrupted in this manner. When the technologically enhanced information exchange is disseminated more frequently and in more depth, interpersonal communication via face to face conversation may become even more strained.

Acceptance of technology tends to fall into three categories. Category one are the early adopters. They relish the freshness and challenge of new technology, even if there are a few warts. Category two are the people that could go either way. If they see a reason to use the technology, they will. If there is no clear benefit, they won't. Category three are the laggards that will never use the technology no matter what. Perhaps they don't see a need, don't have the desire or trust, or just plain do not want to. The functionality of the technology, the way it is delivered and by whom has a profound impact on cultural acceptance – regardless of whether it is a valid option. Just because it originated from a particular person, region, or culture may make acceptance by another person, region, or culture impossible.

The discussion of cultural acceptance of technology is not just between cultures between countries and religions, it is also between ages within a culture. For example, it is often considered acceptable for young people in the United States to simultaneously hold texting conversations between many different people on different topics. However, it is considered rude if the same young person is holding a verbal conversation with an older person and accepts or sends a text message. Age can play as large a factor as other cultural factors. It is also similarly a factor in terms of readiness to accept a technology – and the way the technology is accepted.

The barriers to achieving true global smart device market penetration are not technically based; rather, they have cultural roots. Mistrust, religious beliefs, past history, unscrupulous behavior, and fear of being bypassed by technology are but a few of these roots. Even if these barriers can be adequately dealt with, preventing cultural dominance and obsolescence have faced mankind since the beginning of time and technology will not solve that which is unsolvable.

CONCLUSION

The use of embedded mobile, RFID, and augmented reality in mobile devices shows great promise for advancing everyday personal and business commerce. The combination of these technologies into one device can provide a tool that could accomplish a large percentage of daily personal and corporate tasks, as well as providing great convenience. Industries such as retail, education, and health care are on the cusp of being transformed into something far different that what we have today. RFID has revolutionized the way things are tracked and secured. Embedded systems are found in medical devices, consumer electronics, automobiles, and many other applications. Augmented reality is increasingly being leveraged in mobile applications such as smart phones, often using some combination of the device's compass, GPS, and camera. Each of these technologies has been in existence for over 50 years, but recent developments in com-

munications technology and handheld devices has enabled uses that weren't though possible just a few years ago. Another factor leading to the rapidly increasing penetration of embedded mobile, RFID, and augmented reality is that the applications serve a useful purpose and in many ways can improve quality of life. Challenges to further integration include privacy and security issues, legal and ethical issues, and cultural and technological issues. But the largest challenges involve people, trust, and people's tendency to being averse to change.

REFERENCES

Bimber, O., & Raskar, R. (July, 2005). *Spatial augmented reality: Merging real and virtual worlds*. Wellesley, MA: A K Peters/CRC Press.

Bureau of Economics. (April, 2010). *Credit Card Accountability Responsibility And Disclosure Act Of 2009. Report on emergency technology for use with ATMs*. Federal Trade Commission.

Collins, J. (May, 2006). *RFID's impact at Wal-Mart greater than expected*. Retrieved on November 28, 2010, from http://www.rfidjournal.com/article/print/2314

Gartner.com. (November, 2010). *Gartner outlines 10 mobile technologies to watch in 2010 and 2011*. Gartner Newsroom. Retrieved on November 7, 2010, from http://www.gartner.com/it/page.jsp?id=1328113

Gogolin, G. (2010). The digital crime tsunami. []. Elsevier.]. *Digital Investigation*, 7(1-2), 3–8. doi:10.1016/j.diin.2010.07.001

Gogolin, G. & Jones, J. (2010). Law enforcement's ability to deal with digital crime and the implications for business. *Information Security Journal: A Global Perspective, 19*(3), 109-117. Doi:10.1080/19393555.2010.483931 Top of Form

King, R. (2009). Augmented reality goes mobile. *Businessweek*. Retrieved on November 22, 2010, from http://www.businessweek.com/technology/content/nov2009/tc2009112_434755.htm

Lapeer, R., Rowland, R., & Chen, M. S. (2004). *PC-based volume rendering for medical visualisation and augmented reality based surgical navigation*. Eighth International Conference on Information Visualisation (IV'04).

Morris, J., & Koopman, P. (June, 2003). *Software defect masquerade fault in distributed embedded system*. DNS 2003. FastAbs.

Patel, N. (March, 2008). *RFID credit cards easily hacked with an $8 reader*. Retrieved on November 28, 2010, from http://www.engadget.com/2008/03/19/rfid-credit-cards-easily-hacked-with-8-reader

Rau, E. (2005, December). Combat science: The emergence of operational research in World War II. *Endeavour*, *29*(4), 156–161. doi:10.1016/j.endeavour.2005.10.002

Roberti, F. (2005). *The history of RFID technology*. Retrieved on November 22, 2010, from http://www.rfidjournal.com/article/view/1338

ADDITIONAL READING

ACM http://www.acm.org

Ciaramitaro, B. (2010). *Virtual Worlds and E-Commerce: Technologies and Applications for Building Customer Relationships*. Hershey, PA: IGI Global. doi:10.4018/978-1-61692-808-7

Computer Fraud and Security http://www.compseconline.com

Computer Law and Security Report http://www.compseconline.com

Computer World http://computerworld.com

Digital Investigation. Elsevier. Amsterdam, The Netherlands.

Government Security News http://www.gsn-magazine.com

Government Technology http://www.govtech.net

Homeland Defense Journal http://www.homeland-defensejournal.com

IEE. http://www.ieee.org

Information Security Journal: A Global Perspective. Taylor & Francis. New York, NY.

Journal of Information Technology http://www.palgrave-journals.com/jit

Lomas, N. (June, 2009). *RFID could be in all cell phones by 2010*. ZDNET. Retrieved on the World Wide Web September 4, 2010. http://www.zdnet.com/news/rfid-could-be-in-all-cell-phones-by-2010/315292

Parr, B. (2009). *Top 6 Augmented Reality Mobile Apps*. Mashable.com. Retrieved on the World Wide Web September 4, 2010. http://mashable.com/2009/08/19/augmented-reality-apps

Privacy and Security Law Report http://www.bna.com/products/corplaw/pvln.htm

Privacy Journal http://www.privacyjournal.net

SIAM Journal on Computing http://www.siam.org/journals/sicomp.php

SiC-Seguridad en Inform·tica y Comunicaciones http://www.revistasic.com

Stewart, J. (2008). *CISSP Study Guide*. John Wiley and Sons.

Technology in Government http://www.itbusiness.ca/it/client/en/techgovernment/home.asp

TechWeb: The Business Technology Network http://www.techweb.com

TopLine Quarterly Newsletter http://www.gocsi.com/awareness/topline.jhtml

Virus Bulletin http://www.virusbtn.com

KEY TERMS AND DEFINITIONS

Augmented Reality: reality involves the use of computer generated or enhanced sensory input such as audio and visual components to enhance the perception of reality.

Bluetooth: a wireless technology for transmitting data over short distances. Bluetooth uses short wavelength radio transmissions that are generally limited to about 10 meters or less, although there are ways to increase this distance.

Embedded System: a computer systems designed to perform a specific task or dedicated function.

EReader: Electronic Reader. An electronic device whose primary purpose is to display text and other book or media related information.

Exploit: a software, hardware, virtual, or social engineering feat that takes advantage of a vulnerability to compromise security.

Firmware: The software that controls embedded systems.

Infrared: a technology used to communicate between devices that are in close proximity. Infrared requires a clear line of sight in order to communicate.

MP3: A digital audio encoding format commonly used in portable music players and smartphones.

M-RFID: Mobile Radio Frequency Identifier. Services provided by mobile devices equipped with an RFID tag over a telecommunications network.

RFID: Radio Frequency Identifier. Communication between a tag and a reader using radio waves.

Risk: the likelihood that a specific threat will manifest itself.

Threat: an indication or warning of a potentially undesirable outcome.

Vulnerability: a security flaw caused by an absence or weakness of a countermeasure or safeguard.

Chapter 10
The Impact of Mobility on Social and Political Movement

Nabil Harfoush
Manara International Inc., Canada

ABSTRACT

The strength of social and political movements is often correlated with the cost and risks of organizing the effort. Reaching large numbers of people to inform them of a movement's goals, and the ability to recruit supporters, has historically relied on mass media, both printed and electronic, along with traditional canvassing, public assembly, and public speaking. This has naturally favoured economic and political elites who had easier access to media channels, and who controlled in many cases the rights to public assembly and free speech. The emergence of affordable communications in general and mobile communications in particular, is bringing radical change to this balance of power. This chapter explores some of these changes and suggests directions for future research in this area.

INTRODUCTION

Social and political movements usually develop when a small group of individuals committed to a set of ideas and goals that resonate with the needs, aspirations and culture of a large group of people succeeds in communicating these ideas and goals directly or indirectly to a majority of this group. In contemporary times, groups with

political or economic power traditionally relied for such communication on mass media in printed and electronic forms (newspapers, radio, TV).

In Western-style democratic societies groups with less political and economic power, and therefore less access to mass media, have relied more on traditional canvassing, public assembly and the right to free speech to spread their message. In some cases, they created events specifically

DOI: 10.4018/978-1-61350-150-4.ch010

designed to attract media attention in order to gain access to mass media channels and further expand the reach of their communication. Green Peace boats challenging whaling ships on the high seas or truck drivers blocking highways in and around Toronto are examples of such events.

In authoritarian environments, mass media and telecommunications are directly controlled by the government and the freedom to assemble and speak is tightly controlled if at all permitted. Core groups committed to ideas and goals different from those in control are therefore compelled to use other more covert forms of communication: word of mouth, rumours, clandestine flyers, secret assembly, and distribution of content through parties outside the control of the local powers. In certain cases non-authorized assembly and demonstration has been used to further signal to others the existence of the movement's core and its ideas and goals. The demonstrations of the opposition during the Iranian elections are a typical example.

In general, the centralized nature of media and telecommunications has naturally favoured economic and political elites who had easier access to media channels, and who controlled in many cases the rights to public assembly and free speech. Less powerful groups (opposition, dissidents, minorities, suppressed majorities etc.) are forced to rely on less effective communications tools in a higher risk context. The rapid advances in information and communication technologies (ICT) over the past decade, the wide-spread adoption of the internet as a distributed communications medium, and the emerging and rapidly evolving mobile communications, are bringing radical changes to this balance of power. In this chapter we explore the impact of mobile communications on social and political movements. Using examples from a variety of fields, we look at the effects of dropping communications cost, individual nature of mobile devices, and the peer-to-peer communications model on the forming and spreading of social and political movements.

We also look at the impact of distributed mobile content creation and discuss emerging thinking about the links between people-based movements and the resilience of communities, new models for economic development challenging traditional investments practice, and the concepts of a crowd-sourced science.

BACKGROUND

Over the past decade the cost of communications driven by technological advances has dramatically decreased. Communications came to be accepted as a major driver of economic development and became essential components of the declared developmental targets by global assemblies. The 2005 World Summit on Information Society (WSIS) has ten global developmental targets. Target 10 is to "Ensure that more than half the world's inhabitants have access to ICTs within their reach" (ITU, 2010, pp. 193).

Among all the new ICT technologies, mobile cellular communications showed the most spectacular growth in the past decade, reaching a global penetration of 67 per 100 inhabitants in 2009 (Figure 1).

As a consequence of such rapid growth, supporting the development of mobile communications became the focus of many global efforts. The United Nations agency for telecommunications, the International Telecommunications Union (ITU), recommended for example in its 2010 mid-term report "Monitoring the WSIS Targets" the adoption of a policy of "expanding mobile network coverage in developing countries, particularly in rural areas" (ITU, 2010, pp. 207-208).

ENABLING THE ORGANIZATION

One of the immediate consequences of the greater availability of affordable communications in general and mobile communications in particular is

Figure 1. Global ICT development, 1998 – 2009. (© 2010 ITU. Used with permission)

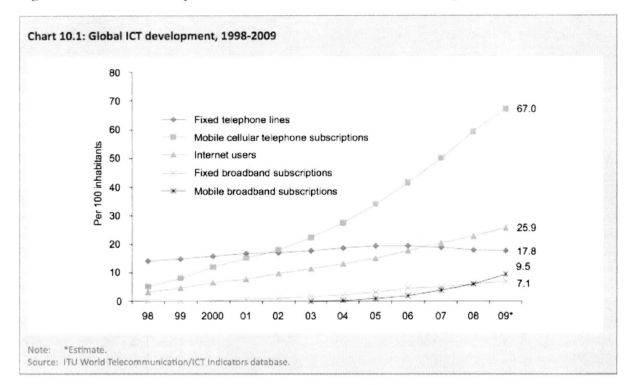

Chart 10.1: Global ICT development, 1998-2009

Note: *Estimate.
Source: ITU World Telecommunication/ICT Indicators database.

a reduction in difficulty and cost of organization in both permissive and non-permissive environments.

In December 2008 a small group of individuals in Toronto, Canada, concerned with urban poverty and looking for a way to channel the season's spirit into action, came up with the idea of organizing a Christmas party in support of the Daily Bread Food Bank charity. Linking the Christmassy theme to the city of Toronto, Ontario, the event was dubbed "hohoTO". The challenge was how to organize the event at such a late date when most venues were already reserved for Christmas and end-of-year corporate and private events. Members of the core group put out the idea on their personal streams on Twitter. It resonated with many of the individuals in those networks, who spread the word further. Soon event-related traffic on Twitter became significant enough to attract the attention of a variety of organizations. The staff and management of The Mod Club donated their venue and volunteered their time for the event. Word about the event continued to spread rapidly over other social networks and ICT tools (text messages, email etc.) and the event was sold out. Corporate managers sensing that an extraordinary event was in the making started offering sponsorships and donations. From concept to full implementation including all logistics, catering, bar, security, ticket sales and access control, all using mobile communications and social networks, the event was fully organized in a few days by a small group of individuals and ended up raising an astonishing $25,000 for the Daily Bread Food Bank, way beyond the organizers initial goal of $10,000. A spin-off event in the summer of 2009 pulled in more than $13,000 for the charity. These early successes consolidated the community and helped establishing a permanent online presence for the event (web page at http://hohoto.ca/, Facebook, Twitter etc.). In December 2009 a second "hohoTO" event was held that raised in one night over $50,000 plus 1,000 lbs of food at the door, again surpassing the organizers goal of $40,000.

Another example, this time from a non-permissive environment, are the June 2009 elections in the Islamic Republic of Iran. When the results of the elections announced a landslide victory by incumbent President Ahmadinejad, opposition parties claimed serious irregularities and mobilized their supporters to protest. The Iranian government facing an unprecedented scale of protests banned all foreign media from covering such protests. The information void created by the ban was promptly filled by citizens with mobile phones, who reported protest-related events and clashes using SMS, Twitter, Facebook and their phone cameras. In the early stages of the protest movement this crowd-based reporting was successful in bypassing the media reporting ban by the government. Opposition forces realizing its reach started using the new medium for coordinating their actions, informing their members of government forces locations and actions, and spreading their news, photos and videos to sympathetic groups and governments outside the country asking for their support. The breach of the government ban on reporting had a major impact on the further development of events. Traditional media anxious to cover the Iranian protest movement jumped on the content provided through the mobile citizens and amplified it through their mass broadcasting channels. This brought the protest events in Iran to the forefront of people's attention in many countries. Governments interested in a change of regime in Iran moved rapidly to leverage this broad movement for their own purposes. A CNN blog reported that the U.S. State Department "is working with Twitter and other social networking sites to ensure Iranians are able to continue to communicate with each other and with the world" (Labott, E., 2009). For example the State Department asked Twitter to reschedule planned systems maintenance to avoid service interrup-

tion for those reporting the events in Iran. Twitter rescheduled its maintenance to coincide with 1:30 AM Iranian time, a period of least activity. Other major social networks responded as well. Facebook released an early version of its site in Persian (Parr, B. 2009) although the translation was not completed. Google also hastily launched Farsi support for their Google Translate tool. As Och (2009), Principal Scientist at Google, wrote at the time: "And we're launching this service quickly, so it may perform slowly at times. We'll keep a close watch and if it breaks, we'll restore service as quickly as we can"

The Iranian government was aware of the potential political dangers posed by social networks and mobile media-enabled citizens. Already on May 23, 2009 they had banned Facebook "to prevent Mousavi supporters from using it for his presidential campaign prior to Friday's poll" (AFP, 2009). The government had also cut service to the main mobile telephone network in the capital Tehran and blocked Facebook and YouTube sites on June 13, 2009 immediately preceding President Ahmadinejad's televised appearance to declare victory in the elections (AFP, 2009). However, the citizen-based coordination and information efforts of the Iranian opposition, and their adoption and amplification by major social networks and foreign governments, was so significant that it prompted the Iranian government to unprecedented measures. "One day after the election, Iran closed the internet down entirely for half an hour, then slowly loosened its grip, as the authorities struggled to gain control" (Leyne, 2010). Cellular service was sporadically shut down. Ultimately the government mounted an active counter-measures campaign that included more sophisticated Internet filtering, blocking opposition communications, tracking down opposition supporters through their use of their mobile devices, using Internet content for interrogation and indictment of arrested opposition members, and an alleged "Iranian Cyber Army" that hacked web sites supportive of the protesters including Twitter

(Leyne, 2010). In response, those supporting the Green movement created proxy servers to mask the internet addresses of Iranians accessing social media networks and blogs. Every time Iranian authorities identified and blocked one of those proxy servers, new ones would be set up. In a short period of time the conflict morphed from an internal confrontation between the Iranian government and its citizen-based Green movement to a full international cyber conflict between pro- and anti-Iranian forces across the globe.

Since those events took place the debate has raged on the role of social media in Iran's political movements. Some called it the Twitter Revolution in reference to the heavy use of Twitter during the post elections crisis. Other contested any major impact of Twitter on the citizen movements *inside* Iran and pointed out that three of the most prominent Twitter accounts commenting on events in Tehran were in fact based in the United States, Turkey, and Switzerland (Esfandiari, 2010). The relative novelty of online social networks, the focus of traditional media on the impact of these networks on their own industry, and the strong interest of advertising agencies in using social media for marketing purposes may explain the excessive attention given to social networks when looking for the main enabler of the Iranian opposition movement. A closer look at why Twitter was considered by many as the medium of the movement, reveals deeper linkages to other enablers: "Its free, highly mobile, very personal, and very quick" (Grossman, 2009). Given the ubiquity of mobile communications globally, it is not surprising that their role in enabling grass-roots movements to self-organize is often overlooked or is hidden within other systems that they enable. In the case of the Iranian elections, Twitter played an important role in spreading information about events sometimes within the Green movement but mostly outside of the country's borders. It should be pointed out, however, that the information distributed by Twitter and other social networks was originally captured and transmitted mostly using

mobile phones. Without the distributed capacity of citizens' mobile cellular phones it would have been much more difficult to get any information about the unfolding events.

A better model for understanding the enablers of self-organized political movements may be to consider the combination of mobility and interconnected social networks. The first element, mobility, provides the advantage of distributed collection of information, which is difficult to suppress. Authorities could shut down the cellular telephone service for a period of time, but such action remains quite a radical measure that most authorities would prefer to avoid in order not to project an image of waning control. Even if cellular telephone service is shut down, as was the case in the Iranian elections, the photos, videos, and text captured on the mobile devices remain available and can be further disseminated using other networks and media.

The second element, social networks, is an assembly of interconnected and overlapping networks. Information in one person's message on Twitter is relayed to that person's followers, each of whom has their own network of followers. Information can also easily cross from one social network to another. A tweet can be simultaneously sent to one's account on Facebook or LinkedIn, thus reaching another layered network and propagating there. As a result the information captured by distributed mobile users is amplified and can spread very rapidly across a multitude of these networks. The dual effect of distributed capture of information and the interconnected nature of that information's distribution over multiple networks makes centrally organized suppression efforts highly ineffective.

EMPOWERING INDIVIDUALS

One of the factors differentiating mobile communications from other communications is that mobile communications devices are usually more personal than family or community devices such as television sets, home computers, or corporate laptops. The individual nature of the relationship with mobile devices makes for more personalized and private interaction with the device. It has enabled the establishment of new peer-to-peer communications channels using cellular network services like SMS and internet-based Instant Messaging (IM) applications such as Microsoft's Live Messenger, Yahoo! Messenger or the BlackBerry Messenger. Among young adults and teens the overwhelming preference is for mobile communications tools. A recent study by the Pew Internet & American Life Project has shown that in America "cellphone texting has become the preferred channel of basic communication between teens and their friends and cell calling is a close second." (Lenhart, 2010). The intensity of this communication is not to be underestimated. The number of text messages sent by a teen daily is 50 messages for one half of American teens and 100 messages for one third of them (Lenhart, 2010).

As cellular phones integrate music, photo and video capabilities, the scope of this communication has also expanded to include multimedia in addition to traditional text and voice. The impact of combining the individual nature of mobile devices with multimedia recording capabilities goes well beyond the entertainment value that initially drove the adoption of these tools. It has opened new avenues of self-expression and participation that would have been impossible with collectively-owned devices (such as landline phones, home TV, home and work computers). One example of the maturing of self-expression using mobile phones is the documentary titled *156 Turns* about the 2010 Pikes Peak International Hill Climb motorcycle race in Colorado Springs that was entirely shot using an Apple iPhone 4.

It is a relatively small and natural step to go from self-expression for entertainment or artistic purpose to self-expression for the purpose of a cause. In 2006 a technician of Comcast, a major U.S. cable company, fell asleep at a customer's home while on phone hold for a long time with the technical department of his company. Using a mobile device the customer video taped the technician sleeping on his couch and created a satirical video that he then uploaded to YouTube. The video resonated with many other customers' dissatisfaction with Comcast's service and spread virally. As of September 2010, the original video posted was viewed over 1.5 Million times. It created a public relations nightmare for Comcast and forced the company to review its customer service and to create "Comcast Cares", a new online customer service group dedicated to monitoring the company's reputation on the Internet and engaging rapidly with irate customer postings promising to expedite resolution of their issue.

A more poignant example is the video of Saddam Hussein's execution, which was captured clandestinely on a mobile phone and leaked to the rest of the world through the Internet. "The unprofessional and undignified atmosphere of the execution drew criticism around the world from nations that both oppose and support capital punishment" ("Execution of Saddam," 2010).

Video clips shot with mobile phones and capturing political events, exceptional weather activity, police actions or natural disasters are now common and almost universal across borders and continents. The above few examples illustrate how mobile phone technology is reducing by several orders of magnitude the cost of mobilizing support for a cause, be it political, environmental, civic, or humanitarian, and how it is empowering individuals to participate more effectively in economic, social, political and cultural issues of interest to them.

The ability to easily build and maintain a personal list of friends or colleagues and to effectively and rapidly exchange with members of such a personal network multiple formats of information of common interest, has led these networks to become durable, resilient, and of increasing personal and social value to their members. Whether it is police action witnessed, a grievance with an employer, frustration with service, protest against government, a new fashionable item, or the buzz about a new song, movie or idea, these peer-to-peer networks have become a prime channel for spreading the word about it. The combination of the following factors is making these peer-to-peer channels a natural and fertile ground for the emergence and rapid spread of social trends and movements of many kinds.

• high adoption rate among young adults and teens, a demographic with rebellious tendencies;
• the self-selecting of individuals with similar interests and inclinations for membership of these networks; and
• the enablement of self-expression for a personal or collective cause,

EMPOWERING MOVEMENTS

In the previous section we have seen how multimedia-enabled mobile phones combined with the distribution power of social networks are enabling the emergence of individual initiatives that quickly garner attention and support from an increasing number of people, rapidly becoming significant enough to attract the attention of traditional main-stream media and to influence the actions of organizations and governments. In this section we will consider the effects of mobile phone technology on how emerging movements or existing organizations reach out and mobilize support.

The personal relationships that users have with their mobile devices are allowing organizations and movements a more granular segmentation of the target population. Whereas a message using television may have had to be adapted to the whole family, mobile phone channels allow the targeting of a more narrowly defined demographic. For example, youth within a family and their parents could be targeted with distinct messages. Such

segmentation can obviously be based on other criteria than generation. The American Red Cross had experimented since 2008 with targeting various new segments of donors with less disposable income asking for smaller donations ($5 to $10) and had collected pledges of more than $190,000 through a program called Text2help. After the 2010 Haiti earthquake and facing a massive need for donations, the American Red Cross in collaboration with The Mobile Giving Foundation launched a similar campaign on a larger scale targeting various segments of users of mobile phones with requests to donate $10 by texting the word "Haiti" to 90999. By January 17, 2010 the organization had "received more than $9 million in donations from more than 900,000 mobile phone users" (Maestri, 2010). A similar SMS-based campaign by Haiti-born musician Wyclef Jean's Yele Haiti Earthquake Fund raised $2 million in the same period. The scope and speed of the results attained in these two campaigns demonstrated the power of tapping into the long tail of the donating population, by targeting many individuals through a personal channel and making it extremely convenient to donate a small amount commensurate with average spending on a typical mobile phone bill. Interestingly enough, these SMS-based campaigns were fuelled and amplified through social networks such as Facebook and Twitter, "where users are urging one another to make donations using cellphones" (Reuters, 2010). This confirms the positive feedback loop created by the combination of mobile devices with social networks.

Nowhere was the hypothesis about the combined power of mobile and social network put more to the test than during the 2008 U.S. elections campaign of President Barack Obama. In a *Washington Post* interview, Scott Goodstein, Obama's Director of External Online Organizing and Mobile, talked about his view of text messaging: "To me texting is the most personal form of communication" (Harfoush, 2009). The text messaging program initially rolled out by Goodstein in June 2007 built on experiences gathered from previous political campaigns on the use of SMS. Limbo Inc., a company specialized in mobile entertainment that had run SMS advertising campaign in the early stages of the campaigns of Democratic presidential candidates Hillary Clinton and Barack Obama, reported that their SMS advertising campaign changed voting intentions for 28% of recipients. The voting intention changed significantly for 6% and a little for 22% of recipients and was explained through a combination of increased awareness and changes in the perception of the candidate. In addition the campaign increased interaction of recipients with the candidate across other media, both paid and unpaid ("Limbo reports mobile," 2008).

The Obama campaign text messaging program built on such results, but went on to innovate in a number of ways that made it remarkable in both methods an achievements.

First, it put emphasis on having users opt in to receive text message updates from the campaign. Those who chose to provide their zip code received information relevant to their neighbourhood. Subscribers could always end their subscription to the campaign's updates. This provided a sign of respect to the community and helped built the necessary trust for further engagement and participation.

Second, the program made a serious attempt to build a two-way conversation with subscribers. Messages sent by the campaign did not only ask for specific actions in support of Obama, it also asked for feedback about issues. Goodstein wrote: "We will use text messaging to ask for your opinions and advice and give you the ability to request information from the campaign" (Harfoush, 2009, pp.119). Campaign staff and volunteers regularly scanned incoming messages, responded to supporters' questions, and watched for early signs of important views or movements forming within their subscriber base. They monitored for any individuals unhappy about the messages they are receiving. Interesting ideas and contributions were posted to other social networks feeding into the amplification of the campaign's message.

Third, the program understood perfectly the power of enlisting the support of the subscribers' personal social networks. Leveraging the ease of forwarding text messages to friends and family, campaign messages asking for volunteers at a specific rally or donations by a certain date were sent out enough in advance to provide sufficient time for supporters to forward these messages to others.

Fourth, text messaging was used for major announcements made by Obama and not relegated to the role of a secondary channel. The announcement of Joe Biden's selection as vice presidential candidate went first to subscribers of text messaging before being distributed to mainstream media. This strengthened the trust building and sense of community among supporters and incited many to increase their efforts in support of the campaign. It is worth noting that the text messaging program was not aimed only at youth. It made a serious effort to study how this medium applies to various demographics. Overall, the text messaging program gathered three million subscribers who received on average 5 - 20 messages per month (Monte, 2009). On election day, battle ground subscribers received three or more text messages (Harfoush, 2008).

The mobile strategy of the Obama campaign was not limited to text messaging. Goodstein was aware of the critical importance of mobile phones. He said that "262 million Americans are using mobile phones. That's roughly 84% of the total population. (&) I think it's a must for campaigns to be using mobile technology. It's the only device that's truly with people for 15 to 24 hours a day" (Bukheit, 2010). So in addition to its personal nature, the mobile phone offers an extended time period for communications. It is no surprise then that the campaign's mobile strategy included designing Obama Mobile, a web site dedicated to delivering campaign content specifically formatted for mobile devices, and that it also made use of a mobile application discussed in the next section.

THE FUTURE OF CAMPAIGNING

Perhaps the best example of how mobile technology can help an organization gain significant efficiencies and empower its members and supporters, is the iPhone application launched by the Obama campaign in October 2008, just over a month before election day. Coded by volunteers in less than three weeks, the free application was installed by 95,000 supporters (Harfoush, 2009, pp.120). This application did not receive the analysis and attention it deserves probably because of its late arrival in the campaign. However its design and functionality provide a glimpse into the future of campaigning (Figure 2).

The iPhone application included the following features.

Figure 2. Obama's campaign iPhone application. (© 2008 Rahaf Harfoush. Used with permission)

Call Friends

Phone calls have been known to be one of the most effective ways of connecting with potential voters and volunteers in a political campaign. The massive size of the task has led most campaigns to one of two solutions: Recruiting and training a corps of volunteers and allocating them to temporary call centers established at the campaign's headquarters to make outgoing calls to the target population. Alternatively campaigns outsourced the outgoing call function to specialized call center companies that use their own staff to implement the calls.

The Call Friends feature of the iPhone application offered a third alternative. It enabled supporters to become mini call centers using their own personal mobile devices. More interestingly, it made use of the supporter's personal network; it accessed the user's address book and automatically organized its contacts by battle ground states, thus prioritizing personal contacts according to central strategic needs of the campaign. Users would call their friends according to the prioritized list. Illustrating the value of using personal networks, these calls coming form a friend would be better received than a call from an unknown person calling from the campaign's office. The application went beyond this. It actually tracked who has been called and offered an easy way for users to document the outcome of the call. As a result, not only were annoying double calls avoided, but the information on the calls' numbers and outcome was collected by back-end servers providing campaign organizers with almost a real-time view of the political landscape of a certain region and the performance of the campaign there. The aggregation of the many mini-call centers in fact provided a distributed calling center capability at a fraction of the cost of traditional solutions.

News

The News function provided a single aggregated source for all campaign press releases and statements. The format was optimized for mobile devices; each news item had a short summary with a link to the full text of the item. Users could switch with a single click between local and national press items. They could also easily email an item to anyone in their address book. With the overload of information during the campaign and the various attacks and rumours spread by competitors, this feature helped respond very rapidly to such attacks and rumours and allowed Obama's positions to be available to supporters without the delays and distortions experienced through the traditional media channels. It also provided supporters with immediate access to a reliable and trustworthy repository of Obama's positions on a wide range of issues.

Media

Similar to the News function this feature provided access to a library of photos and video clips related to Obama activities and speeches. The higher fidelity of the visual media leveraged the candidate's charisma. The feature allowed users to easily email a photo or video clip to anyone in their address book. In order to facilitate finding the right photo or video clip for a specific topic or discussion, items in this repository were searchable by tags. A link was provided to Obama's YouTube channel, where not only campaign generated content could be found but also a wide range of supporter-produced contents, such as the video song "Yes We Can" by *will.i.am*, that provided emotional and entertaining connections to the candidate.

Issues

This Issues feature provided the most comprehensive information source on Obama's position on a wide range of issues, again formatted for mobile users: each issue was briefly explained; Obama's position on the issue was listed in bullet points; his voting record on this particular issue was also provided; and a link to his speeches that touched on this issue was also provided. Supporters seeking more detailed information could download PDF documents that addressed the issue in depth. The availability of such relevant information played an important role in the campaign's efforts to persuade undecided voters. Supporters felt confident and well informed with that information at their finger tips. True to the campaign's principles in creating a two-way conversation with supporters, users of this feature could fill a form providing their thoughts, ideas, and feedback about their persuasion efforts back to the campaign.

Local Events

Using the GPS functionality of the iPhone this features allowed supporters to find out events organized by the campaign within a set distance from their location. Later events organized by supporters were also included.

Get Involved

Also using the GPS functionality this feature provided users with the nearest volunteers operations centre and made them aware of the various volunteering opportunities available.

SYNERGY BETWEEN INDIVIDUAL AND ORGANIZATION

The combined use of mobile communications tools (SMS, iPhone application, mobile web site) with Internet-based social networks as demonstrated by the 2008 Obama campaign has yielded interesting results. The campaign raised a record amount of $630 Million, 67% of which was pledged online. Youth turnout in the 2008 Primary Elections set a high record. The scope and depth of supporters mobilization and engagement reached unprecedented heights. Although many other factors contributed to the success of Obama's campaign and culminated in his election as president, there is general consensus that mobile technology and social networks have played an important role in this success and will have an increasing role in future campaigns of all kinds, political, social and commercial, at least in three areas.

First, the capability of capturing information in audio and visual formats by distributed mobile devices combined with information amplification through social networks has created a fertile environment for the emergence of social and political trends and movements. This fertile environment makes it easier for individuals to rally rapidly around new ideas, preferences, or causes, thus enabling easier formation of broad movements.

Second, both emerging movements and existing organizations can now reach more people faster at less cost and can deliver their messages more efficiently through the finer segmentation enabled by the personal nature of mobile devices. They gain additional benefits through leveraging the personal social networks of their supporters.

Third, mobile technologies are enabling the distribution of all major functions of a campaign management office including organizing, volunteer recruitment, fundraising and communications. This distribution taps naturally into crowd-sourced resources, ideas and initiatives, leading to substantial reductions in cost and to increasing the agility and resilience of the campaign's organization.

The benefits from these increased capabilities are not limited to organizations. The wide spread of mobile and social technologies is creating an increased awareness and confidence at the level of the individual as well. In January 2009, when US Airways Flight 1549 had to make an emergency landing in the waters of New York City's Hudson river, the first report, picture, and video of the evacuation of the passengers did not come from any news channels or professional reporters. They came instead from citizens equipped with mobile phones and digital cameras. The words "citizen journalism" are now accepted broadly and many event organizers formally offer citizen journalists who blog or tweet full accreditation and media privileges at their events.

CITIZEN NETWORKS

With the increased sense of ability, citizens are organizing more complex collective efforts to take into their hands functions traditionally handled by various levels of government or international organizations. Take for example Ushahidi, a web site originally developed by Kenyan citizen journalists to monitor and display on a map reports of violence following the 2008 elections. Reports could be submitted through the Internet or through regular mobile phones using SMS. The initiators of the web site were surprised by the 45,000 users in Kenya alone that flocked to their site. That same year they shared their code with civics in South Africa, who used it to monitor incidents of xenophobic violence. Realizing the need for such a citizens platform, the founders rapidly evolved from an ad hoc group of volunteers to an organization aligned with the open source development movement, involving a growing network of volunteer software developers from across Africa but also from Europe and the United States.

The new organization promptly developed a new framework and released the Ushahidi platform, which even in its alpha version was tested by 11 organizations within Kenya and 4 deployments outside including Gaza, India and Pakistan. Since then "(t)he Ushahidi platform has been used to monitor elections in India, Mexico, Lebanon and Afghanistan. It has been deployed in the DR Congo to track unrest, Zambia to monitor medicine stockouts and the Philippines to track the mobile phone companies" (Ushahidi, 2010). More recently, the platform has been used to map citizen-generated data during the post-earthquakes crisis in Haiti and Chile. It was also used in a Sudanese civil society initiative to monitor the elections in Sudan.

While many of the deployments of Ushahidi are in a political or major disaster context, the ubiquitous availability of the mobile technology, the rising awareness and sense of ability of individuals, together with the integration and visualization capabilities of platforms like Ushahidi are driving many initiatives outside of those main areas. In the Atlanta metro area crimes can be tracked on a map of the city at the Atlanta Crime Maps web site (http://crime.mapatl.com/main) using the Ushahidi platform. Data is provided by the Atlanta Police Department or is reported by citizens through a web interface accessible from computers or smart phones. In Washington (DC) after the heavy snowfalls of the 2009/2010 winter a web site called "Snowmaggedon – The Clean Up" was created in collaboration with The Washington Post. It uses the Ushahidi platform to display areas in urgent need of snow removal. Citizens are guided to action in 3 simple steps: (a) report a problem (car stuck, driveway or sidewalk problem etc.) or a solution (available plow or snow-blower etc.); check for reports near you; (c) connect with neighbors (using the comments section in the report form) and organize a cleanup party.

This rapid adoption of collaborative platforms enabled through the combination of mobile communications with social platforms that have data integration and visualization capabilities is thus enabling a global movement for monitoring the ac-

tions of incumbent powers, whether governments (Kenya, Sudan, Gaza), corporations (Philippines phone companies, gas price monitoring) or illegal groups (monitoring human trafficking). This new level of scrutiny is having an impact on these powers, mostly in curbing excesses and forcing a more inclusive approach in their actions for those open to change and at least a more cautious action for those who resist it.

A NEW DEFINITION OF RESILIENCE

The demonstration of the effectiveness of self-organizing crowds in disaster situations small and large, as shown on the various examples of Ushahidi deployments, is fostering a new school of thought in the emergency preparedness and disaster relief communities, which argues that "next generation resilience relies on citizens and communities, not the institutions of state" (Edwards, 2009, pp.1). Edwards observes that traditional command and control centers that are mandated with handling emergency or disaster situations, usually tend to build their own "network of participation" from scratch every time depending on the nature of disaster and its location, rather than using existing citizen networks that have local knowledge and capabilities and could act immediately based on long-established trust between their members. The problem is that most of the time authorities don't know about these citizen networks. Even when they know, they usually do not trust such networks because they don't understand their reach and capabilities, but mostly because they are used to a command and control structure that usually does not allow for decentralization and empowerment. However, citizen networks are often well organized, efficient and armed with tremendous local knowledge that authorities lack. Edwards writes about three such networks in the UK: The Farm Crisis Network (FCN), The South West ACRE Network of rural community councils, and faith-based communities.

The purpose of the FCN was to support farmers and farming communities with pastoral and practical help. In 2001, when farming communities were hit hard with the foot and mouth disease, government authorities handling the crisis, both central and local, ignored local expertise and knowledge of geography, supply chains, and local resources. This isolated the community and wasted the opportunity to use locally available knowledge and resources to help address the crisis. "In contrast to the government, the FCN adopted a more nuanced approach, reaching out to farmers and the wider community and supporting thousands of households with pastoral and practical help" (Edwards, 2009, pp.65). Similarly, a community partnership of faith-based organizations, the Islington Faiths Forum (IFF), emerged from the chaos following the 1987 deadly fire at King's Cross station in London for the purpose of organizing a better response by these communities in case of emergency. IFF received some support from the local emergency planning official, who paid for high-visibility jackets, green flashlights for the car, and most importantly a mobile phone for coordinating emergency work. The value of such self-organized communities became visible during the aftermath of the coordinated terrorist attacks on London's public transport system in July of 2005. Organizing their activities like traditional command and control systems do, IFF was able to roll out services rapidly around King's Cross Station and coordinate response from various IFF communities including local Muslims.

The fact that IFF needed external financial support to acquire a mobile phone illustrates the cost barrier that faced self-organizing communities in the late eighties and early nineties regarding communications in a disaster situation. The explosive growth in the availability and affordability of mobile communications worldwide has removed such barriers and enabled a larger number of self-organizing communities that are concerned with their preparedness and resilience to emergencies to organize themselves. Where central or local government authorities understood the value of such communities for emergency preparedness and disaster relief, the integration of the capabilities of such citizen networks with traditional command and control centers is creating the foundations of a more resilient and agile nation, in which the definition of resilience is not limited to the ability to recover from a disaster but rather extended to include the ability to adapt, collaborate, and learn collectively.

At the dawn of the twenty first century human civilization is facing several serious threats ranging from radical climate change to dwindling natural resources and deadly diseases. Our civilization relies for its survival on highly complex and interdependent systems such as food and energy supply chains, energy distribution grids, global telecommunications, and long-range transportation. Our reliance on these complex systems is creating a brittle society. These critical systems are the first to be disrupted in emergencies and disaster situations. In some cases disasters destroy or disable not only parts of that critical infrastructure, but also the central command and control capacity itself as was demonstrated in the Haiti earthquake. Crisis after crisis have demonstrated the need for local action and for integrating local knowledge and capabilities into both pre-disaster preparedness and post-disaster relief and recovery efforts. The rapid spreading of mobile communications is accelerating the emergence of self-organizing citizen networks that are addressing various aspects of this need and has defined citizens and communities rather than the institutions of state as the cornerstones of next-generation resilience. The requirement by this new model of resilience to enable and empower many overlapping citizen networks and to integrate them with centralized emergency command and control centers brings with it significant shifts in the distribution of power and authority, making this process a serious challenge for most if not all forms of governments.

SOCIO-ECONOMIC DEVELOPMENT

Traditional development theory focuses efforts on basic needs such as food, shelter, clean water, sanitation, primary healthcare and employment. As a consequence, investments and development aid funding were aimed at these areas and were dispensed according to these priorities. Decisions about such investments or aid were several levels removed from the communities in need of development and were usually made by external development "experts" and government officials. While some progress was made, this model has not proven very effective or efficient. This is best illustrated through the United Nations Millennium Development Goals (MDG), which were established by a world leaders summit in 2000 and set eight goals to be achieved globally by 2015. In September 2010 the United Nations held a high-level summit to review mid-term progress on MDG. As Ban Ki-moon, Secretary General of the United Nations writes in a report highlighting gaps in the implementation of MDG:

"Delivery of official development assistance is slowing down. The Gleneagles commitments to doubling aid to Africa by 2010 will not be met. The Doha Round of multilateral trade negotiations remains stalled. Debt burdens have increased, with a growing number of developing countries at high risk or in debt distress. And rising prices are hampering access to medicines, while investment in technology has weakened" (MDG, 2010).

With this backdrop and the effects of the global economic crisis, mobile communications technology for developing regions was considered a luxury irrelevant to development goals and therefore at the bottom of the development investments priority list. This traditional development policy has many critics.

One of the most interesting and relevant critics is Iqbal Quadir, who argues that traditional aid in fact only empowers the authorities of the recipient countries and removes their incentives for empowering their people (Quadir, 2006). Quadir was born and raised in Bangladesh. In the early nineties, he was practicing as an investment banker in New York and collaborating with colleagues through a computer network. When the network failed one day, it reminded him of an episode from his childhood, where in the absence of communications, he had to walk 10 miles to a pharmacy, only to find it closed and walk back having wasted a day for nothing. He realized that communications is not only a basic human need but also a major contributor to productivity. He set out to create a mobile service in his native Bangladesh but was rejected several times by various investment and aid funding sources.

Inspired by Grameen Bank's micro financing concepts pioneered by fellow Bangladeshi and Nobel Peace Prize recipient Muhammad Yunus, Quadir founded Grameen Phone, a joint venture between Telenor, Norway's largest telecommunications service provider, and Grameen Telecom Corporation, a non-profit sister company of Grameen Bank. One of the first programs Grameen Phone offered was the Village Phone Program, started in 1997, which used micro loans to help more than 210,000 people, mostly women living in rural areas, to acquire a mobile phone and use it to provide service to others. The program supported universal access to telecommunications service in remote rural areas: People lacking the means to own a phone could gain access to communications through the services offered by these so-called Village Phone operators, who had an opportunity to earn a living providing this essential service.

Grameen Phone became remarkably successful. It is now the leading telecommunications service provider in Bangladesh serving by December 2009 over 23 million subscribers. It created direct permanent or temporary employment for over 5,000 people; more than 100,000 people depend indirectly on the company for their livelihood. More importantly, Grameen Phone "provides telephone access to more than 100 million rural people living in 60,000 villages and generates revenues close to $1 billion annually" ("Iqbal Quadir", 2010). The company has become one of the largest taxpayers if not the largest in Bangladesh.

The success of Grameen Phone demonstrated that mobile communications can create significant economic development even in very poor areas. Economic development in turn has always a social impact. A 2008 Deloitte study commissioned by Norway's Telenor on the impact of mobile communications in Serbia, Ukraine, Malaysia, Thailand, Bangladesh and Pakistan found both direct and indirect social impact. Among the direct impacts found were increased social cohesion. With increased migration of some family members from rural areas to urban centers and even overseas countries, mobile communications are helping maintain important connections between family members. In addition to the universal access provided through small service provider (such as the Village Operators of Grameen Phone) to users who can't afford their own dedicated phone line or cellular phone, the study also found that mobile communications provide a basis for sharing Internet access particularly in areas where Internet access is not easily accessible (Deloitte, 2008, pp. 5).

The study found that in 2007 across the countries studied mobile communications contributed between 3.7 and 6.2% of GDP. It also found a signicant correlation between mobile penetration and economic growth rates: "(...) a 10% increase in penetration will, holding other factors equal, lead to a 1.2% increase in long-term growth" (Deloitte, 2008, pp. 8-10). While hard to quantify in monetary terms, improvements in productivity are correlated with improved living standards and hence with social change. The ability to earn an income and become more financially independent improves significantly the status of the earners, particularly in the case of women. Similarly, the ability to improve business performance for self-

employed and small business owners, leading to higher disposable income, improves their social status.

The social impacts of mobile communications that are correlated with economic development are sometime grouped under the label "tangible" social impact. They usually include four categories (Bhavnani, Won-Wai Chiu, Janakiram, & Silarszky, 2008):

Entrepreneurship and job search: Mobile communications facilitate carrying out a number of business functions, mostly informational and transactional ones. This has an impact both on the supply and demand sides of the market. Entrepreneurs and business owners can run their operations more efficiently. Job seekers have easier access to the labour market. There is much anecdotal evidence to support this view particularly from a study by Chipchase (2006) cited by Bhavnani (Bhavnani, Won-Wai Chiu, Janakiram, & Silarszky, 2008, pp.16).

Information asymetry: Using mobile phones to arbitrage over price information from potential buyers and optimize transactions has been shown to reduce price variations. In a 2007 study of fishermen in India, the introduction of mobile phones has reduced the mean coefficient of price variation from 60-70% to 15%. The study of 300 sardine fishing units concluded that "the use of mobile phones: (a) increased consumer surplus (by an average of 6%); (b) increased the fishermen's profits (by an average of 8%); (c) reduced price dispersion (by a decline of 4%) and reduced waste" (Bhavnani, Won-Wai Chiu, Janakiram, & Silarszky, 2008, pp.16).

Market inefficiencies: Economists see the lack of affordable access to relevant information and knowledge as a critical impediment to efficient markets. As mobile communications are proving to be an efficient tool to provide access to such information and knowledge, they are gaining prominence in the efforts to reduce poverty particularly in rural areas of developing countries where three quarters of the developing world's population lives. One example is the Palliathya help line in Bangladesh, which in it's pilot phase offers helpline services to the people living in 4 villages. The services aim at preventing exploitation by middlemen, provide employment opportunities particularly for rural women, reduce information gaps, save cost and time, and strengthen access of service providers to rural people (Bhavnani, Won-Wai Chiu, Janakiram, & Silarszky, 2008, pp.17).

Transport substitution: with the availability of mobile communications the need for face-to-face meeting and physical presence is reduced and with it the cost of transportation and the unproductive time spent commuting. Mobile workers in urban and rural settings benefit most from such reduction in cost and downtime. For urban workers, whose job involves much travel (taxi drivers, plumbers, sales people etc.) mobile communications brought time savings of 6%; "56% of businesses in South Africa identified reduced travel as a beneficial impact of the mobile phone" (Bhavnani, Won-Wai Chiu, Janakiram, & Silarszky, 2008, pp.17).

SOCIAL CAPITAL

There is also a range of "intangible" benefits from mobile communications, "which are difficult to value, may not have direct economic benefit, but will certainly enhance and promote the growth of culture, society and societal ties" (Bhavnani, Won-Wai Chiu, Janakiram, & Silarszky, 2008, pp.18). One of the most important intangible benefits could be the so-called social capital or social cohesion, which encompasses the relationships and norms among a collective that enable members of that collective to pursue shared objectives more effectively. It is now generally accepted

that social capital contributes to productivity and that it generates advantages to the collective that accumulates it.

A 2005 study of communities in South Africa and Tanzania found links between mobile usage in rural communities and three types social capital:

• as an amenity & shared commodity
• to mediate strong links (with family and friends and other community members) and
• to mediate weak links (with individuals 'outside' the community, e.g. businessmen, government officials, tradesmen, etc.) (Bhavnani, Won-Wai Chiu, Janakiram, & Silarszky, 2008, pp.19).

HEALTHCARE

Another area where mobile communications is having a significant impact is healthcare. "There are 2.2 billion mobile phones in the developing world, 305 million computers but only 11 million hospital beds" according to Terry Kramer, strategy director at British mobile phone operator Vodafone (Perez, 2009). This high penetration by mobile phones across populations of developing countries offers an unprecedented opportunity to reach large numbers of individuals with relevant and timely health information. More importantly, communications can bridge the urban rural divide of healthcare: healthcare providers (doctors, nurses etc.) are usually concentrated in urban areas, while the majority of the population in the developing world lives in rural areas without access to these resources. For example, "in India (&) there are 1m people that die each year purely because they can't get access to basic healthcare. The converse angle to that is that 80% of doctors live in cities, not serving the broader rural communities where 800 million people live" (Perez, 2009).

Because mobile technology is a more efficient way of providing telecommunications services in rural areas, it can connect those that desperately need healthcare with the large pools of healthcare providers, who live in the urban centers. As Grameen Phone has demonstrated, providing mobile communications services can be very lucrative even in the rural areas of poor countries. The interest in providing mobile healthcare and health education services is therefore rapidly rising among national governments, international organizations, and service providers. Vodafone, a major international provider of mobile communications services has joined the Rockefeller Foundation and the United Nations Foundation in creating the mHealth Alliance, whose objectives are to advance the use of mobile technology for healthcare in the developing world and to guide governments, NGO's and other mobile firms on the use of mobile technology for healthcare.

The process is advancing rapidly. In a recent report prepared for the UN Foundation and Vodafone Foundation and titled "mHealth for Development: The Opportunity of Mobile Technology for Healthcare in the Developing World" over 50 mobile health initiatives across 26 countries were listed (Vital Wave Consulting, 2009). These projects cover a wide spectrum of healthcare areas:

Education and Awareness: SMS messages are used to deliver information about testing and treatment methods, availability of healthcare services, and disease management directly to user's phone. Grameen Phone for example works closely with the health authorities in Bangladesh to raise awareness about immunization and uses SMS to alert subscribers to current immunization campaigns. The United Nations uses SMS to distribute food vouchers to the over one million Iraqi refugees in Syria. In addition to such one-way campaigns several interactive campaigns have been deployed. For example in India, South Africa, and Uganda interactive campaigns are being used to promote AIDS education and testing, provide information about other communicable diseases like TB, or promote maternal or youth reproductive health issues. While other media have been previously used for similar purposes, the personal nature of the mobile device and the privacy of communications with the individual have made the mobile channel the most popular of all communication media and has lead to higher participation rates than other media.

Remote Data Collection: Public health policies and government actions rely critically on information about the subject of the policies or actions. Data collection across rural areas has its challenges, particularly where many people are not able to reach a secondary or tertiary healthcare facility even in the case of severe illness. Maintaining the quality and consistency of the data collected is also challenging. Even at the pilot stage, data collection using mobile devices are providing health officials at all levels with better information to assess the effectiveness of healthcare programs and optimize the allocation of resources. A mobile data collection project deployed in Uganda resulted in 25% savings in the first 6 months plus higher job satisfaction amongst health workers. Other parties could also benefit from getting information from masses of people on mobile phones in developing countries. For instance, pharmaceutical companies could collect data on how people respond to drug treatments, use mobile phones to ensure that drugs that are reaching people in developing countries are authentic, not counterfeit, or have not been tampered with en route.

Remote Monitoring: mobile devices are used to monitor patients' compliance with medication regimen, monitor health condition, and maintain caregiver appointments. In a project in Thailand TB patients were given mobile phones that healthcare workers (themselves former TB patients) call daily to remind them to take their medication. Compliance reached 90%. While remote health monitoring projects using mobile technology are relatively limited in developing countries, they are taking off rapidly in the developed world, particularly for monitoring patients with chronic conditions at home. This rapid adoption in the developed countries will undoubtedly drive rapid adoption of the technology in the near future in the developing world, although the focus of the monitoring would be different.

Communication and Training for Healthcare Workers: the acute shortage of healthcare workers in the developing world is exacerbated by the difficulty to update the knowledge of those scarce resources while they are facing excessive demands for service. Providing health workers with access to relevant sources of information via mobile devices reduces the effort and costs of such training. Mobile devices enable healthcare workers to better communicate among themselves in a region, thus better coordinating resources and referrals, for example referring a patient to a hospital without knowing if a bed is available for that patient.

Disease and Epidemic Outbreak Tracking: Communicable diseases often brake out in pockets. The earlier these pockets are identified, the faster they can be addressed, and the smaller the spread of the disease. The ability of mobile devices to capture data in a widely distributed way and at low cost, plus the availability of data integration and visualization platforms like Pachube are providing powerful new solutions for the rapid capture and transmission of data and the early detection of any outbreaks. Mobile outbreak tracking solutions are being used in Peru, Rwanda, and India. A recent success story is that of EpiSurveyor, an open-source software allowing health workers to easily create data collection forms adapted to their specific needs. The World Health Organization (W.H.O.) has adopted this software as an electronic data collection standard and it is being used by the health ministries of 20 countries in Sub-Saharan Africa (Butcher, 2009).

Diagnostic and Treatment Support: The effectiveness of treatment is highly dependent on the proper diagnostic of the problem. With the severe shortage of healthcare resources locally in most developing and rural areas, the challenges for diagnosing are significant. In developed countries the trend has been to implement telemedicine networks using high-bandwidth networks and expensive terminal equipments. In the developing world, where such telemedicine network may only be feasible to link the main healthcare facility of the country with a foreign country's high-profile facility, the trend emerging is to use mobile based solutions to enable diagnostics and support treatment. Solutions include equipping health workers mobile devices with step-by-step diagnostic support, built-in calculator for drug dosage, and reference materials. The ability to capture images with mobile devices and to transmit these along with vital signs to a remote center, where specialized resources are available, reduces the need for transportation and improves the speed of the process.

As with any new solution there are of course challenges facing such mobile health applications. Among the technological ones are the autonomy provided by the battery of the mobile device, when these are used in remote areas with inexistent or intermittent power supply. This challenge is being addressed at many levels. Some of the mobile service providers are designing the power generation for their points of presence to have some excess capacity that can be used for the surrounding communities. Some of the local inhabitants of remote villages and communities are self-organizing as well: they collect all the mobile devices that need recharging and take them collectively to another location where they can be charged. New grid-independent power charging equipments are appearing every day: solar-, wind-, and motion-based devices are becoming increasingly available, although many of them are not yet reaching the mass-production stage that would make their cost affordable for large scale deployments in developing countries.

The majority of the challenges, however, are non technological. They mostly are manifestations of limited institutional capacity. For example, integrating data collected through mobile devices into the decision making process of the health authority is not always a simple process. As with many projects and new technologies it is difficult to quantify social benefits and express such quantification in monetary terms. However, with the increasing pressures to contain rapidly escalating healthcare costs and to find new effective healthcare delivery mechanisms, and with the continuing rapid decrease in the cost of digital communications and wireless networks, the Moor's law driven increase in processing and storage performance of mobile devices, and the rapid maturing of mobile platforms (smart phones, tablets etc.), it is highly probable that the penetration of mobile devices in the world will continue to increase and that mobile healthcare applications will continue to accelerate along with their multi-faceted impact on a variety of health and social related issues.

EDUCATION

Given the strong uptake of mobile technology among youth, it was obvious that this technology would have an impact on all aspects of their social lives: between peers, within families, and of course at school.

At the peer group level the impact of mobile communications has been extensive, as communications amongst group members is central to the identity of the individual. The impact is further amplified by the increased peer group's influence during adolescence. Communications within the peer group has always had a functional and a relational aspect. The functional side is mainly the coordination of group activities. In previous generations its primary form was physical presence at pre-arranged locations or later the family's landline telephone. The personal mobile device brought independence of location and private communications, which contributed to an increase of the influence of the peer group.

The relational aspect goes beyond communications for the purpose of coordinating actions to creating and maintaining strong relationships within the peer group without or with less adult interference. The importance of the relational level is illustrated by the expectations among youth when using mobile text communications: received text messages must be answered within 15 – 30 minutes, beyond which timeframe an apology is necessary (Campbell, 2005).

Negative effects of mobile technology include ostracizing those without mobile devices, which is creating a new divide; decreasing inter-personal skills by hiding behind the technology in emotionally charged situations, and cyber bullying. The higher capability of coordinated action has also brought by new cultural byproducts such as using mobile phones to "gatecrash" parties in large numbers, organizing "rave" parties where these are still possible, or organizing a "flash mob", which Wikipedia defines as "a large group of people who assemble suddenly in a public place, perform an unusual and pointless act for a brief time, then disperse".

At the family level the uptake of mobile communications was mainly driven by parents's perceptions of improved security for their children. It lead to renegotiating freedoms between children and parents, providing options to extend curfews and places where children are permitted to go, while maintaining some form of parental control. While some researchers see the mobile phone given by parents to a child as an intrusion into young people's lives, the majority of researchers believe that mobile technology has undermined the authority of parents, who don't really know who their children friends are nor have easy access to monitor communications of their child with such friends (Campbell, 2005).

The impact on the school was more significant and challenging. Young people's increased reliance on mobile communications within their peer group is making them reluctant to turn their devices off while in class. Surveys found that 86%

of Italian 9- and 10-year old who owned mobile phones and 66% of New Zealand students who took a mobile phone to school kept them on during class. In addition to the disruption or loss of attention that unobtrusive text communications can bring to the classroom, mobile communications are also blurring the lines between the school and the outside as students can immediately communicate with parents, who in turn contact the school (Campbell, 2005). The built-in image and video capture capabilities of new mobile phones is creating opportunities for misuse leading to new challenges, some of which quite serious as illustrated by the continuous string of gay youth suicides in U.S. schools.

The challenges posed to the school system by mobile communications are within the broader context of the changes brought by the digital revolution and the transition from industrial and post-industrial economy to a knowledge-based economy. These changes are driving a rapid evolution of the fundamental structures in all contemporary institutions: political, economic, industrial, commercial and social. The educational institutions that had developed over the past few centuries to supply the needed skilled manpower to these various institutions are no exception. They needed to adapt to these changes as well. The signs for that transformation, albeit slow and fragmented, can be seen everywhere and at every level of education from K-12 to post secondary and graduate education. In addition to rethinking the rigid traditional structures of knowledge domains and opening up to more interdisciplinary collaboration even across unrelated knowledge areas, this transformation required more often than not embracing new technology rather than resisting and banning it from the educational process.

Mobile communications technology being the most widely adopted communications technology both in developed and developing countries, particularly amongst young people, it is only natural that many of the new effort for the transformation of education engage that technology and seek to

learn how it can be harnessed for educational purposes. These efforts are well underway and cover almost every aspect or level of education. The following provides a sample of the transformation underway.

Targeting young children aged two to five years Fisher Price, a major brand of children toys by Mattel, launched in early 2010 three new applications for the iPhone. One of the applications trains children to identify animals. The two other apps provide an electronic version of popular toys (Little People Farm and Chatter Telephone). While not many 2-5 years old would have an iPhone, many of their parents do. The value of the apps to them is to provide reasonably useful distraction for young children, for example while waiting at the pediatrician. For pre-schoolers the company has developed the iXL Learning System. Featuring resemblance to the iPad tablet device, iXL is a tiny computer complete with speakers, a touch screen, big colored button to interact with the device, and a collection of apps that includes Story Book, Note Book, Game Player, Art Studio, Music Player, and Photo Album (Caleb, 2010).

St. Mary's City School in Ohio is an example of developments happening at the primary school level. Students from third to sixth grade have their own mobile learning devices and the school is planning to equip seventh graders next year. Another example is The Bishop Strachan School in Toronto, Ontario, where every student is equipped with an iPad. As part of its Mobile Learning project Abilene Christian University in Texas is giving 2,000 college students and their teachers Apple iPhones and iPod Touches and telling them "Go mobile, go digital?" The project is trying to find out whether always-on, always connected, personal digital devices and social networks will transform teaching and learning. While some of the push for equipping students with personal mobile devices may be supply-side driven, there can be no doubt that the availability of powerful mobile devices for all students within a learning collective and their teachers is focusing and accelerating the efforts to adapt existing content and pedagogy to the new medium and to develop new content and methods that are specifically designed to take advantage of these devices.

In the middle school range the challenges with math education have been known for a while. According to statistics by the U.S. Department of Education only 31% of eight graders score at or above "proficient" level on standardized math tests. In some school districts high-school-algebra failure rates approach 50% (Svoboda, 2009). There are multiple initiatives exploring new solutions for this problem. Verizon Wireless has partnered with educators in Texas to implement a smart-phone driven math curriculum for fifth graders in the town of Keller. K-Nect is another initiative funded by Qualcomm to distribute cell phones for math instruction. Launched in 2008 the pilot is testing the concept at Southwest High School in Jacksonville, NC and five other schools in North Carolina, where a high percentage of students receive free and reduced-price lunches. The results have been remarkable: "(...) students in the Project K-Nect group scored higher on state Algebra I proficiency tests than their nonconnected counterparts did. At Southwest High, every student in one Project K-Nect class notched a 100% proficiency rating in algebra; students in a non-Project K-Nect class with the same teacher averaged 70% proficiency" (Svoboda, 2009).

Meanwhile at Purdue University students use an application called Hotseat to participate interactively in a collaborative learning process. The application provides near real-time feedback during class and enables professors to adjust the course content and improve the learning experience. Students not only can post their own questions about instructor-framed content of lectures but can also vote on other students' questions that they would like answered. They can also comment on such questions and answers and contribute their own thoughts to the conversation. Students can post from their web-enabled mobile phones, by text messages, through their Twitter or

Facebook accounts, or by logging into a web site. Thus, Hotseat leverages mobility and the power of social networks to create a virtual collaborative classroom that transcends the traditional classroom both in space and time.

The impact of mobile technology on education is not limited to the formal education processes. Similar transformation can be observed in adult learning and training. London-based design shop BERG recently released The Michael Thomas Method, a mobile app for learning language through listening and speaking (http://www.michelthomasapp.com/). Based on deep observations of mobile users' behavior and their lifestyle the app was designed with one big "pause" button that allowed users to control accurately the snippets of learning they could absorb while living their fast-paced lives. Lynda.com, a leading provider of software training videos online, is enabling anytime, anywhere access to their tutorials with a new iPhone app. Users can access their history and the content they have been watching from several supported mobile devices picking up exactly where they left regardless of which device they use to access their account.

There have been also a convergence between mobile gaming and learning, which is not limited to the Fisher-Price examples discussed earlier for early childhood education and entertainment but rather extend to youth and adults. A new category called "edutainment" has been created and this trend continues unabated bringing gaming to mainstream education and reaching even bastions of the traditional higher education system. In January of 2010 the makers of the popular mobile game Foursquare signed an agreement with Harvard University for using the game to encourage students to explore the university's campus and surrounding places of interest.

CROWD-SOURCED SCIENCE

The transformation of the educational systems through participation of larger numbers of individual users each equipped with a powerful mobile device is enabling the emergence of a new type of science. Initially, the availability of a broad distributed basis of people and devices capable of capturing information in multiple formats and transmitting these rapidly to central locations that can aggregate and visualize the data collected for a variety of purposes was primarily used for the collection of measurements and observations. Some of these initial applications used human-based observation and reporting while others used automated sensor data collection from mobile devices carried by people.

Project NOAH (which stands for Networked Organisms and Habitat) is a human-based data collection project. It provides a common technology platform that nature lovers can use to explore and document local wildlife and research groups can use to harness the power of citizen scientists. For nature lovers the platform allows users to select coarse categories for the organism they are observing, add some descriptive tags and submit the information through their mobile phone. The submission is automatically date/time stamped and the information along with some location details is stored in a central species database. Users can then check most recent sightings based on their location and can get additional species information that is usually complemented by local experts' knowledge.

The project also launches periodically so-called missions based on the needs of partner research groups and organizations to help gather important data, that otherwise would be very difficult if not impossible to collect. "Missions can range from photographing specific frogs or flowers to tracking migrating birds or invasive species" (NOAH, 2010). Recent missions included monitoring environmental damage from the oil spill in the Gulf of Mexico, reporting fox and grey squirrels, mushroom mapping, vegetable varieties for gardeners, and the lost ladybug project.

Oil Reporter, developed by Intridea for the Crisis Commons group, is an Android and iPhone application that "enables trained citizen journalists to use their mobile phones to capture and upload quantitative and qualitative data, as well as geo-tagged photos and videos to help in the recovery efforts" from the oil spill following the explosion and sinking of the Deepwater Horizon offshore drilling platform in the Gulf of Mexico in April 2010. Users of the application are asked a series of questions about their observations to collect details consistently. They can also attach photos or videos that are automatically geo-tagged and date/time stamped. Government agencies wishing to conduct their separate data collection using the application can segregate their data through an organizational ID. The application was developed within few days using the Appcelerator's native mobile application rapid development platform.

Another example of crowd-sourced human-based collection project is University of California's statewide effort to map roadkill using citizen observers. Volunteers use their mobile phones to capture GPS coordinates, photos and species information of killed animals and upload them to a central site, where the information is used to populate a Google map visualizing the kills.

Most of these citizen science projects use the various sensors built into mobile phones such as GPS, accelerometer, and camera in increasingly complex and sophisticated ways. Computer scientists at the university of California's School of Engineering have developed an Android application called Visibility, which uses participants' cameras to analyze and measure air pollution. Users simply point their phone camera towards the sky and snap a photo. Using data from the phone's GPS and compass the app then determines the location and direction of the photo, and compares the results with pre-established models of sky luminance. This in turn leads to a measurement of air quality, which is impacted by natural sources (e.g. salt entrainment, wind-blow dust) and manmade sources (exhaust, mining activities, etc.) (Perez, 2010).

The other type of collection that uses automated collection of data from mobile devices carried by people is best illustrated with two projects. The first is Google Maps traffic functionality, which crowd-sources traffic information from drivers' and passengers' GPS-enabled phones, including their location, speed, and geographic concentration to build real-time maps of traffic in cities and on highways. The drivers are not necessarily aware or participating in the data collection activities, rather Google relies on the fact that a statistically a percentage of the drivers (or their passengers) are using mobile phone with location services enabled.

Another automated data collection project is at the core of start-up Root Wireless Inc. The company plans to map the cellular signal strength of all major mobile service providers in the U.S. (Verizon, AT&T, Sprint and T-Mobile). These service providers make many claims about the reach, signal quality and reliability of their networks, but there have been no independent and impartial sources to confirm or refute such claims. An on-line poll of 2,300 web visitors to CNET.com showed that "26 percent of respondents have returned a cell phone because of poor service, and that 46 percent think there's not enough information on service quality when buying a handset" (Richman, 2009). The company's application debuted with a beta testing period on CNET.com for eight specific U.S. markets. An ambitious expansion plan calls for 4,000 volunteers – 1,000 per carrier – in each major market and fewer in smaller markets. Ultimately, the company wants to have 200,000 volunteers constantly monitoring the mobile networks nationwide. Contrary to the Google Maps traffic example, participants in the signal strength monitoring must download a small application to their mobile phone, and hence are aware of the data collection activity. The application runs in the background and the company claims that it consumes little battery power and bandwidth and that any personal data is stripped before being recorded in the central database.

Whether monitoring urban air quality or cellular signal strength, a widely distributed sensor data collection network is superior to current data collection systems in scope, accuracy, reliability and resilience. As important as air quality is, its traditional assessment is usually based on very few measurement locations. Traditional signal strength measurement equipment is bulky and expensive and needs to be trucked to the various locations. In comparison, crowd-sourced sensing networks can span thousands of measuring locations at a fraction of the cost of traditional methods. The resilience of a distributed system is also usually much higher than that of systems consisting of few nodes.

As crowd-sourced data collection has developed and proved a viable alternative to traditional collection systems, tasks assigned to the crowds started evolving from data collection only (whether manually or automated) to more valuable tasks involving human intelligence and skills.

One of the earliest and now evolved examples of crowd-sourcing human intelligent tasks is Amazon's Mechanical Turk. It is an online exchange for tasks requiring human intelligence to be carried out. In fact, the units of work made available through the site are called HITs for Human Inteligence Tasks. Like any exchange Mechanical Turk has both a supply and a demand side. On the demand side individuals or organizations seeking to crowd source some tasks have a way to present their tasks in a standard format easy to search and to distribute. On the supply side people wishing to offer their skilled labor to work on HITs can register and describe their interests and abilities in standardized formats that enable the platform to offer them HITs more in line with their interests and compatible with their skills. There are many qualification processes within the system itself to ascertain the level of knowledge and expertise in a particular field. Task offers range from simple opinions expression about a product or a service, to detailed testing of a new web site, to translating snippets of text or longer texts, to transcribing audio recordings that can be over an hour long. Remuneration ranges accordingly from 2 cents to over 15 dollars per task. For people living in developing countries, where the exchange ratio of local currency against the U.S. Dollar is high, even tasks of the 2 cents category can be quite appealing, particularly given the ability to easily pick and choose tasks and to work on them at one's own pace. The exchange acts as an impartial party between the task offerers and the task takers. It accepts prepayment from task offerers to guarantee payment and pays task takers upon completion by crediting their Amazon accounts.

The mobile communications revolution enabled the unprecedented access of billions of people around the world to crowd sourcing platforms like Mechanical Turk creating new value on both sides of the crowd sourcing equation and validating the concept itself as a viable production mode. One of the most prominent examples of this combination is txteagle, a company that has developed a platform enabling "approximately 2 billion mobile phone subscribers in 80 countries to earn money or airtime by work on a phone or desktop computer" (txteagle, 2010). Similarly to Mechanical Turk, the platform's focus is on business process outsourcing (e.g. forms processing, translation, audio transcription, fact checking) but also on local knowledge gathering. With access to millions of people distributed across the developing world, it becomes possible to gain reliable real-time information on local conditions, product penetration or current prices in specific geographies.

Txteagle has built a powerful network of alliances with over 220 mobile operators and with financial service providers interested in accessing the millions of people, who don't use banks. Compensation is monetary or through crediting airtime for the participan's mobile phone. The

airtime compensation has been integrated into the billing systems of txt eagle's mobile service provider partners creating the capability of instant compensation to a large number of workers anywhere in the world. This payment through local mobile service provider partners shields the companies outsourcing through txteagle from the complexities of paying a globally distributed workforce in over 50 currencies.

The company has also taken the qualification of its "workforce" to new heights. Each worker undergoes a rigorous application and certification process and must complete specific training sessions and sample tasks before being admitted to perform client tasks. Using its patent-pending statistical quality management system, txteagle is inverting the traditional work-labor paradigm: instead of bringing labor to the task, it delivers the task to a distributed workforce that can operate without the traditional managerial oversight. Work quality is evaluated by sophisticated statistical algorithms and workers are compensated based on the usefulness of their input, which creates incentives for self-improvements and continued learning, while eliminating the costs associated with traditional workforce management.

Earlier in this chapter we looked at Ushahidi as a platform for aggregating data collected from many mobile devices and visualizing it for various purposes. Another way of looking at Ushahidi is as crowd sourcing platform for various types of information: occurrence of violence in Kenya, military action against civilians in Gaza etc. Sometimes two different crowd sourced systems can be overlaid to provide a more powerful solution. After the Haiti earthquake of 2010, a flood of SMS and Twitter messages, mostly in Creole, arrived at a rate of more than one message per second, which overwhelmed first responders. A group of volunteers realized that it was critical to organize this information into categories that can be used by the military and rescue and relief workers on the ground.

Within days a crowd-sourced effort was mounted. The first step was to assign a unique code (4636) as a destination for all messages related to the crisis. This allowed all such messages to populate a queue. Because the Creole language has many varied spellings, it was not possible to automate the translation of messages. FrontlineSMS, the organization working initialy on this idea, decided therefore to crowd-source the translation. Initially, volunteers started translating the Creole messages into English, which enabled a much larger number of volunteers at Tufts University to parse, categorize and geo-tag messages so they can be stored in appropriate databases. Samasource, an organization specialized "in providing digital work opportunities for the people who need them most" that include data entry, video captioning, business listing verification, and book digitization, suggested using Haitians, who are on the ground and have lost their livelihoods, instead of volunteers for the translation, thus offering them an opportunity to earn money in the process. The cellular network proved to be the most resilient among the various communications networks in Haiti's crisis and allowed not only thousands of people to request help (some from under the rubble), but also for volunteers to assist and for Haitians to be actively involved in the rescue and relief efforts.

It became rapidly clear that most messages fell into one of two categories: request for help and search for family and friends. The information related to requests for help was used to populate Ushahidi maps that helped relief workers understand the urgency, scope and location of pressing needs and allocate resources more efficiently. Messages seeking to locate family and friends were routed to a database that allowed people to find information on the persons they were searching for more easily. The crow-sourcing effort in effect provided invaluable assistance to rescue and relief efforts by streamlining the flow of information coming from the scene of the disaster, rapidly establishing a centralized information resource for finding missing persons, and in the process creating revenue earning possibilities for people, who had lost their livelihoods through the disaster.

Crowd sourcing is on the verge of going mainstream, expressing the emergence of a new collaborative production mode. In almost every new crowd sourcing initiative or project mobile is playing a major role. Even fields as specialized as investigative journalism, which is arguably one of the foundations of the Fourth Estate in any democratic system, are not immune to this trend. The 2010 "We Media PitchIt Challenge" by We Media and Ashoka's Changemakers sought "the most innovative ideas inspiring a better world through media". The winner of the challenge's $25,000 prize in the category "non-profit" was announced in January 2010. It was Capital News Connection's submission titled: "AssignIt - Mobile Crowd-sourcing Apps for Hyper-Local Investigative Journalism". The submission proposed "teaming citizens with professional journalists at every step of gathering, vetting and distributing local news". Item number one in the submission's implementation plan included the "(d)evelopment, design/user experience and QA of interactive mobile apps" (Wittstock, 2010).

This rapid emergence of mobility in crowd sourced activities is related to the current clash between the traditional hierarchical organization of production and the emerging self-organizing peer to peer production modes. The vanguard research on self-organized networks came once more from the computer networking technologies and their potential military applications. The concept of increasing network reliability through decentralization of network functions is not new. It was behind the packet-switching concepts of ARPANET and then the ITU's X.25 and related standards, which led ultimately to today's Internet. But while that search for increased resilience took place in a context of mainly fixed communications nodes each serving multiple users, today's environment is characterized by a large number of smaller, even individual, mobile nodes. The question emerged of whether such small and mobile nodes could form ad hoc networks as they came in proximity to each other using non-regulated spectrum bands such as those of Blue Tooth or IEEE's 802.x series of standards, and whether such networks could operate meaningfully without the support of a fixed infrastructure and central database servers. This field of research, initially termed ad hoc networks and later more precisely called MANET (for mobile ad hoc networks) took off rapidly and is today present in numerous higher education and research institutions around the world.

The current MANET research topics are mostly about the network protocols for the mobile discovery of local resources in peer-to-peer wireless networks. MOBI-DIK (MOBIle DIscovery of Knowledge about local resources) is an initiative at the Computer Science Department of the University of Illinois at Chicago, who forsees a wide range of applications for such ad hoc mobile networks:

- **Social networking:** In large professional or political associations and social gatherings, MANETs could facilitate face to face meetings based on matching profiles or shared interests.
- **Transportation:** MANETs could help share location-relevant information on traffic jams, parking spots, available taxicabs and ride sharing opportunities.
- **E-commerce:** MANETs could match buyers and sellers in a mall or facilitate peer-to-peer trading of products (music, tickets, maps, etc.) or location-relevant knowledge (special offers, opening hours, security alerts etc.).
- **Emergency response:** MANETs could help rapidly replace the fixed communications infrastructure destroyed or disabled by the emergency.

- **Assets management:** Devices and appliances equipped with MANET capabilities could automatically form into networks oand transitively relay status information or alerts.

If you think such application are far-fetched or the domain of science fiction, think again. In September 2007 the Swedish company TerraNet AB demonstrated a mesh network of mobile phones without the need for cell sites. Each handset in such a network uses any other handset within range as an additional node, with which voice and data can be exchanged over the handset's own radio frequency. If a handset number is dialed, the calling handset checks whether the destination number is within range of any handset in the mesh network, and if so the call is relayed.

A similar concept is used by the One Laptop Per Child (OLPC) Association, who aims at providing children in developing countries with rugged energy-efficient laptops costing under $100 per unit. OLPC's XO laptops have built-in wireless and can form into mesh networks when in proximity of each other without the need for a central hub; they can share data and applications without a central database server. The XO design inspired the wave of netbooks that flooded the market in the last few years, but it remains the only one that can dynamically join mobile ad hoc networks.

In July 2009 TerraNet successfully completed a product demonstration in Salinas, Ecuador, where government officials "walked along the streets while talking on their TerraNet modified handsets" (TerraNet, 2009). One participant in the demo was able to stay on the phone for 20 minutes without a single failed handover across the mesh network. While commercial MANET implementations are still rare and relatively of a small scale, the implication of a scaled up implementation of this peer-to-peer technology is tremendous. What would be the impact if MANET was available at a large scale in a situation similar to the Iranian protests? What if the authorities could not block protesters communications by shutting down cellular service because protesters were not relying on fixed infrastructure owned by the government or a large corporation? What would the future of telecommunications providers be if all handsets in an urban environment were MANET capable?

Obviously, incumbent economic and political powers are worried about such developments. Handset manufacturers have shown interest in TerraNet's technology but they have to tread cautiously, as their major customers currently are the national and multi-national telecommunications service providers, who are keen to protect their investments in fixed infrastructure and their current revenue streams. Some technologies that can facilitate mobile peer-to-peer systems development and deployment, for example data broadcasting, are well understood and technologically easy to implement, but have yet to be deployed by cellular service providers.

A glance at any mobile peer-to-peer computing bibliography shows rapid advances are being made in mobile ad hoc network database technology and towards applications in transportation, e-commerce etc. However, despite the research and development momentum behind the MANET concept and its derivatives there are significant technological hurdles to be overcome. Some problems are complex and difficut to solve. For example the problem of routing messages between two devices that are out of each other's transmission range, using mobile devices as intermediaries, is still not fully resolved after twenty years of work on the issue. Other hurdles include "resource constraints on the mobile device, security and privacy, variable and/or disconnected network topology, and heterogeneity of devices" (Wolfson, 2009).

Despite these hurdles the mobile device with its increasing capabilities and its ubiquitousness is rapidly becoming the focus of distributed computing research and development. In October 2010 the First International Workshop on Issues

in Computing over Emerging Networks is being held in Delhi, India, with many of the papers addressing various aspects of resource-constrained mobile peer-to-peer networks. Contrary to the perceptions of some, mobility is not a second-rate access to the Internet, when the primary connection is not available. As Oberman wrote in a blog once, "(i)t is simply a new, portable and lightweight way to approach the Internet, which in turn, will completely reconceptualize the way in which we think about the Internet" (Oberman, 2006). It has already brough papable change in several important areas, and will continue to be a powerful driver behind several changes that have the potential of radically changing the structures of our current economic, political and social systems.

REFERENCES

Anonymous. (2009, June 13). *Mobile phones, Facebook, Youtube cut in Iran*. Retrieved from http://www.google.com/hostednews/afp/article/ALeqM5jSPlmVgh-SfeEO9WhpOVG6Slnu0w

Bhavnani, A., Won-Wai Chiu, R., Janakiram, S., & Silarszky, P. (2008). *The role of mobile phones in sustainable rural poverty reduction*. ICT Policy Division, Global Information and Communications Department (GICT). Washington, DC: World Bank. Retrieved from http://siteresources.worldbank.org/extinformationandcommunicationandtechnologies/Resources/

Bukheit, C. (2010, August 16). *How state political candidates are using mobile texting* [Web log message]. Retrieved from http://www.nonprofit-mediaworks.com/2010/08/16/how-state-political-candidates-are-using-mobile-texting/

Butcher, D. (2009, May 08). *Mobile software reinventing healthcare in developing world* [Web log message]. Retrieved from http://www.mobilemarketer.com/cms/news/software-technology/3204.html

Campbell, M. (2005). The impact of the mobile phone on young people. *Proceedings of the Social Change in the 21st Century Conference*. Brisbane, Australia: Queensland University of Technology-Centre for Social Change.

Cox, J. (2010, March 23). *Can the iphone save higher education?* [Web log message]. Retrieved from http://www.networkworld.com/news/2010/032310-iphone-higher-education.html

Edwards, C. (2009). *Resilient nation*. London, UK: Demos.

Esfandiari, G. (2010). The Twitter devolution. *Foreign Policy*. Retrieved from http://www.foreignpolicy.com/articles/2010/06/07/the_twitter_revolution_that_wasnt

Execution of Saddam Hussein. (2010). *Wikipedia*. Retrieved September 23, 2010, from http://en.wikipedia.org/wiki/Execution_of_Saddam_Hussein#Execution_proceedings

Grossman, L. (2009, June 17). Iran protests: Twitter, the medium of the movement. *Time*. Retrieved from http://www.time.com/time/world/article/0,8599,1905125,00.html

Harfoush, R. (Designer). (2008). *Yes we did*. [PowerPoint slides].

Harfoush, R. (2009). *Yes we did: An inside look at how social media built the Obama brand*. Berkeley, CA: New Riders.

Iqbal Quadir. (2010). *Wikipedia*. Retrieved September 24, 2010, from http://en.wikipedia.org/wiki/Iqbal_Quadir

ITU. (2010). *Monitoring the WSIS targets: A midterm review*. Geneva, Switzerland: International Telecommunications Union.

Labott, E. (2009, June 16). *State department to Twitter: Keep Iranian tweets coming* [Web log message]. Retrieved from http://ac360.blogs.cnn.com/2009/06/16/state-department-to-twitter-keep-iranian-tweets-coming/

Lenhart, A., Ling, R., Campbell, S., & Purcell, K. (2010). *Teens and mobile phones*. Informally published manuscript. Washington, DC: Pew Internet and American Life Project, Pew Research Center. Retrieved from http://www.pewinternet.org/topics/Teens.aspx

Leyne, J. (2010, February 11). *How Iran's political battle is fought in cyberspace* [Web log message]. Retrieved from http://news.bbc.co.uk/2/hi/8505645.stm

Limbo. (2008, February 08). *Limbo reports mobile advertising changes voters' attitudes and behaviors* [Web log message]. Retrieved from http://www.limbo.com/presscenter?pr=pr20080204.html

Maestri, N. (2010, January 16). *US texting raises $11 million for Haiti* [Web log message]. Retrieved from http://in.reuters.com/article/idINIndia-45435720100116

MDG Gap Task Force (Ed.). (2010). *The global partnership for development at a critical juncture – Mdg Gap Task Force report 2010*. New York, NY: United Nations.

Mobile Behavior. (2010, February 16). Mobile in education: Mattel announces fisher-price ixl learning system [Web log message]. Retrieved from http://www.mobilebehavior.com/2010/02/16/mobile-in-education-mattel-announces-fisher-price-ixl-learning-system

Monte, L. (2009). *The social pulpit – Barack Obama's social media toolkit*. New York, NY: Edelman.

NOAH. (2010). *Networked Organisms And Habitats home page.* Retrieved from http://www. networkedorganisms.com/about

Oberman, J. (2006, March 08). *What some people do not get about the mobile buzz: Mobile at politics online day one* [Web log message]. Retrieved from http://personaldemocracy.com/content/ what-some-people-do-not-get-about-mobile-buzz-mobile-politics-online-day-one

Och, F. (2009, June 18). *Google translates Persian* [Web log message]. Retrieved from http://google-blog.blogspot.com/2009/06/google-translates-persian.html

Parr, B. (2009, June 18). *Facebook releases Persian translation for #iranelection crisis* [Web log message]. Retrieved from http://mashable.com/2009/06/18/facebook-persian/

Perez, S. (2009, February 20). *Mobile phones to serve as doctors in developing countries* [Web log message]. Retrieved from http://www.readwrite-web.com/archives/mobile_phones_to_serve_as_doctors_in_developing_countries.php

Perez, S. (2010, September 22). *Android users crowd-source air pollution analysis* [Web log message]. Retrieved from http://www.readwriteweb.com/archives/android_users_crowd-source_air_pollution_analysis.php

Purdue University. (2010). *Hotseat: Enabling collaborative micro-discussion in and out of the classroom.* Retrieved from http://www.itap.purdue.edu/tlt/hotseat/

Richman, D. (2009, October 12). *Crowdsourcing digital signal strength* [Web log message]. Retrieved from http://www.msnbc.msn.com/id/33239992/ns/technology_and_science-wireless/

Svoboda, E. (2009, November 1). *Cellphonometry: Can kids really learn math from smartphones?* [Web log message]. Retrieved from http://www.fastcompany.com/magazine/140/cellphonometry.htm

Talks, T. E. D. (Producer). (2006). *Iqbal Quadir says mobiles fight poverty.* Retrieved from http://www.ted.com/talks/iqbal_quadir_says_mobiles_fight_poverty.html

The_Role_of_Mobile_Phones_in_Sustainable_Rural_Poverty_Reduction_June_2008.pdf

Txteagle. (2010). *Overview.* Retrieved from http://txteagle.com/?q=about/overview

United breaks guitars. (2010). *Wikipedia.* Retrieved September 06, 2010, from http://en.wikipedia.org/wiki/United_Breaks_Guitars

Ushahidi. (n.d.). Retrieved from http://www.ushahidi.com/media/Ushadidi_1-Pager.pdf

Vital Wave Consulting. (2009). *Mhealth for development: The opportunity of mobile technology for healthcare in the developing world.* Washington, DC: UN Foundation and Vodafone Foundation.

Wittstock, M. (2010, January 19). *Assignit - Mobile crowd-sourcing apps for hyper-local investigative journalism* [Web log message]. Retrieved from http://www.changemakers.com/node/68605

Wolfson, O. (2005). Mobi-dic: Mobile discovery of local resources in peer-to-peer wireless networks. *Bulletin of the IEEE Computer Society Technical Committee on Data Engineering, 28*(3). Retrieved from http://cs.uic.edu/~boxu/mp2p/deb-wolfson.pdf.

Wolfson, O. (2009). *Foreword to the book. Mobile peer-to-peer computing for next generation distributed environments: Advancing conceptual and algorithmic applications* [Web log message]. Retrieved from http://www.cs.uic.edu/~boxu/mp2p/foreword.pdf

KEY TERMS AND DEFINITIONS

Crowdsourcing: the act of outsourcing tasks to a large group of people or community rather than to suppliers, contractors and employees. The mass collaboration capabilities brought by web 2.0 technologies have pushed crowdsourcing to the main stream of various business and social activities.

Edutainment: A form of entertainment designed to educate as well as amuse.

Flash Mob: a large group of people who assemble suddenly in a public place, perform an unusual and pointless act for a brief time, and then disperse. The term is generally applied only to gatherings organized via telecommunications, social media, or viral emails, but not to events organized by public relations firms, protests, and publicity stunts.

Microfinance: the provision of financial services to low-income clients who traditionally lack access to banking and related services. It is not limited to credit and can include other financial services such as savings, insurance and fund transfers.

Moore's Law: a long-standing rule in computing hardware by which the density of transistors on an integrated circuit doubled approximately every two years. As a consequence the capabilities of digital devices are improving at roughly exponential rates: processing speed, memory storage capacity, sensors etc. The law is named after Intel co-founder Gordon E. Moore, who described the trend in a 1965 paper.

Social Movements: a type of group action. They are large informal groupings of individuals and/or organizations focused on specific political or social issues, in other words on carrying out, resisting, or undoing a social change.

SMS: Short Message Service, a text communication service most frequently used by mobile phone users to exchange short text messages (up to 160 characters) using standardized communications protocols. It is claimed t be the most widely used data application in the world.

Social Networks: a social structure made up of individuals or organizations called "nodes", which are connected by one or more specific types of interdependency, such as friendship, kinship, common interest, financial exchange, dislike, sexual relationships, or relationships of beliefs, knowledge or prestige. More recently the term refers more frequently to social structures relying on web 2.0 communications technologies for their connections, such as Facebook, LinkedIn or Twitter.

Chapter 11
Next Generation Multimedia on Mobile Devices

Mikel Zorrilla
Vicomtech Research Centre, Spain

Juan Felipe Mogollón
Vicomtech Research Centre, Spain

María del Puy Carretero
Vicomtech Research Centre, Spain

David Oyarzun
Vicomtech Research Centre, Spain

Alejandro Ugarte
Vicomtech Research Centre, Spain

Igor García Olaizola
Vicomtech Research Centre, Spain

ABSTRACT

The multiplatform consumption of multimedia content has become a crucial factor in the way of watching multimedia. Current technologies such as mobile devices have made people desire access to information from anywhere and at anytime. The sources of the multimedia content are also very important in that consumption. They present the content from many sources distributed on the cloud and mix it with automatically generated virtual reality into any platform. This chapter analyzes the technologies to consume the next generation multimedia and proposes a new architecture to generate and present the content. The goal is to offer it as a service so the users can live the experience in any platform, without requiring any special abilities from the clients. This makes the architecture a very interesting aspect for mobile devices that normally do not have big capabilities of rendering but can benefit of this architecture.

INTRODUCTION

Multimedia Content landscape is changing very quickly. Social networks are enabling a prosumer and collaborative attitude from the community. Social networks and other popular sites let users create and share information with other people. Their popularity in recent years has increased the active participation and content production of the web because these systems often contain blogs, photo galleries and other means for sharing digital content (Bonhand et. al., 2007). In addition consumers are demanding multimedia content access from everywhere and at anytime through any device, being the use of mobile devices the most popular and increasing trend of the last years.

DOI: 10.4018/978-1-61350-150-4.ch011

The term SaaS (Software as a Service) is being used to refer to an abstraction devoted to offer distributed services over the cloud. SaaS can be a possible replacement to traditional software, where buyers obtain a permanent license and install and maintain all necessary hardware, software and other technical infrastructure (Choudhary, 2007).

Taking this term as a reference, the term SaaS can be extended to MaaS (Multimedia as a Service) where the user wants to consume multimedia content in "service mode", adapting the content properties and delivery to users' needs and context. Moreover, the term can be further extended to everything available on the Internet as XaaS (Everything as a Service). This fact allows universal access and enables the creation and combination of new contents and services.

There are also some emerging technologies, such as 3D content. Integrating these technologies into multimedia content, the user can enjoy a multimedia experience where he/she can find natural interfaces, virtual and augmented reality environments and automatically generated content personalized to each user.

With the aim of supporting this multimedia experience, Quality of Experience (QoE) is a crucial fact which, depending on the context of the content consumption, decides the most efficient Quality of Service (QoS) values.

Therefore, the main goal of this chapter is to propose an architecture that allows the access to the next generation multimedia on any device without needing a dedicated hardware in the client side where mobile devices can benefit of this architecture.

BACKGROUND

What is Multimedia Technology?

According to Vaughan (1993), "multimedia is any combination of text, graphic art, sound, animation, and video that is delivered by computer". In addition, Vaughan explains that when users are allowed to control what and when these elements are delivered, it is called interactive multimedia. Besides, providing a structure of linked elements through which the user can navigate, interactive multimedia becomes hypermedia.

In common usage, the term "multimedia" refers to an electronically delivered combination of media including video, images, audio, text and 2D or 3D animations that can be accessed interactively. Nowadays the expression multimedia is frequently linked to natural interaction because it provides different ways to access the content and thanks to interactive multimedia, users can decide the means to inform themselves. The scope of multimedia is very extensive. The influence of multimedia content from education to business makes people be attracted to more catchy and interactive information.

THE EVOLUTION OF THE MULTIMEDIA CONTENT

Most of the problems of the multimedia are related with how to manage large quantities of data so the compression of the multimedia is a crucial point in order to reduce the storing and transferring data.

The main access to video content occurred when the VHS won the Videotape format war. The first Video Cassette Recorder appeared in 1972 by Philips. However, the first system that succeeded was Sony's Betamax in 1975. Immediately the Japanese enterprise JVC presented Video Home System (VHS) and in 2000 Philips developed the Video 2000. Instead of developing a single format with the benefits of the different systems, a format war started and VHS became the most popular. This made possible to have a video recorder in most of the houses and popularize it.

Very quickly the Compact Disc became very popular using an optical disc to store digital data. The CD-ROOM that was created by Philips and Sony in 1979 became a popular media to distribute multimedia applications and contents. At the moment there is a new format war with the high definition media format over an optical disc, between Blu-ray Discs and HD-DVD.

The Internet and its increasing bandwidth have influenced the way multimedia is consumed. Internet brought many advantages to people providing them access to any kind of information instantly. Web pages are not only composed by text but also by images, audio and video. In addition several pages include animations in order to attract the visitor's attention. The variety of media offering information has been the key to the success of multimedia content.

Web 2.0 has brought collaborative content creation where people share information through different manners such as blogs, wikis, social networks, etc. In addition, new technologies and devices like mobile phones have given society access to this information everywhere and at anytime thanks to ubiquitous technologies.

Nowadays, new current technology brings the opportunity to access information at anytime and anywhere and using any device. For example, people can get information outside their homes using mobile devices. Mobile technology is being developed very fast and users can have nearly the same experiences with their mobile devices than while using their personal computers. Multimedia content is frequent at mobile devices. For example people can communicate with others thanks to MMS (Multimedia Messaging System) adding images, videos or music to the written text.

Finally, in recent years, talking about Web 3.0 new formats and methods of interaction are being integrated on the Internet. As an example 3D graphics are demanded by users. Virtual worlds and social networks are more frequent in our daily lives and they integrate multimedia to the information that they bring. Users and developers are demanding more immersive interaction. What is more, the Law demands universal accessibility in order for everybody to have the same rights.

This is the reason why natural interfaces and personalization as well as virtual and augmented reality are fundamental technologies for the next generation multimedia.

Limitations in Current Multimedia Consumption

Multimedia offers the possibility of using different types of content as a complement to the given information. Thanks to multimedia, users can read, watch, listen and view this information and interact with it. However there are some disadvantages that should be resolved in order to allow a more immersive interaction.

Although multimedia content gives the opportunity of having an interactive experience with the content or the information, the user is often forced to interact in a certain way. The interaction is not always natural and the user has to adapt oneself to interact with the system. That is a drawback when users have a handicap and they have no choice to select the most appropriate interaction mode for them. Even though it has been said that multimedia is often linked to natural interaction, there are still some limitations to be resolved in order to offer natural interaction to each person.

Another main deficiency is that multimedia applications usually require specific hardware. There are still technological issues to achieve ubiquitous and context aware multiplatform publication infrastructures. To reach this hardware independent many-to-many approach (from many content sources, to many end user devices), a new architectural design is needed.

With the aim of improving the explained limitations, the purpose is to offer a more immersive experience while users interact with multimedia content. There are several new technologies that can be integrated as multimedia content that allow the user to have interaction. Virtual and augmented reality, personalization and natural interfaces are technologies that can be integrated with multimedia content that let users have a more immersive experience while they interact with any content, system or device.

MULTIMEDIA EXPERIENCE

Collaborative Content Creation

Collaborative content can be created by several ways (Traunmüller, 2010), for example as wikis which are knowledge collections built by collaborative editions; as blogs where people create notes open to be commented by others; or tagging that makes sharing information possible.

Besides, social networks include collaborative content. When a computer network connects people, it is called social network (Wellman, 1997). A social network is a set of people or a social entity connected by a set of socially-meaningful relationships.

Apart from the business behind social networks, their purpose is to communicate or interact with other people in addition to share information with them. Well known social networks like Facebook, YouTube or MySpace allow communication with others by multiple media, where users can share information or/and their thoughts or opinions by text, pictures, videos, music, etc. As Bonhand et. al., (2007) explains, some of the most successful sites let users create and share content, and users can connect and communicate with others and thus provide a richer and more interactive user experience.

Virtual worlds are also social networks, for example Second Life or Imvu. Thanks to these 3D social networks people can communicate with others in a more immersive way. Developers of virtual worlds are integrating multimedia content to offer multimedia experience.

The role of multimedia content on the collaborative Internet is very important. It has allowed people to share their information through different media, enriching the users' experience. But also

Internet has helped building (to build) collaborative multimedia applications (Nicol et. al., 1999).

Wrapping Multimedia Content in Services (MaaS)

According to Turner et. al., (Turner et. al., 2003) SaaS (Software as a Service) "is one of a demand-led software market in which services are assembled and provided as and when needed to address a particular requirement".

The use of this term is being extended on the Internet community. Developers of software and business professionals have a tendency to associate the term SaaS with business software as a probably lower-cost method for businesses. Through a well-designed implementation and proper licenses, SaaS provides benefits without associated complexity and high cost to equip devices with applications they might not need.

Taking all these ideas into consideration, the term SaaS can be extended to MaaS (Multimedia as a Service) where users consume multimedia content as a service. For this purpose, it is necessary to adapt the content properties to users' needs as well as to the context. The presentation of multimedia has to be a completely transparent service where the user does not need to know underlying details like which are the sources that have been used to create the content or where the rendering process happened which are giving him an immersive multimedia experience.

Additionally, this term can be extended to XaaS (Everything as a Service) that means everything available through this high layer abstraction. The XaaS paradigm moves the Internet towards the universal access of every digital resource.

Immersive Experience

According to several authors as Dede (2009) shows, immersion is "the subjective impression that one is participating in a comprehensive, realistic experience". Taking this into consideration, providing immersive experience to the users is to allow people to have a realistic experience using new technologies.

The term immersion is often linked to virtual and augmented reality. However, there are other new technologies that allow offering immersive experience to the users possible while they are accessing to multimedia content so as to complement it and enrich users' experience with the content.

Immersive experience complements the access to the multimedia content giving a more complete idea of the content and allowing the users to have other or a complementary perspective of it. That means that not only can the user watch, read and/or listen to the content but users can also feel immerse on it. This enriches users' experience giving them the illusion for example of interacting with an object something that otherwise cannot be possible.

The use of immersive experience can be linked to any scope, not only is it connected to games but it can also be applied to work, education, entertainment, people care, etc. Nowadays, people are demanding new technologies and a more immersive experience in order to make their life easier and comfortable.

The following technologies have been detected as very significant to offer immersive experience to the user: virtual/augmented reality, personalization and natural interfaces.

VIRTUAL/AUGMENTED REALITY

Nowadays technologies like virtual and augmented reality are at their very peak. Although they are more associated to games and entertainment, new applications try to make use of them to enrich their characteristics no matter which is the intention of these applications. But, what are these technologies and what is their purpose?

One of the given definitions of virtual reality (VR) is "the use of computers and human-computer interfaces to create the effect of a three-dimensional world containing interactive objects with a strong sense of three-dimensional presence" (Bryson, 1996). The purpose of VR is to simulate places or situations that can take place in a real life but in a virtual environment.

Azuma (1997) described augmented reality (AR) as a variation of virtual reality. According to his definition, VR technologies "completely immerse a user inside a synthetic environment. While immersed, the user cannot see the real world around him. In contrast, AR allows the user to see the real world, with virtual objects superimposed upon or composited with the real world".

Thus virtual and augmented reality technologies help to improve the experience of the user while they interact with the system, application or device because these technologies allow the users to live a situation that emulates a real one or complement it with additional information.

Features and Advantages

Some authors talk about immersive virtual reality where the user becomes totally immersed in an artificial and three-dimensional world that is completely generated by a computer.

The characteristics of immersive VR can be resumed as:

- Views are three-dimensional perspective and often stereoscopic which enhances the perception of depth and the sense of space.
- The point of view is often egocentric. The views change matching the physical position of the users and the direction of his/her gaze. That provides the immersion of the user in the virtual environment.
- The environment is usually dynamic.
- Multi-sensory techniques are employed to interact with the user. User input can be multimodal. That means using whatever voice, text, gestures, etc.

- Realistic interaction can be possible thanks to gloves or similar devices that provide manipulation and control of virtual objects.

According to Nooriafshar et., al.(2004) the advantages of VR is notable on situations where:

- Access to the real object or environment is difficult or impossible.
- Using the real object is unsafe or has a health hazard for people.
- Experimenting with the real object is too expensive.

An augmented reality system has the following three characteristics (Azuma, 1997):

- it combines virtual and real (reality);
- it is interactive in real time;
- and it is registered in 3D.

Virtual and augmented reality technologies are applied in a wide scope of applications from e-learning systems to surgery training tools.

NATURAL INTERFACES

People's communication is made through gestures, expressions and movements. According to Valli (2004) natural interaction is made through systems that understand these actions and engage people in a dialogue. In addition a principal feature is that it is not necessary that people wear any special device or learn any instructions while they interact.

Natural interfaces are those in which the user can interact with the interface through gestures, speech, touch, vision and/or smell based interaction. In other words, natural interaction involves "actions and sensations that refer to our daily life experiences" (Mignonneau. et. al., 2005).

As Mahoney (2000) explains, the most effective user interface is one that is comfortable and natural. And the most comfortable and natural

interface is one that is nearly invisible. That means that there is no separation between control and the presentation of the information. One of the characteristics of a successful natural interface is the reduction of the cognitive load on people while they interact with it (Valli, 2006).

An example of the advantages of Natural Interfaces is that it can enhance visitors' experience at a museum or exhibition (Alisi, 2005). At their work, the authors show a system which allows visitors to have a natural interaction with works of art. Visitors do not need to wear a special device to interact with the system, tested at a museum in Florence.

Allowing natural interaction between human and objects in a virtual world provides reality (Murakami, 1991). Besides, as Mankoff et. al., (2002) explains natural interfaces are particularly useful in settings where a keyboard and mouse are not available such as a very small or very large display and with mobile and ubiquitous computing.

Not forcing the consumer to interact in a specific way would satisfy user experience. In other words, it would be appropriate that the user could choose the interaction mode that he/she need or prefer. Nowadays, it is frequent to use multimodal interaction which implies visual, touch, gesture and voice interfaces. However it is often difficult to translate all of these manners to mobile devices because of the current technologies. The architecture explained in this chapter solves this problem and makes it possible to access to natural interfaces on any device without a dedicated hardware.

PERSONALIZATION

With the purpose of universal accessibility, it is necessary to personalize and adapt multimedia content to users, not only to their preferences but also to their needs.

There are two terms associated to personalization: customization and adaptation. The first one, customization is the personalization requested directly by the user. Adaptation means personalization that is automatically performed by the interface or the system (Weld et. al., 2003). As an example to view the difference between customization and adaptation is Miis and the game Wii Fit from Nintendo. A Mii is a virtual character that represents the player, who can decide or customize the appearance of his/her Mii. The system (game) by itself adapts the figure of the Mii according to the real weight of the player. In this case the user cannot decide if his/her Mii is slim or fat, it is adapted by the game according to reality.

Customization is not a new concept. There are many desktop user interfaces that allow the user select the tools they prefer on the menus and toolbars. However nowadays, current technology gives the opportunity to access to nearly the same content from different devices or environments. The problem is that the appearance of the users interface can change because of the characteristics of the device or the environment. For example, the different size of the screen between PCs and mobile devices makes the user interface to be different in many cases and the customization made on PC cannot be possible otherwise.

Regarding users' preferences, it is common to think about colors, fonts, size, etc. However, the preferences can involve the order of the content, users' tastes or the place from where people access to the multimedia content. That means that for example, if the user is at home, s/he may prefer access to the content through audio. Conversely, if s/he is outside, possibly the best way could be the access through reading text so nobody else can listen to the information.

As for users' needs, a simple rule that any content should fulfill is that everybody should have access to it. That means that people who have a handicap could have the same rights and/or benefits than the rest of the society. For that reason, multimedia content has to take into account possible handicaps. In addition the information should be given in different formats, such as video, audio, and text so it can be accessible from

different ways. Moreover, contents have to be adapted to assistive technologies to transmit the information to the users. In this way, everybody can access to the information and interact with new technologies.

Thereby personalization should be linked to natural interfaces so that people are able to interact with the interface or the system in a natural way according to their skills and capabilities or their desires. By this way, users could access and interact with multimedia content in a natural way according to their personal interface.

PROPOSED ARCHITECTURE

Architecture to address the problem of generating and presenting the next multimedia content is presented in order to get the content from many sources in combination with virtual reality and present it in any platform. This solution is device independent and generates and presents multimedia content without a dedicated hardware on the client side.

Due to the lack of support to render computer graphics in most of the platforms, a solution of rendering the multimedia content (augmented reality) in a server and delivering the video using streaming technologies has been adopted.

This architecture only needs a few render capabilities at the client side, such as an RTSP video client that most of the devices have, and it delegates all the technological challenge to the server.

Another main characteristic of the architecture proposed is its device independence. That means that each type of device that is connected to The Internet (no matter the connection medium: 3G, Ethernet, Wireless, etc) should be able to receive the content. It is not dependent of the codecs that are supported on the device, its screen size, the bandwidth of the network connection, the needs of the person that is accessing and watching the content, etc. This is possible thanks to the server makes all the work as well as differences among different devices. Thus, it takes parameters from the context, the device and the user to select the stream that fits for that connection (video codec, bit rate, resolution, complexity of the augmented reality content, etc).

Implemented Architecture

Current implemented architecture (see Figure 1) is based on three main parts;

- a Web Service that communicates with a Client device via HTTP,
- a Gstreamer RTSP server that is connected to these Web Services,
- the Client device that communicates with the Web Service to start a communication and receives the content via a RTSP stream maintaining an interaction via.

Client

The Client device does not require advanced rendering capabilities. It only needs to be able to render an RTSP stream that should be already

fitted by the Server depending on the device. This point makes the mobile phones and other mobile devices ideal for this architecture because they give the user the possibility to consume content anywhere but they normally do not have capabilities to render 3D computer generated content. Thanks to this solution, the user has access to a high characteristic multimedia content anywhere and with a common device.

Web Service

The Web Service server is responsible for receiving the requests from the Clients. It includes different calls that can be used to request new content and interact with it. For each devices connection it starts a new slot with the content generation, rendering and codifying it, to send to the user.

Several services have been implemented as a Web Service that can be called by the Client. For example a Web Service called *prepareConnection* provides the RTSP URL that gives the stream to the client. This call receives different arguments to receive information about the needs of the user that is asking the connection, the context and the device which is using in order to adjust

Figure 1. Implemented architecture's diagram

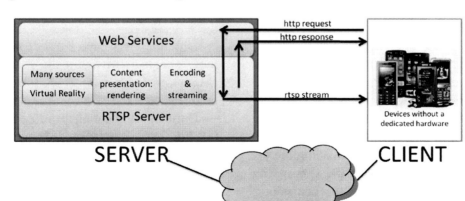

the parameters of the stream to the best values of the situation. This call returns to the client the RTSP URL that can be understood and showed by the device.

Other services are developed in order to let the user set values about the content which is being watched at that moment and the way to interact with it in real time. This makes it possible that during the receiving of the stream the user is able to call these services and watch immediately the changes in the stream.

RTSP Server

This module presents the content to the user composing a content created from many sources on the cloud and adding 3D computer generated content. It renders it following the user's needs and encodes the stream according to the clients device needs.

It is based on Gstreamer and has the capability to modify the properties of each component during running time, providing interactivity with the different devices.

There are two main parts inside this RTSP Server:

- A Gstreamer module that provides the integration of the 3D computer generated content to the main stream and allows that content to mix with other multimedia content on the cloud in order to get the desired Augmented Reality. Its properties can be modified at running time, so it is possible to interact with that content in real time and change its parameters.
- Encoding and streaming module. This module is responsible of encoding the already generated media in the appropriate way for each device and the connection context.

Architecture's Workflow

In Figure 2 the sequence diagram of the architecture is presented from a device connection request to the interaction with the media stream.

Validation of the Implemented Architecture

A scenario for the validation of the mentioned architecture has been developed. Therefore a 3D computer generated multimodal virtual character was used with the intention of being users' assistant while they are navigating on web pages.

PHP files were developed where HTML web pages are created and an RTSP player is embedded in. PHP asks the Web Service to prepare a Virtual Character with specific characteristics (female or male, eyes color, age, voice, etc) and the Web Service answers a RTSP URL with the video stream. That is shown in the embedded player together with the other entire HTML page. During the navigation of the user, PHP calls more services included in the Web Services such as *speakAvatar*, sending the text and the language of the sentence that wants to hear as parameters. Other services are implemented to provide animation to the Virtual Character (face and body expressions, voice changes, etc). This guides the user during the experience of the navigation into the web pages.

Different tests have been done to consume these web pages including the Virtual Character stream from PCs and from mobile devices, such as mobile phones and PDAs with satisfactory results.

During the validation the researchers realized that the architecture proposed could be used as a preliminary good solution. However some points that needed to be improved were checked. On one hand, the server needs to render and codify an avatar for each client, so one server can provide service to a limited value of clients. A distributed server farm has to be design in order to have a scalable capability to answer to a lot of clients

Figure 2. Implemented architecture's sequence diagram

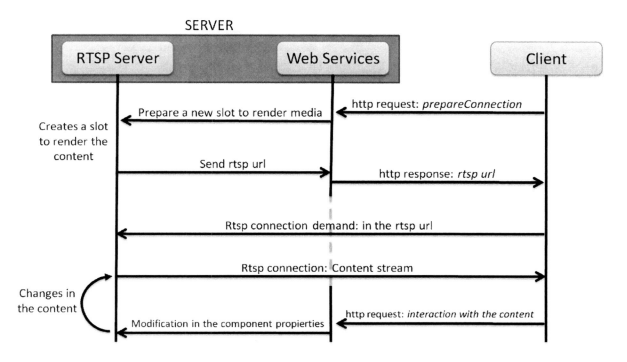

in a real implementation. On the other hand, the interaction reaction time must be improved to be able to watch immediately the requested changes in the stream in our device. A possible solution for these points is given in the next section as a scalable proposal of architecture.

Scalable Architecture Proposal

As it was explained in the previous section two main issues were detected to improve the implementation developed of the architecture: the scalability on a real implementation and the response time of the interaction requests.

In this section an extended architecture is proposed, taking into account the detected drawbacks in the implemented architecture (See Figure 3).

For this scalable proposal there is a completely new module called Processing Manager. This module integrates the Web Services shown

on the implemented architecture but does not include any RTSP server. This module receives the Web Service requests and manages these requests to redirect the processing requested to the appropriate RTSP Server available on the cloud. So there will only be a Processing Manager to reply all the requests, but there will be unlimited RSTP Servers to render and encode all the multimedia content. The Client will communicate at the beginning with the Processing Manager, but afterwards it will be redirected to an RTSP connection with one of the RTSP Servers. The different modules are explained with more detail.

Processing Manager

This module has to manage the client's requests by forwarding them to the RTSP Server. The Processing Manager should have to know which processing capabilities that request needs and

Next Generation Multimedia on Mobile Devices

Figure 3. Scalable architecture's diagram

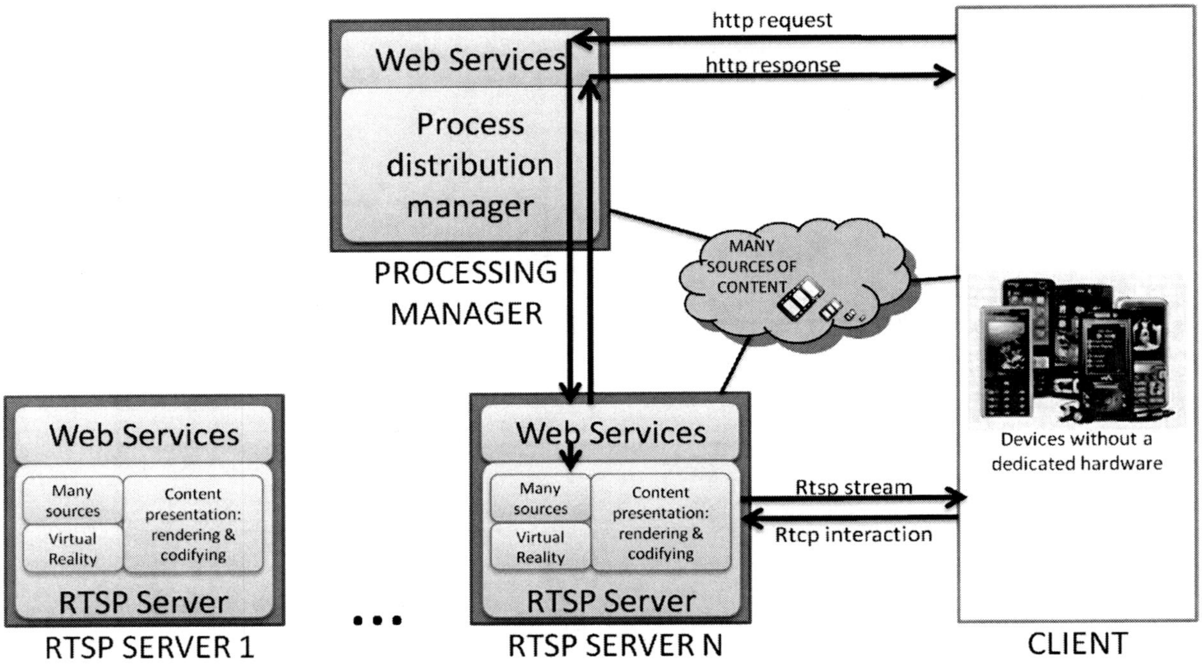

know which RTSP Server is available to render that process. This module has to manage the number of RTSP Servers on the cloud and their processing status in order to be able to forward the requests.

RTSP Server

This module includes the Gstreamer-based RTSP Server presented on the implemented architecture with a new layer of Web Services that make it possible for it to communicate with the Processing Manager. This way it is possible to add as much RTSP Servers as necessary on the cloud and only informing the Processing Manager about its availability they can start to work, even in run time. This new Web Service layer would not be public and it would only be known by the Processing Manager.

The RTSP Server receives new connection requests from the Processing Manager and informs the RTSP URL where to give the stream. The Processing Manager is the one that informs the

Client the address of the RTSP stream but after that any other communication occurs between Client and RTSP Server. The RTSP Server also communicates the Processing Manager their processing status after each modification in order to notify their availability for further processing.

The interaction channel can also be directly opened between the Client and the RTSP Server using Real Time Control Protocol (RTCP). That way the response time for the interaction is improved. The RTCP has been developed to provide a feedback of the Quality of Service (QoS) of the RTP communication, but it can also be used to transmit additional information over the stream.

Client

The changes shown on the scalable architecture does not have any effect on the client. It asks the Processing Manager for a new connection and the only difference is that the RTSP URL received is not located in the same physical server even if it

178

belongs to one of the other RTSP Servers. Once the connection is established, the client should interact with the content using the RTCP protocol. In Figure 4 a sequence diagram of the presented scalable architecture is shown.

Benefits of the Architecture

Two architectures have been presented here; an implemented one and an improved architecture that tries to fix the weaknesses detected on the implementation architecture. However both of them pursue the same goals. First of all, they multimedia content as a service on the cloud similar to the Multimedia as a Service (MaaS) concept previously mentioned. The client does not have to worry about codecs, does not need specific hardware on the client's device to render advanced multimedia content (powerful video cards, etc), it only has to ask for the service and the Server provides the best context adapted content.

FUTURE RESEARCH DIRECTIONS

This architecture provides a multimedia consumption service to users that allow them to access to a multimedia world using any device. A dedicated hardware is not necessary on the client and this brings several benefits to the users. For this purpose, all the data processing for the rendering of the automatically and semi-automatically generated multimedia content must be on the server's side.

An inversion of server farms is needed in order to give a massive service to the user. That is why the atomization of the render process into different and distributed servers is very important. This is the main point in the future research in order to improve the architecture.

Furthermore it is necessary to research the best way to integrate the interaction orders from the client to the RTSP Server over the RTSP stream. An adaptation of the RTCP protocol for this use could provide a very good response time providing the user a better experience. Thus it will not be important anymore whether the render of the content is being done locally or remotely.

Figure 4. Scalable architecture's sequence diagram

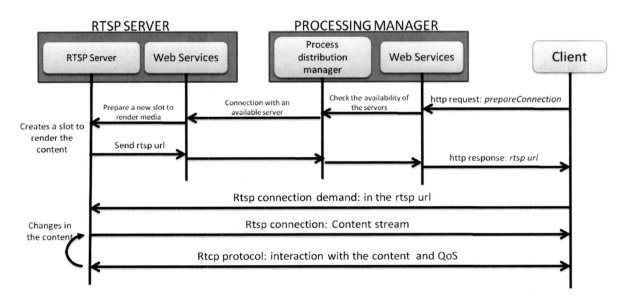

CONCLUSION

This chapter makes a revision of the multimedia content, analyzing its evolution and evaluating the consumption limitations it nowadays has. A necessity of including immersive experience has been detected in order to empower users' experience while they interact with multimedia content.

In order to integrate immersive experience, new technologies are needed. Not only virtual and augmented reality technology has been detected, but also personalization and natural interfaces. The multimedia content can be the interface providing automatically generated and personalized content with an efficient interaction channel. All of these technologies can improve users' experience and allow them to consume next generation multimedia content from many sources through any device.

A new architecture to present the next generation multimedia has been presented. The main characteristic of this architecture is that it does not require a dedicated hardware in the client side. As a result, less powerful devices such as mobile devices can get a profit of this architecture.

To conclude, next generation multimedia is appearing and industry is offering solutions to enrich user's experience and give them a service to consume it from any sources, anytime and in any platform.

REFERENCES

Alisi, T. M., Bimbo, A. D., & Valli, A. (2005, July). Natural interfaces to enhance visitors' experiences. *IEEE MultiMedia*, *12*(3), 80–85. doi:10.1109/MMUL.2005.52

Azuma, R. (1997). A survey of augmented reality. *Presence (Cambridge, Mass.)*, *6*(4), 355–385.

Beier, K. P. (2004). *Virtual reality: A short introduction*. Retrieved September 2, 2010, from http://www-vrl.umich.edu/intro/

Bryson, S. (1996). Virtual reality in scientific visualization. *Communications of the ACM*, *39*(5), 62–71. doi:10.1145/229459.229467

Choudhary, V. (2007, January). Software as a service: Implications for investment in software development. *International Conference on System Sciences*, (p. 209)

Dede, C. (2009). Immersive interfaces for engagement and learning. *Science*, *323*(5910), 66–69. doi:10.1126/science.1167311

Mahoney, D. P. (2000, February). Innovative interfaces. *Computer Graphics World*, *23*(2), 39-42, 44.

Mankoff, J., Abowd, G. D., & Hudson, S. E. (2000). OOPS: A toolkit supporting mediation techniques for resolving ambiguity in recognition-based interfaces. *Computers & Graphics*, *24*(6), 819–834. doi:10.1016/S0097-8493(00)00085-6

Mignonneau, L., & Sommerer, C. (2005). Designing emotional, methaforic, natural and intuitive interfaces for interactive art, edutainment and mobile communications. *Computer Graphics*, *29*, 837–851. doi:10.1016/j.cag.2005.09.001

Murakami, K., & Taguchi, H. (1991). Gesture recognition using recurrent neural networks. *Proceedings of the SIGCHI Conference on Human Factors in Computing Systems: Reaching Through Technology*, (pp. 237-242). New Orleans, Louisiana, United States.

Nicol, J. R., Yechezkal, S. G., Paschetto, J., Rush, K. S., & Martin, C. (1999). How the Internet helps build collaborative multimedia applications. *Communications of the ACM*, *42*(1), 79–85. doi:10.1145/291469.291474

Nooriafshar, M., Williams, R., & Maraseni, T. N. (2004). *The use of virtual reality in education*. The American Society of Business and Behavioral Sciences (ASBBS) Seventh Annual International Conference, Cairns, Queensland, Australia, 6th-8th August.

Traunmüler, R. (2010). Web 2.0 creates a new government. *Electronic Government and the Information Systems Perspective. Lecture Notes in Computer Science, 6267,* 77–83.

Turner, M., Budgen, D., & Brereton, P. (2003, October). Turning software into a service. *Computer, 36*(10), 38–44. doi:10.1109/MC.2003.1236470

Valli, A. (2004). *Notes on natural interaction.* Retrieved August 20, 2010, from http://www.citeulike.org/user/eckel/article/4324923

Valli, A. (2006). The design of natural interaction. *Multimedia Tools and Applications, 38*(3), 295–305. doi:10.1007/s11042-007-0190-z

Vaughan, T. (Ed.). (1993). *Multimedia: Making it work.* Berkeley, CA: Osborne/McGraw-Hill.

Weld, D., Anderson, C., Domingos, P., Etzioni, O., Gajos, K., Lau, T., & Wolfman, S. (2003). Automatically personalizing user interfaces. *Proceedings of the International Joint Conference on Artificial Intelligence, (IJCAI03),* Acapulco, México.

Wellman, B. (1997). An electronic group is virtually a social network. In Kielser, S. (Ed.), *Culture of the Internet* (pp. 179–205). Mahwah, NJ: Lawrence Erlbaum Associates.

Chapter 12
Transforming eMaintenance into iMaintenance with Mobile Communications Technologies and Handheld Devices

Nalin Sharda
Victoria University, Australia

ABSTRACT

iMaintenance stands for integrated, intelligent and immediate maintenance; which can be made possible by integrating various maintenance functions, and connecting these to handheld devices, such as an iPhone, using mobile communication technologies. The main innovation required for developing iMaintenance systems is to integrate the disparate systems and capabilities developed under the current eMaintenance models, and to make these immediately accessible by ubiquitous and intelligent computing technologies –such as Digital Ecosystems and Cloud Computing– connected via wireless networks and handheld devices such as the iPhone. A Digital Ecosystem is a computer-based system that can evolve with the system that it monitors and controls, and can be embedded in the system's components, thereby providing the ability to integrate new functionality without any downtime. Cloud Computing can provide access to additional software services that are not available in the local Digital Ecosystem. This chapter will show how these computing paradigms can provide mobile computing and communication facilities required to create novel iMaintenance systems.

DOI: 10.4018/978-1-61350-150-4.ch012

INTRODUCTION AND HISTORICAL PERSPECTIVE

Maintenance is an essential aspect of any system; because, without the required maintenance the system's performance can deteriorate; and this is at the lower end of the consequence spectrum, while on the higher end of the consequence spectrum, the system can fail catastrophically.

Early Maintenance Methods

"Prior to the Industrial Revolution, generally held to have begun in England in about 1750, maintenance consisted of individual craftsmen such as carpenters, smiths, coopers, wheelwrights, masons, etc. repairing the buildings, primitive machines and vehicles of the day" (Sherwin, 2000). In those days failures were repaired primarily by making a new part, or repairing the old one. As there was no concept of 'remove-and-replace' spare parts, these repairs took a long time. Most early manufactured systems were made to last as long as possible through repair oriented maintenance.

A case in point is Lord Nelson's flagship HMS Victory, which was 40 years old in 1805, when it took part in the battle of Trafalgar. It was kept either in active service or in reserve until 1860, and then as a hulk until 1922. Towards the end, almost all of its original structure would have been replaced. Such a maintenance policy was economically viable in those days, because skilled labour was cheap as compared to the value of the system components (Sherwin, 2000).

However, in the modern world the cost of labour often exceeds that of the system components; therefore, one of the main objectives of the current maintenance strategies is to reduce the Mean Time Taken to Repair (MTTR). This is achieved by using fast diagnostic techniques and 'remove-and-replace' spare parts. The concept of preventive maintenance has also become commonplace to reduce the Mean Time Between Failures (MTBF).

Diagnosis and Repair

In 1785 Thomas Jefferson noted that musket parts could be interchangeable, if made accurately enough; this led to the idea of 'remove-and-replace' spare parts. Good quality maintenance however, still requires skilled craftsmen; presently, such craftsmen are either rare or too expensive in the well-to-do countries. Consequently maintenance, particularly of items that require high level of skills, is becoming a problem (Sherwin, 2000).

"Shortages of competent repairers and their consequent higher wages have affected maintenance policy and subsequently the design of industrial, commercial, and especially domestic machinery" (Sherwin, 2000). Simple household systems such as toasters are treated as throwaway items in rich counties, but are repaired in less well-to-do countries. This has also changed the fundamental design philosophy behind goods such as motor cars. A car that has longer MTBF and lower MTTR is preferred even at a higher cost. Much of this improvement (in MTBF and MTTR) is achieved by introducing electronics and computer technology under the bonnet of the car for faster diagnostics. However, older, and simpler designs can be used where the cost of repair by skilled craftsmen is still affordable (Sherwin, 2000).

Preventive Maintenance

Preventive maintenance models are used to avoid unpredicted system breakdown, and thus, loss of production. One of the methods for selecting the frequency of preventive maintenance and the choice of parts to be replaced is based on historical data. Usually such preventive maintenance schemes are time-based, and do not consider the current health of the product. Such schemes are inefficient and cost more to a customer (Lee, et al., 2006). An improvement over such time-based preventive maintenance methods is degradation analysis, which measures the changes in physical characteristics of the system components to predict the likelihood of failure.

Thus, the corrective action for the cost inefficiency of simple time-based maintenance is: condition-based maintenance (CBM), in which the need for maintenance is deduced from the current degradation in the system parameters, and their evolution with time. Further advancement on the traditional CBM methodology is based on gathering the product degradation information by using on-line sensing techniques, thereby minimizing the system downtime (Lee et al., 2006).

Evolution of the eMaintenance

These on-line monitoring techniques, combined with other Information and Communication Technology (ICT) systems have led to the evolution of the eMaintenance paradigm. eMaintenance systems have been evolving for over a decade for many industrial systems.

In recent years ICT systems have broken free from being tethered to a wired network, as wireless networks now provide excellent communication bandwidth at a very reasonable cost; which, at times, is less than that for setting up a wired network.

Handheld devices such as the iPhone have now put powerful computers in the palms of our hands. These devices can access a large repository of information through the Internet, and provide access to almost any other device connected to the Internet. Cloud Computing has done away with the need for maintaining one's own high powered computers, by using surrogate servers. The concept of Digital Ecosystems is making use of wireless communication technologies to create systems that can evolve with time and grow as the system they monitor and control grows.

All of these technologies can be combined to develop iMaintenance systems; systems that provide immediate, intelligent and integrated maintenance. The main aim of this chapter is to present an overview of technologies involved,

and a framework suitable for the development of iMaintenance systems. Specific objectives include:

1. Report on the evolution of maintenance and eMaintenance systems.
2. Discuss various technologies that can be used to enhance eMaintenance systems into iMaintenance systems.
3. Present a model for transforming eMaintenance systems into iMaintenance systems.
4. Discuss the benefits of the proposed iMaintenance systems.
5. Present future research directions and conclude.

TRADITIONAL MAINTENANCE

According to Lee et al. (2006), "A lot of machine maintenance today is either reactive (fixing or replacing equipment after it fails) or blindly proactive (assuming a certain level of performance degradation, with no input from the machinery itself, and servicing equipment on a routine schedule whether service is actually needed or not)".

However, both of these methods are wasteful and costly. While on the surface it might seem that machines fail unpredictably, in reality, machines go through a gradual and measurable process of degradation prior to the failure. Often this degradation is hidden from the naked eye, and not even sensed through hearing before the machine fails; therefore the human user feels that the machine failed suddenly, whereas it had been deteriorating incessantly.

To overcome these problems many sophisticated sensors and computerized components, with built-in diagnostic capabilities have been developed in recent years; these components are capable of regular monitoring of the components, and delivering data about the machine's wear status and/or current performance (Lee et al., 2006). These sensors, when used in conjunction with various computer and communication technologies lead to the eMaintenance paradigm.

However, even with the help of sensors and computers, maintenance remains a complex process; it included many complex decisions, such as: selection of the component to be monitored; sensor installation; data acquisition, transformation and analysis; followed by decision making; planning of maintenance operation, reporting to the operators and management; as well as keeping track of spare parts (Han & Yang, 2006).

In many complex industrial systems there can be as many as 10,000 hardware components, their monitoring devices and related software systems, which add to the complexity of the complete maintenance process chain (Han & Yang, 2006). Therefore the transformation of traditional maintenance processes into an eMaintenance system requires a clear definition and systematic architecture for its development.

E-MAINTENANCE

The term *e-Maintenance* has been in use from about early 2000, and since then, it has become a common term in the maintenance-related literature (Muller, Crespo-Marquez, & Iung, 2008). However, there seem to be many definitions and many viewpoints about eMaintenance theory and practice. Some engineers may consider e-Maintenance as a concept, other as a philosophy or phenomenon (Muller et al., 2008), however, it needs formal models for successful implementation.

Some of the concepts that eMaintenance embodies are included in the following equation in Table 1 (Muller et al., 2008):

Equations 1 and 2 highlight the fact that eMaintenance is a multifaceted idea even at a conceptual level. Therefore, inevitably, is has many physical manifestations as well. However,

Table 1. eMaintenance concept equation

E-main-tenance	= Excellent maintenance (1)
	= Efficient maintenance (do more with fewer people and less money) (2)
	+Effective maintenance (improve RAMS metrics)
	+Enterprise maintenance (contribute directly to enterprise performance)

the essential ingredient of eMaintenance is that it integrates information and communication technologies (ICT) with some maintenance strategy and its implementation plan. (Li et al., 2005).

In today's technology focused world, the eMaintenance paradigm has become essential to provide a link with other ICT-based methodologies for supporting production using eManufacturing, and, at the top level, for supporting business using eBusiness techniques (Yoshikawa, 1995).

Considering its various perspective, e-Maintenance can be viewed as (Li, Chun, Nee, & Ching 2005):

- A maintenance strategy, i.e. a methodology for maintenance management
- A maintenance plan, i.e. a well defined set of tasks
- A maintenance type, e.g. an implementation of condition-based maintenance (CBM)
- A maintenance support system, providing resources and services for maintenance

Further explanation of each of these viewpoints is expounded in the following section, as presented by Muller et al. (2008), based on extensive literature review.

E-Maintenance as a Maintenance Strategy

E-maintenance can be merely defined as a maintenance strategy where the tasks are managed electronically using real-time equipment data obtained thanks to digital technologies (i.e. mobile devices, remote sensing, condition monitoring, knowledge engineering, telecommunications and Internet technologies) (Tsang, 2002). From this point of view, e-maintenance is interpreted as a maintenance management process (Hausladen, & Bechheim, 2004), which deals with the expansion of the volume of data available. This definition is refined by Baldwin (2004) or by Moore and Starr (2006) in the following way: "E-maintenance is an asset information management network that integrates and synchronises the various maintenance and reliability applications to gather and deliver asset information where it is needed when it is needed".

E-Maintenance as a Maintenance Plan

E-maintenance can also be seen as a maintenance plan, which meets the needs of the future e-automation manufacturing world in the exploration of the approaches of CBM, proactive maintenance, collaborative maintenance, remote maintenance and service support, provision for real-time information access, and integration of production with maintenance (Ucar & Qiu, 2005). The implementation of an e-maintenance plan requires a proactive e-maintenance scheme, i.e. an interdisciplinary approach that includes monitoring, diagnosis, prognosis (Muller, Suhner. & Iung, 2007), decision and control processes.

E-Maintenance as a Maintenance Type

Generally speaking, e-maintenance is the symbol of the gradual replacement of traditional maintenance types ((Han & Yang, 2006), by more predictive/proactive types. Regular periodic maintenance should be advanced and shifted to the intelligent maintenance philosophy to satisfy the manufacturer's high reliability requirements (Tao, Ding, & Xion, 2003). Hence, Koc and Lee (2001) referred e-maintenance (system) as predictive maintenance (system), which provides only monitoring and predictive prognostic functions (Lee, 2001).

E-Maintenance as a Maintenance Support

Last but not the least, e-maintenance can be referred as a maintenance support. For example, Zhang, Halang, & Diedrich (2003) consider that e-maintenance is a combination of Web service technology and agent technology, which provides a way to realize intelligent and cooperative features for the systems in an industrial automation system. Crespo-Marquez and Gupta (2006) define e-maintenance as a distributed artificial intelligence environment, which includes information processing capability, decision support and communication tools, as well as the collaboration between maintenance processes and expert systems.

Muller et al., (2008) propose a definition of eMaintenance by combining the different perspective presented by various authors and also taking into account the European Standard (EN 13306:2001) for maintenance terminology. This definition also considers eMaintenance as a component of the eManufacturing (Lee, 2003). Based on the above imperatives Muller et al., (2008) proposed the following definition of eMaintenance:

Maintenance support, which includes the resources, services and management necessary to enable proactive decision process execution. This support includes e-technologies (i.e. ICT, Web-based, tether-free, wireless, infotronics technologies) but also, e-maintenance activities (operations or processes) such as e-monitoring, e-diagnosis, e-prognosis, etc.

An alternative view of eMaintenance is presented in Figure 1, along with its relationship to eBusiness and eOperations, as defined by Lee & Ni (2004). This view predicates that eMaintenance comprises Condition-Based Monitoring (CBM) and use of data obtained from CBM for Predictive Prognosis. This eMaintenance system is connected to sensors and controllers connected to the various products and machine parts that make up the entire system.

In this model eMaintenance is encompassed in the bigger eOperations system that includes Remote Diagnostics and Asset Management; and

may include the facilities for simulation, optimisation and decision making based on their results. These two aspects are enclosed in the wider perspective of eBusiness, which included eCommerce for effective cash flow, information flow for customer relationship management, and materials flow using supply chain management. All of these systems need to use ICT systems. While a wide variety of ICT systems are available for each of the functions, often the challenge is to integrate these into one package.

By now, eMaintenance is recognised as an important methodology for the industry to adopt, and many companies have adopted it. Even entire countries have invested in further research and development of eMaintenance systems to gain efficiencies in the maintenance of industrial systems, leading to increased productivity (Han & Yang, 2006).

A more detailed model of generic eMaintenance architecture is shown in Figure 2. While an eMaintenance system aims to minimise human

Figure 1. E-maintenance related to e-operations and e-business (Lee & Ni, 2004)

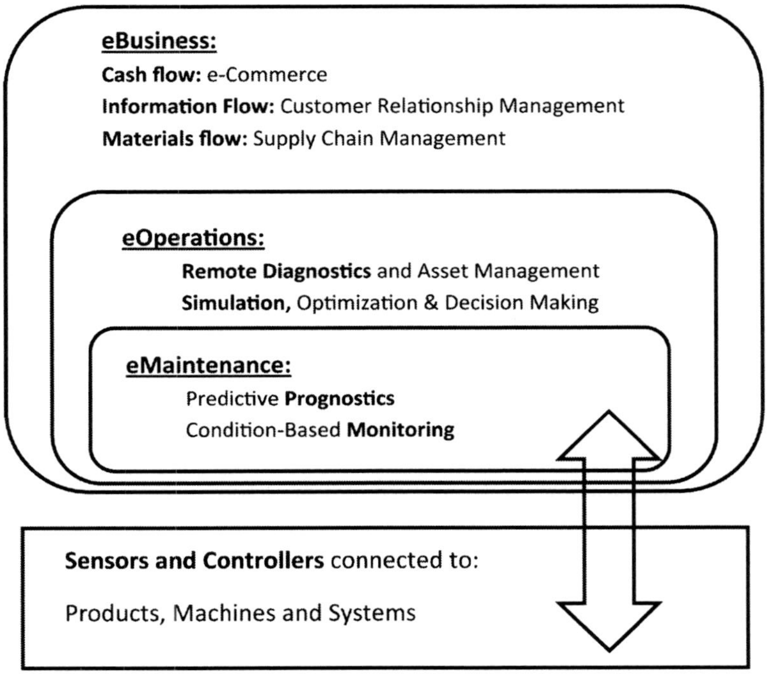

intervention, it does not eliminate it. The local (human) experts need to participate in creating the maintenance strategy. This strategy is applied to the machine components for real-time condition monitoring. The results of the condition monitoring are stored in a local database. The fault diagnosis system uses the data from the database to create degradation predictions. These degradation predictions in turn drive the maintenance strategy. In a sense the human expert provides the oversight to what looks like a feedback loop.

The eMaintenance centre is connected to the main monitoring loop and the database via a data link. This link will be the main component that will allow us to transform eMaintenance into iMaintenance. More on this transformation will be presented in later sections.

Another area where mobile technologies provide great opportunity for advancing eMaintenance to iMaintenance is the links between the sensors and the real-time monitoring modules, over and above the link between the eMaintenance centre and the user interface.

ICT Frameworks for eMaintenance

Many new ICT frameworks have been proposed for developing eMaintenance systems (Koc & Lee, 2001). Yu, Iung, & Panetto (2003) have given an integrated framework based on multi-agents for developing eMaintenance system; which also uses case-based reasoning (CBR) for solving service scheduling problems.

PROTEUS is a project sponsored by the French, German and Belgian governments; its aim is to establish a European generic software platform with the aim of encouraging and supporting the development of web-based eMaintenance practices (Szymanski et al., 2003).

However, any system with a fixed architecture is likely to become out-of-date with time. A new concept is emerging now to overcome this problem of obsolescence: that of Digital Ecosystems.

Figure 2. Local maintenance strategy linked to e-maintenance (Han & Yang, 2006)

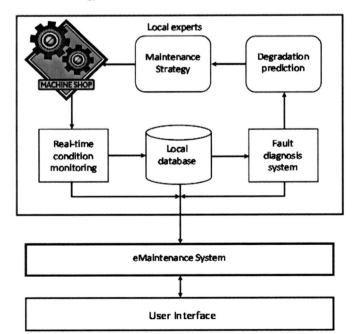

DIGITAL ECOSYSTEMS

A digital ecosystem is an ICT technology based system, which can evolve with time and even blend into its application environment (Dini et.al., 2005; Nachira, 2005).

From the stand alone digital electronic control systems and large computers of the mid-twentieth century era, the modern computers have evolved into powerful desktop computers and hand-held devices such as the Personal Digital Assistant (PDA), and mobile phones, e.g. iPhone; making computational power inexpensive and widely available, right in the palm of our hands (Sharda, 2009a).

Furthermore, computers can now be connected to each other using wired as well as wireless connections including the Internet. These connections can change and reconfigure as the application demands. These features have led to the concept of Digital Ecosystems, i.e. digital computer and communication systems that can evolve (Sharda, 2009a). Therefore a Digital Ecosystem (DE) is a digital system that has the following features:

- Evolves with time
- Adapts to the user and application needs
- Distributed geographically
- Embedded within the application environment

A DE infrastructure is likely to be:

- Pervasive digital environment
- Populated by intelligent digital components
- Capable of evolving
- Adapting to local conditions

The above listed features of Digital Ecosystems point to these having the following properties (Sharda, 2009b).

- These DEs need to be extensible in space and time; being extensible in space implies

that an inner DE can communicate with the outer DEs with the help of wired and / or wireless communications infrastructure. In the context of eMaintenance it implies that the eMaintenance DE needs to communicate with the eOperations DE and the eBusiness DE, as shown in Figure 3.

The communication bus shown on the right hand side, in dotted lines, represents the communication path that allows the various nested DEs to communicate with each other and the external ICT infrastructure called Cloud Computing. Further details of Cloud Computing are presented in a later section.

To be extensible in time a DE needs to use lessons learnt from the living world in terms of how they evolve. The manifestation of this evolutionary character in DEs can be incorporated by using 'plug-and-play' components that are automatically programmed to work interactively with existing devices. This will make it possible to bring in a new device into the system such that it interoperates with the existing system without any downtime to reinstall and reconfigure exiting devices and their software interfaces.

Therefore DEs will need to embody a high degree of learning ability to enable them to learn from the historical data collected for condition-based monitoring and predictive prognosis.

To achieve the above mentioned properties these DEs will need to have (Sharda, 2009b):

- No central control, no plans defined in advance
- Built on complex hierarchies
- Fault tolerant: No central point of failure
- Diversity and autonomy (recursively)
- Adaptation to the local conditions
- Selection and evolution

Figure 3. E-maintenance DE communicating with e-operations and e-business DEs

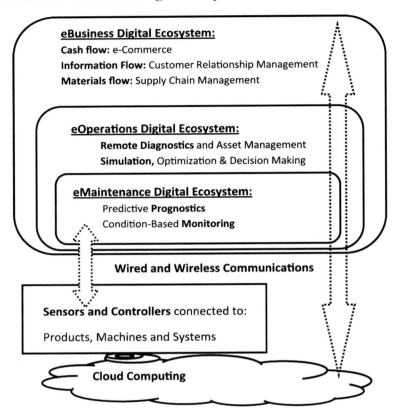

DE Enabling Technologies

The main enabling technologies required for transforming eMaintenance into iMaintenance are the following. A brief description of each is given here, and a more detailed explanation for the most important ones follows.

- **Wireless Sensor Networks:** Sensors for condition-based monitoring.
- **Plug and Play Hardware / Software:** This will allow one to expand a system without the need for shutting it down or reconfiguring it.
- **Web Services:** Software components that are available on the World Wide Web, and can be downloaded by a system as and when required.

- **Mobile Phones 3G-4G:** Third Generation (3-G) mobile phones already allow many advanced communication features, such as video conferencing. With the fourth generation (4G) mobile technologies much higher Quality of Service (QoS) will be available.
- **Cloud Computing:** Provides computing power outside the boundary of a user's own infrastructure.

To be able to develop DEs that will lead to the transformation of eMaintenance into iMaintenance paradigm, one of the essential ingredients will be Wireless Sensor Networks (WSNs). These WSNs will form the core of the system that links various sensors and controllers that feed in the data for the condition-based monitoring and predictive prognosis.

Wireless Sensor Networks

Wireless Sensor Networks (WSNs) are a collection of sensors connected with wireless networking technologies, to monitor the physical environment in which these are deployed. The parameters monitored by these sensors can include temperature, sound volume, vibration, pressure, motion, and wear in machine parts for maintenance systems (Sharda, 2011).

Development of WSNs was originally motivated by military applications, such as, incognito surveillance in a battlefield. Over the years, these WSNs have found many industrial as well as civilian application, for example: industrial process monitoring, machine health monitoring, environment and habitat monitoring, healthcare, home automation, and traffic control (Kay & Mattern, 2004).

As the name implies, in a WSN each node contains one or more sensors. Many WSN nodes are designed as generic systems, in which a range of sensors can be plugged in a motherboard, as and when required. Each note includes a radio transceiver which forms the link between the actual sensor and the other wireless communications components in the WSN. The WSN node is controlled by a low power microprocessor. Of course, there needs to be some power source on each node; this could be a battery, or a solar cell for outdoor deployment (Kay & Mattern, 2004).

"The evolution of WSNs started in the 1980s with the advent of the microcontroller chips, and has progressed incessantly since then to the present day, and, of course, will continue in the future. Sarigiannidis (2006) divided the evolution of WSNs into three generations, based on the information presented by Chong and Kumar (2003). The classification shown in Table 2 is based on the same", (Sharda, 2011). However, the proposed third generation WSNs are yet to become widespread. IN the current environment extended versions of the second generation WSNs are likely to be used in iMaintenance systems.

Plug and Play Hardware / Software

Plug & Play hardware and associated software allow one to expand a system without the need to shut it down, or reconfiguring it. The concept is similar to the Universal Plug and Play (UPnP)

Table 2. Three generations of Wireless Sensor Networks (Chong and Kumar, 2003)

Parameter	Generation-1	Generation-2	Generation-3
Period	1980's to1990's	Early 2000s	Mid 2000s onwards
Size	Large shoe box and above	Pack of cards to small shoe box	Dust particle to coin
Weight	Kilograms	Grams to Kilograms	Milligrams to grams
Node architecture	Separate sensing, processing and communication	Integrated sensing, processing and communication	Integrated sensing, processing and communication
Topology	Point-to-point, & star	Client server & Peer-to-peer	Peer-to-peer
Power supply	Large batteries	AA batteries	Solar, ambient temperature/pressure
Lifetime	hours, days and longer	days to weeks	months to years
Deployment	Vehicle-placed or air-dropped individual sensors	Hand placed	Embedded or "sprinkled", and left behind.

feature included in operating systems such as Windows XP.

According to Fout (2001), "With the addition of Device Plug and Play (PnP) capabilities to the operating system it became a great deal easier to setup, configure, and add peripherals to a PC. Universal Plug and Play (UPnP) extends this simplicity to include the entire network, enabling discovery and control of networked devices and services, such as network-attached printers, Internet gateways, and consumer electronics equipment".

Furthermore, the UPnP model is more than a simple extension of the older Plug and Play peripheral model. It is designed to support zero-configuration, thereby achieving "invisible" networking, and when required, automatic discovery from faults for a range of devices manufactures by a wide range of vendors (Fout, 2001).

One of important feature of UPnP is that a device can dynamically join a network. To achieve this, it first obtains an IP (Internet Protocol) address, then it conveys its capabilities to other devices, and finally it learns about the presence and capabilities of other devices. It does all this automatically; requiring no manual configuration. All devices on the network can then communicate with each other, using peer-to-peer networking techniques.

According to Fout (2001): "The scope of UPnP is large enough to encompass many existing, as well as new and exciting scenarios including home automation, printing and imaging, audio/video entertainment, kitchen appliances, automobile networks, and proximity networks in public venues". While at the time of UPnP's launch in 2001, the concept of Digital Ecosystems was not prevalent, it can now be recognised that UPnP or its extensions will be an important factor in creating self forming and evolving DEs. In turn, these DEs can be used for developing the iMaintenance systems.

One of the advantages of UPnP is that it uses the standard TCP/IP (Transmission Control Protocol /Internet Protocol) standard as the communication mechanism. This allows it to fit in seamlessly into existing networks that use the TCP/IP standard, and most networks do. Furthermore, UPnP is distributed and open network architecture. It is independent of any specific operating system (OS), programming language, or physical medium (Fout, 2001).

WEB SERVICES

Web Services are software components that are available on the World Wide Web (WWW), and can be downloaded by a system as and when required.

Booth et al. (2004) describe the purpose of web services as: "Web services provide a standard means of interoperating between different software applications, running on a variety of platforms and/or frameworks". The Web Services Architecture (WSA) documentation provides a common definition of a Web service, and how it is placed within a larger Web services framework. The WSA provides a conceptual model for understanding and using Web services and the relationships between its various components.

Web service is defined by Booth et al. (2004) as:

A Web service is a software system designed to support interoperable machine-to-machine interaction over a network. It has an interface described in a machine-processable format (specifically Web Services Description Language (WSDL)). Other systems interact with the Web service in a manner prescribed by its description using Service Oriented Architecture Protocol (SOAP) messages, typically conveyed using Hyper Text Markup Language (HTTP) with an Extensible Markup Language (XML) serialization in conjunction with other Web-related standards.

The key feature of the Web services architecture is that it provides the foundation for interoperability; thus, it defines how various elements of the global Web services network are linked to ensure interoperability between a range of Web services available on the Internet.

Types of Web Services

Web Services are essentially a new way of creating and running Web applications. "They are self-contained, self-describing, and modular applications that can be published, located, and invoked across the Web" (Ravuthula, 2002). Web Services perform many functions of varied complexity, from simple data requests to complicated business processes. An important aspect of Web services is the ability of other users to discover their presence; for example, even when a Web Service is discovered, requested, and deployed by one application, other applications (and even other Web Services) can discover and invoke the same Web service (Ravuthula, 2002).

This is possible as the Web services architecture connects Service Providers and Service Requestors. Service providers need to maintain information about their services in a registry, which is accessible by others. Service requesters, or consumers, search these registries for the required services. Once a required service is found, it can be invoked and thus used. This process requires a Service Broker (also called Registry Provider), which is a database of various services registered with that Registry (Ravuthula, 2002).

To create iMaintenance systems we need to create eMaintenance Service Brokers, so that various developers can create specific maintenance oriented Web services and register these. These Web services may be created by the manufactures of the various monitoring hardware device, by iMaintenance service providers, by technical societies, or even individual developers.

One of the crucial elements of the iMaintenance paradigm would be the provision of user interface through mobile devices such as PDAs and advanced mobile phones, such as the iPhone.

Information Transmission Over Mobile Phones

The ability to access information and issue commands for initiating maintenance actions via mobile phones is what is going to transform eMaintenance into iMaintenance. What type of information is transmitted for monitoring and executing control functions is the main issue when using mobile services. Of course, the size and the resolution of the mobile phone screen can be an issue in some cases, especially if complex images have to be viewed.

Transmission of multimedia information becomes more difficult when it has to be transmitted over mobile networks (Sharda, 2006), as compared with transmission over wired networks. This is because, in general, mobile networks have lower bandwidth than that for wired networks. Where WSNs are used for linking the sensors to the condition-based monitoring system, information transmission is even more challenging due to further lower bandwidth, less memory, lower processing speed and battery power than that available over the standard infrastructure based mobile networks. As the WSNs use ad-hoc routing protocols, transmission of multimedia information becomes even more difficult (Sharda, 2011).

However, most severe challenges are encountered when the information being transmitted comprises still or moving pictures. The current Third Generation (3G) mobile phones provide adequate bandwidth, and allow many advanced communication features for still images transmission and even video conferencing. With the fourth generation (4G) mobile technologies much higher Quality of Service (QoS) will be available for image and video transmission.

The scenario for using mobile systems for maintenance is rather positive, as most maintenance systems transmit data for machine to machine communications. Image transmissions will be required mainly if one end sends image of the offending part, which needs to be viewed and interpreted by a human maintenance engineer. However, in the future, some image processing-based diagnostic systems may require image transmission even for machine to machine communications.

The issue of having enough computing power or the lack of it, especially in mobile devices, can be overcome with the help of Cloud Computing.

Cloud Computing

The availability of computing power outside the boundary of a user's own infrastructure is termed Cloud Computing. This term is used because computing power is being accessed via the Internet, and the Internet has been represented as a cloud for many years. Therefore, Cloud Computing refers to delivering hosted services over the Internet.

Cloud Computing services can be classified into three categories:

- **Infrastructure**-as-a-Service (IaaS)
- **Platform**-as-a-Service (PaaS) and
- **Software**-as-a-Service (SaaS)

As per Kim (2009): "Perhaps the simplest working definition of cloud computing is being able to access files, data, programs and 3rd party services from a Web browser via the Internet that are hosted by a 3rd party provider and paying only for the computing resources and services used"

The architecture of a computing cloud includes the following key modules:

- User interaction interface,
- System resource management module with a services catalog, and
- Resource provisioning module.

The system resource management module includes a large network of servers running in parallel. "Often it also uses virtualization techniques to dynamically allocate and deallocate computing resources" (Kim, 2009).

As per Kim (2009), some of the advantages of Cloud Computing include:

1. The 3rd party provider owns and manages all the computing resources (servers, software, storage, and networking) and electricity needed for the services. The users only need to "plug into" the cloud. The users do not need to make a large upfront investment on computing resources; the space needed to house them; electricity needed to run the computing resources; and the cost of maintaining staff for administering the system, network, and database.

2. The users can increase or decrease the level of use of the computing resources and services flexibly and easily.

3. The users pay most likely much less for the services, because they pay only for the computing resources and services they use, and the subscription-based or payper- use charges are likely much lower than the cost of maintaining on-premises computing resources. If the users are to maintain on-premises computing resources, they also need to make the worst-case plan to account for the occasional or seasonal peak needs.

4. The users can in practice access the cloud for services anytime from anywhere.

The above mentioned advantages make cloud computing an excellent way of providing the computing power that may be lacking when eMaintenance is transformed in to iMaintenance. Many vendors, including some large and some small vendors, are now in the Cloud Computing services market. The bigger vendors include Yahoo and Google.

AN EXISTING IMAINTENANCE SYSTEM

An existing iMaintenance system (2010) called i-Maintenance Solutions is already available as a commercial product for building maintenance. Figure 4 shows a graphical view of the same.

Table 3 presents the key features and benefits of the iMaintenance system shown in Figure 4, marketed by iMaintenance Solutions™. While the benefits listed in Table 3 pertain to property management, these could well be applied to the maintenance of a large factory.

This is a complete property maintenance management system. While it is different from the industrial maintenance systems being described, it is interesting to see that it has many benefits that are similar to the iMaintenance paradigm being proposed.

Some of the main benefits touted by the vendor, iMaintenance Solutions™, include the following (iMaintenance system, 2010):

- **Saves you money:**
  - ○ Streamlines your operations
  - ○ Increases overall efficiency
  - ○ Increases tenant retention

Figure 4. A graphical view of the iMaintenance system (2010) for building maintenance

- **Saves you time:**
  - ○ Reduces incoming trouble calls
  - ○ Schedules maintenance events
  - ○ Tracks maintenance & service requests
- **Increases your productivity:**
  - ○ Automates time-sensitive requests
  - ○ Automates tracking and follow-through
  - ○ Instant reporting of open & closed maintenance requests
  - ○ Instant access to historical data

The so listed Property Manager Benefits are similar to the benefits that would flow to the maintenance manage in a factory. He will be able to get an automatic response from the system that the Maintenance Request was received. It will allow him to allocate one person to the job, and CC it to other stakeholders to keep them informed. Where spare parts are required, he will be able to notify the Vendor. Furthermore, he will be able to tracks the performance of the Vendor, and follow up the Maintenance Request to completion. The benefits under the three other categories: Owner Benefits, Tenant Benefits, Vendor Benefits, would similarly, have benefits related to the industry maintenance scenarios.

iMaintenance with Digital Ecosystems and Cloud Computing

Form the discourse so far we can deduce that iMaintenance systems would require ubiquitous and intelligent computing technologies. Figure 5 presents a collage of technologies required for creating the proposed iMaintenance systems. The key challenge is in integrating the various technologies developed thus far. Some of the monitoring systems currently used as stand-alone devices need to be turned into Smart Systems, i.e. systems that have the intelligence to work with other Plug-and-Play technologies, so that

Table 3. Key features of the iMaintenance system (2010)

iMaintenance Solutions™ Overview
Maintenance requests are time sensitive and lawsuit friendly
NEVER AGAIN will you put you or your company at risk
NO LONGER can someone claim you failed to return a call or perform
iMaintenance Solutions is the **"missing link"** that keeps the Maintenance Requests on track, keeps Owners and Tenants well informed and Vendors on task.

Property Manager Benefits:	**Owner Benefits:**
• Automatic system response that the Maintenance Request was received • Designates one person to open the Maintenance Request and disburse it • CC's owners, keeping them informed • Notifies the Vendor and keeps the Vendor on task • Tracks the performance of the Vendor • Follows the Maintenance Request to completion • Aids in Tenant retention - reduces the number of complaints about Maintenance Requests	• Receives a copy of the Maintenance Request by e-mail • Can edit and update their personal and property data • Knows which Vendor received the Maintenance Request • Can offer information pertinent to the Maintenance Request • Stays informed from the report of the Maintenance Request to its completion • Knows when the invoice will be paid
Tenant Benefits:	**Vendor Benefits:**
• Can submit Maintenance Request by e-mail • Can check e-mail for responses from Company • Can edit and update personal data • Knows which Vendor was assigned to the Maintenance Request • Retains control of access to their unit • Stays informed from reporting the Maintenance Request to its completion	• Receives the Maintenance Request by e-mail • Receives unit/tenant access information • E-mail clearly states work to be performed • Can ask for clarification of work to be performed • Submit an estimate for approval by e-mail • Invoice for payment via e-mail

Figure 5. Creating iMaintenance systems with digital ecosystems and cloud computing

these can become a part of a Digital Ecosystem for iMaintenance.

Some of research directions that can be explored include:

- Development of Smart Systems for monitoring and control
- Development of Web services for maintenance
- Creation of Service Providers for maintenance Web services

- Development of generic iMaintenance frameworks that can be adapted to specific industries.

CONCLUSION

We are entering the era of inexpensive, and yet powerful mobile computing. Over the last decade the concept of eMaintenance has evolved and a wide variety of maintenance models and eMaintenance architectures have been proposed. Now these eMaintenance systems can be advanced further to create the iMaintenance paradigm, by using ubiquitous and intelligent computing technologies such as Digital Ecosystems that form the local infrastructure, and Cloud Computing to provide the ICT facilities not available within the local infrastructure. Further research into the development of Smart Systems, Web Services and Web Service Provides targeted towards maintenance functions is desirable.

REFERENCES

Baldwin, R. C. (2004). *Enabling an e-maintenance infrastructure*. Retrieved from www.mt-online.com

Booth, D., et al. (2004, February 11). *Web services architecture*. W3C working group note. Retrieved from http://www.w3.org/TR/2004/NOTE-ws-arch-20040211

Chong, C. Y., & Kumar, S. P. (2003). Sensor networks: Evolution, opportunities, and challenges. *Proceedings of the IEEE*, *91*, 1247–1256. doi:10.1109/JPROC.2003.814918

Crespo-Marquez, A., & Gupta, J. (2006). Contemporary maintenance management: Process, framework and supporting pillars. *Omega*, *34*(3), 313–326. doi:10.1016/j.omega.2004.11.003

Dini, P., De Wilde, P., Rowe, J., Petrou, M., & Heistracher, T. (2005). *The digital ecosystem research vision: 2010 and beyond*. Position paper FP7 Workshop, 2005. Retrieved from http://www.digital-ecosystems.org/events/2005.05/de_position_paper_vf.pdf

Fout, T. (2001). *Universal plug and play in Windows XP*. Retrieved from http://technet.microsoft.com/en-us/library/bb457049.aspx

Han, T., & Yang, B.-S. (2006). Development of an e-maintenance system integrating advanced techniques. [Elsevier.]. *Computers in Industry*, *57*, 569–580. doi:10.1016/j.compind.2006.02.009

Hausladen, I., & Bechheim, C. (2004). E-maintenance platform as a basis for business process integration. *Proceedings of INDIN04, Second IEEE International Conference on Industrial Informatics*, (pp. 46–51). Berlin, Germany. *i-Maintenance system*. (2010). Retrieved September 30, 2010, from http://www.track-it-systems.com

Kay, R., & Mattern, F. (2004). The design space of wireless sensor networks. *IEEE Wireless Communications*, *11*(6), 54–56. doi:10.1109/MWC.2004.1368897

Kim, W. (2009). Cloud computing: Today and tomorrow. *Journal of Object Technology*, *8*(1), 65–72. doi:10.5381/jot.2009.8.1.c4

Koc, M., & Lee, J. (2001). A system framework for next-generation e-maintenance system. *Proceedings of the Second International Symposium on Environmentally Conscious Design and Inverse Manufacturing*, Tokyo, Japan, 2001.

Lee, J. (2001). A framework for next-generation e-maintenance system. *Proceedings of the Second International Symposium on Environmentally Conscious Design and Inverse Manufacturing*, Tokyo, Japan.

Lee, J. (2003). E-manufacturing: Fundamental, tools, and transformation. *Robotics and Computer-integrated Manufacturing, 19*(6), 501–507. doi:10.1016/S0736-5845(03)00060-7

Lee, J., & Ni, J. (2004). *Infotronics-based intelligent maintenance system and its impacts to closed-loop product life cycle systems.* Invited keynote paper for IMS'2004—International Conference on Intelligent Maintenance Systems, Arles, France, 2004.

Lee, J., Ni, J., Djurdjanovi, D., Qiu, H., & Liao, H. (2006). Intelligent prognostics tools and e-maintenance. *Computers in Industry, 57*, 476–489. doi:10.1016/j.compind.2006.02.014

Li, Y., Chun, L., Nee, A., & Ching, Y. (2005). An agent-based platform for web enabled equipment predictive maintenance. *Proceedings of IAT'05 IEEE/WIC/ACM International Conference on Intelligent Agent Technology,* Compiegne, France.

Moore, W. J., & Starr, A. G. (2006). An intelligent maintenance system for continuous cost-based prioritisation of maintenance activities. *Computers in Industry, 57*(6), 595–606. doi:10.1016/j.compind.2006.02.008

Muller, A., Crespo-Marquez, A., & Iung, B. (2008). On the concept of e-maintenance: Review and current research. *Reliability Engineering & System Safety, 93*(8), 1165–1187. doi:10.1016/j.ress.2007.08.006

Muller, A., Suhner, M.-C., & Iung, B. (2007). Maintenance alternative integration to prognosis process engineering. *Journal of Quality in Maintenance Engineering Journal of Quality in Maintenance Engineering, 13*(2).

Nachira, F. (2005). *What is a digital ecosystem?* Retrieved from www.digital-ecosystems.org/events/2005.05/050518-fnak.pdf

Ravuthula, S. (2002). *Web services applications and security: Part 1.* Retrieved from http://www.developer.com/services/article.php/1550461/Web-Services-Applications-and-Security-Part-1.htm

Sarigiannidis, G. (2006). *Localization for ad-hoc wireless sensor networks.* Master of Science Thesis, The Technical University of Delf, The Netherlands, August 2006.

Sharda, N. (2006). Quality of service issues in mobile multimedia transmission. In Karmakar, G., & Dooley, L. (Eds.), *Mobile multimedia communications: Concepts, applications and challenges.* Hershey, PA: Idea Group Inc.

Sharda, N. (2009a). *A high-level model for and intelligent system to maximise the utilisation of harvested rainwater.* OZWATER 09 – Australia's National Water Conference & Exhibition, 16-18 March, 2009, Melbourne.

Sharda, N. (2009b). *Multimedia and communication technologies in digital ecosystems.* Seventh International Conference on Information, Communications and Signal Processing, Macau, 7-10 Dec 2009.

Sharda, N. (2011). Multimedia transmission over wireless sensor networks. In L.-M. Ang & K. P. Seng (Eds.), *Visual information processing in wireless sensor networks: Technology, trends and applications.* Hershey, PA: IGI Global. (Accepted for publication in 2011).

Sherwin, D. (2000). A review of overall models for maintenance management. [MCB University Press.]. *Journal of Quality in Maintenance Engineering, 6*(3), 138–164. doi:10.1108/13552510010341171

Szymanski, J., Bangemann, T., Thomesse, J. P., Lang, C., Thron, M., Reboeuf, X., & Garcia, E. (2003). Proteus—A European initiative for e-maintenance platform development. *Proceedings of the Emerging Technologies and Factory Automation Conference, IEEE ETFA, 03*, 415–420.

Tao, B., Ding, H., & Xion, Y. L. (2003). IP sensor and its distributed networking application in e-maintenance. *Proceedings of the 2003 IEEE International Conference on Systems, Man and Cybernetics,* vol. 4, Washington, DC, USA, (pp. 3858–63).

Tsang, A. (2002). Strategic dimensions of maintenance management. *Journal of Quality in Maintenance Engineering, 8*(1), 7–39. doi:10.1108/13552510210420577

Ucar, M., & Qiu, R. G. (2005). E-maintenance in support of e-automated manufacturing systems. *Journal of the Chinese Institute of Industrial Engineering, 22*(1), 1–10. doi:10.1080/10170660509509271

Yoshikawa, H. (1995). Manufacturing and the 21st century—Intelligent manufacturing systems and the renaissance of the manufacturing industry. *Technological Forecasting and Social Change, 49*(2), 165–213. doi:10.1016/0040-1625(95)00008-X

Yu, R., Iung, B., & Panetto, H. (2003). A mutli-agents based e-maintenance system with case-based reasoning decision support. *Engineering Applications of Artificial Intelligence, 16,* 321–333. doi:10.1016/S0952-1976(03)00079-4

Zhang, W., Halang, W., & Diedrich, C. (2003). An agent-based platform for service integration in e-maintenance. *Proceedings of ICIT 2003, IEEE International Conference on Industrial Technology,* vol. 1, Maribor, Slovenia, (pp. 426–33).

Chapter 13
Mobile Security

Barbara Ciaramitaro
Ferris State University, USA

Velislav Pavlov
Ferris State University, USA

ABSTRACT

Over the past few years, cyber criminals have expanded their focus from desktop PCs to mobile devices such as smart phones, PDAs, and tablet computers. Unfortunately, even though many mobile devices approach personal computers in functionality, most mobile users are not aware of the degree of security threats in the mobile environment. "As mobile Internet usage continues its rapid growth, cyber criminals are expected to pay more attention to this sector" (Siciliano, 2010, p. 1). There are several security threats related to mobile devices. The most common security threat associated with mobile devices is their propensity to become lost, stolen, or misplaced. Social Engineering is a method used by cybercriminals to trick users into providing personal or financial information, or downloading malicious software. One common social engineering attack against mobile devices involves attempts to collect personal, credit card, and banking information from users. Malware is short for malicious software and refers to a collection of malevolent software tools designed to attack the pillars of information security: confidentiality, integrity, availability, and authentication. Although malicious software and security attacks can occur in a number of ways such as SMS text messaging, the primary mode of infection is through the download of mobile applications such as games. Unfortunately, all mobile devices and all mobile operating systems are subject to mobile malware attacks. As a result, malware has become a prevalent threat to mobile devices.

INTRODUCTION

The three core pillars of information security are confidentiality, integrity and availability. Known as the *CIA Triad*, these elements form the basis for

DOI: 10.4018/978-1-61350-150-4.ch013

developing and assessing all information security efforts. *Confidentiality* means that information is not available to people who are unauthorized to access it. *Integrity* means that changes to the information by unauthorized personnel is not allowed. And *availability* means that information is available to authorized users when and where they need it. The concept of authenticity is frequently included when discussions of information security occur. *Authenticity* means that the user requesting or accessing the data is who he or she states they are. (Stewart, 2008) Threats against any of these security principles form the basis for malicious attacks and cyber criminal activity.

The term *mobile device* can mean different things to different people. Some consider all devices that can be used to connect through wireless networks to be mobile devices. For the purposes of this chapter, mobile devices are defined as the category of handheld devices including cell phones, smart phones, PDAs (Personal Digital Assistants), and tablet devices such as the Apple iPad. Desktops, laptops and netbooks are included in the family of personal computers. These two categories of computing devices have much in common but there are some distinguishing issues related to handheld mobile devices. This chapter will focus on those security issues that are applicable to handheld mobile devices.

Mobile devices are uniquely susceptible to security risks in that they are always on and accessible, and provide several means of communication and connectivity through text and multimedia messaging, voice, and wireless connectivity through Bluetooth and Wi-Fi. Although they offer tremendous benefits with their convenience, functionality and immediate access to data, messaging, downloaded applications, and Web services, mobile devices are also fertile grounds for cyber criminal attacks. Each of these avenues is the means to distribute malicious software to the mobile device.

Users of computer technology commonly translate security threats into the need to protect their computers from hackers and cyber criminals who may try to steal information for nefarious purposes or destroy data for fun. Many users understand the risk of security threats such as viruses and identity theft and have taken steps to protect their personal computers with software to protect against these malicious software attacks.

However, "Viruses, Malware, & Spyware are no longer threats for just PC owners. Malicious software is steadily making its way into our pockets." (Kardos, 2010, p. 1). Over the past few years, cyber criminals have shifted their focus from desktop PC's to mobile devices. Unfortunately, even though many mobile devices approach personal computers in functionality, most mobile users are not aware of the degree of security threats in the mobile environment. "As mobile Internet usage continues its rapid growth, cyber criminals are expected to pay more attention to this sector" (Siciliano, 2010, p. 1). A troublesome trend in mobile devices is indicated by the escalating numbers of mobile malware and the increasing use of Internet capabilities in mobile devices. Together these indicate "a growing malware development community" and an "increasing source of potential attack vectors." (Jansen, 2008, pp. 3-9).

MOBILE SECURITY THREATS

Mobile Malware

Malware is short for malicious software and refers to a collection of malevolent software tools designed to attack the pillars of information security: confidentiality, integrity, availability and authentication. In the computer world at large, some estimate that the number of malware programs exceed that of legitimate software programs (Symantec, 2008). Unfortunately, malware has become a prevalent threat to mobile devices.

Mobile malware is not just a nuisance; it can cause serious harm to the mobile device and its owner. Mobile malware can initiate voice calls or send SMS or MMS messages that are unauthorized by the user. These calls and messages are then charged to the mobile device owners' bill. Mobile malware can steal private information such as contacts, user information such as username and password, businesses communications, and online banking transactions. This private information can then be sold to a third party for profit. Another mobile malware threat is the use of the mobile device to perform illegal operations and create communication channels for attackers. This is usually accomplished by taking over remote control of the device and using it in an organized attack. Mobile malware can destroy information and can install malicious software that can spread to other users through messaging or Bluetooth connection. The use of spyware and key loggers allow cybercriminals to track and record the actions and keystrokes of mobile users, eavesdropping on their behavior and communication. With the growth of GPS services, malware can now track the location of the user. From an enterprise business perspective, mobile devices can provide an entry point to corporate information. Mobile malware can be used to penetrate a corporate network in order to steal or destroy information. (Jain, 2010)

Mobile malware can spread in a number of ways including downloads of games and applications from the Internet or application stores, messaging services such as Instant Messaging, SMS, MMS, and local wireless connections such as Bluetooth. Bluetooth is a technology that allows wireless devices to connect to each other over short distances. Malware can be attached to emails messages or embedded in applications or Web sites. Under most circumstances, the user must act in some way to accept and open the message or consent to the download or game. As will be discussed shortly, social engineering is often used to trick the user into accepting the message or installing the application (Jansen, 2008).

Malware can be classified into several categories. The most well known category is that of *infectious malware* which include *viruses* and *worms*. These malware software products are distinctive in their ability to spread quickly from one mobile device to another as attachments to email, SMS messages, Bluetooth connections, games or other downloaded applications. Both viruses and worms can carry what is known as a *payload*, which is a malicious program designed to steal or destroy data.

A second category of malware is known as *concealment malware* and includes Trojan horses, root kits and backdoors. A *Trojan horse* is a software program designed to look like a desirable game or application but that carries with it malicious software. Its most common method of distribution is through the download of what appears to be attractive ringtones, wallpaper or games. *Root kits* are a type of software that modifies the mobile operating system so that its presence may not be detected. *Backdoors* are installed applications that allow malicious users to bypass the standard authentication processes of the phone and gain access without providing passwords or PIN numbers.

A third category of malware is specifically created for financial gain. These include spyware, keystroke loggers, dialers, and other intrusive applications. These programs are the most common ones used by organized cybercriminals to steal personal and financial information. *Spyware* software is designed to track the user's actions and provide information to the creator. They are often inadvertently downloaded by users from Web sites, attached to SMS or email messages, or hidden in games, ringtones and wallpaper downloads. *Key loggers* work by intercepting the keystrokes of a user then they are entering authentication information, credit card, banking or other personal information. The captured keystrokes are then sent to the creator of the malicious software for future exploitation. *Dialers* are malicious software that takes over control of the mobile device and either places unauthorized voice calls or sends unauthorized SMS or MMS messages.

Lost or Stolen Mobile Devices

The most common security threat associated with mobile devices is their propensity to become lost, stolen or misplaced. In fact, it is estimated that over 8 million phones are lost with 32% of them never recovered (Jansen, 2008). Unless a user specifically establishes security precautions on their mobile device, once it is in the hands of unauthorized users they have the ability to access and use any of the data and functionality provided by the mobile device. The compromised mobile device could then be used to place voice calls, send SMS (Simple Message System) and MMS (Multimedia Message System) messages, access banking information, make purchases, and accumulate other charges related to downloaded applications. The optimum solution to this threat is to install software on the mobile device that is able to erase the contents of mobile devices and deactivate their status when they have been reported as lost or stolen. Lookout (www.mylookout.com) is a software product that provides a set of services for lost or stolen mobile devices. If your mobile device is equipped with GPS, then Lookout can locate your phone. It can also issue a loud audible alarm. Most importantly it can either remotely lock your mobile device or remotely eliminate all of its contents before it falls into the hands of malicious users.

Closely related to the loss of a phone is the security threat of unauthorized access. If a mobile device comes into the hands of an unauthorized user, their first goal will be to access the information and resources stored on the mobile device. The types of information stored on today's mobile devices go far beyond a listing of contacts and phone numbers. Mobile devices now hold personal and financial information as well as records of calls, texts, and locations visited. Unfortunately, very few mobile users implement password or PIN (Personal Identification Numbers) on their mobile device. If they chose to lock their mobile devices, many users rely on the default settings such as 1234 or 0000 to unlock the device. Additionally, a more recent authentication threat appeared with the use of touch screens. Security experts have found that attackers are able to use the fingerprint smudge pattern on the face of the mobile device to deduce the password or PIN (Schwartz, 2010). In

order to protect their device against unauthorized access users should establish a password that is prompted when they activate their mobile device. Additionally, a mobile device should become inactive after 2 to 3 minutes of idle time, once again requiring a password when it is reactivated each time for use. Another recommendation is to encrypt personal and proprietary information stored on the mobile device so that even if it is accessed, it is unreadable in its encrypted form (Trend Micro, 2007).

Mobile Social Engineering

Social Engineering is a method used by cyber-criminals to trick users into providing personal or financial information, or downloading malicious software. "Social engineering is all about catching users off guard and luring them into clicking on malicious links or sharing sensitive information." (Bradley, 2010, p. 1). Some researchers believe that as mobile malware protection against viruses and other malicious software increases through improved hardware and software tools, smarter social engineering will remain a leading issue as it relies on manipulating people into providing information or access (Security Watch, 2009).

One common social engineering attack involves attempts to collect personal, credit card, and banking information from users. There are several variations of this scam. *Smishing* uses SMS texting as the basis of illegal requests for users to provide personal information. The smishing message will provide the user with a phone number to call or a web site to access. On the other end of the phone or web site is a cyber criminal anxious to steal whatever personal information is provided. Smishing is similar to *phishing* attacks which use the same tactics in the email environment. *Vishing* is a similar illegal scan in which voicemail messages are left for the user requesting that they call back or access a web site to provide more information. Unfortunately, even if the user does not provide information on the web site, there is a very high probability that by simply accessing the web site, the user will download malicious software onto the mobile device (Montalbano, 2010).

*Spam i*s another type of social engineering in that mass distributions of email or SMS messages are sent to a broad spectrum of users with the hope that only a small percentage will respond. These messages are considered spam when they are sent regardless of the users' interest in the email content. One serious problem is that some mobile users pay to receive text messages, and spam messages can become quite costly. Additionally, spam often contains malware such as viruses or worms that can be dangerous and destructive to the mobile user (Jansen, 2008). Although laws have been enacted to prevent mobile spam such as the CAN-SPAM Act of 2003, they are not regularly enforced as spam is hard to investigate and difficult, if not impossible, to stop. Only a few states including California and Ohio have initiated laws against mobile Spam. Other countries such as Australia and Hong Kong have been more successful in their regulations against mobile spam. (Wise Geek, 2010).

Although not often considered social engineering, the collection of data from mobile users can fall into this category under some circumstances. Companies will often request information from the user in order to provide a discount coupon or other service. Privacy and ethical considerations require that the user is informed, and consents to, the collection of data and the purpose for which it is collected. However, when collected data is used or sold for purposes without the users' consent, it become a security threat and violates the confidentiality rights of the users.

Mobile Eavesdropping

Mobile eavesdropping and tracking can occur in two ways: direct and indirect. One of the common and direct ways for mobile eavesdropping to occur is through the use of spyware which is usually downloaded as part of a game, message or application. Once the spyware is installed, it tracks, collects, and forward information about the user their transactions to a recipient server or phone (Jansen, 2008). Most mobile devices provide the capability to connect to local wireless networks such as the Wi-Fi connections in coffee shops, book stores, or airports. Wi-Fi refers to a set of standards that guide wireless connections in close proximity. Although Wi-Fi ad hoc connectivity is very convenient for the user, it is rife with security risks and provides the means and opportunity for an indirect eavesdropping or tracking attack (Ibid). It is common for a cyber criminal to create a rogue wireless access point in a public Wi-Fi area in order to intercept and hijack information transmitted from mobile devices (Couture, 2010). "In order to steal your data through a rogue access point attack, a hacker needs only to overpower the local Wi-Fi access point ... with his evil network rather than the public one. You will still be connected to the Internet, except all your personal data will pass through the hacker's computer which acts as a man-in-the-middle." (Best Security Tips, 2007, p. 1). The best advice for mobile users is to refrain from sending any personal, proprietary or financial information when you are using a public wireless network. If it becomes necessary to use a public network, then it becomes essential to encrypt the data before it is transmitted.

Mobile Games and Applications

Although malicious software and security attacks can occur in a number of ways such as SMS text messaging, the primary mode of infection is through the download of mobile applications such as games. "If your device is penetrated, cyber criminals would not only have the ability to cripple your phone's software, but they could also pinpoint your exact coordinates, track all your incoming and outgoing messages, activate your mic (sic) to record any audio in the room, and much more." (Kardos, 2010, p. 1).

One of the causes of malware distribution through mobile games is the lack of thorough malware scanning and review conducted by mobile phone application providers. The Google Android application marked is considered the most vulnerable as Google does not have employees dedicated to reviewing applications submitted to its store for security threats. They rely primarily on their users to inform them of malicious software (Ante, 2010). Although both Apple and Blackberry review mobile applications before they appear in their application store, security threats remain (Ibid). Some believe that mobile application store providers should assume primary responsibility for the mitigation of the future spread of malicious software (Kardos, 2010).

It is also important to note that some mobile applications are designed to provide unauthorized access to the mobile device by disguising its functionality. During a recent conference, the App Genome Project was announced. This project was designed to examine what applications are actually doing underneath the user interface. They found that many free applications do not disclose their full functionality and in many cases developers incorporate malicious code into their offerings. This malicious code is designed to access contacts, browser history, location, transaction logs and phone information (Mahaffey, 2010).

Some mobile applications are able to turn the mobile device into what is known as a *zombie* which carries out the commands of a central hacker. During the RSA Conference in 2010, security researchers demonstrated an experimental mobile botnet comprised of 8,000 iPhones and Android smartphones. Derek Brown and Daniel Tijerina from TippingPoint's Digital Vaccine Group designed an innocently looking weather application, WeatherFist, which provided weather forecast information. The researchers also developed a malicious version of the application, which intended to obtain user's GPS coordinates, read files like contacts and cookies, spread spam, and access Facebook, Twitter, email, and passwords.

The application turns the mobile devices into "zombies" controlled by a command server. The project aimed to raise awareness about mobile botnets. The application was not distributed via the iPhone Apple Store or Android Market. The threat of mobile botnets is real. It spreads far beyond the personal computer. (Higgins, 2010)

Mobile Networks

There are two main competing networks for use with cellular services associated with mobile devices: Global System for Mobile Communications (GSM) and Code Division Multiple Access (CDMA). GSM has been in existence since 1982 and now provides over 3.4 billion connections worldwide (Ananda, 2008). It is considered the world's most popular standard for mobile services with 80% of the global mobile market using this technology (GSM World, 2009). Currently, T-Mobile and AT&T use the GSM technology standard (Fendelman, 2010). GSM is a transmission technology that uses one of four frequencies ranges through which to send and receive information: 850 MHz, 900 MHz, 1800 MHz, and 1850 MHz. GSM supports both authentication and encryption technologies. The security goals of GSM are to provide authentication of users, confidentiality of user data, use of SIM cards (Torrani, 2008). One of the distinctive features of the GSM system is the use of SIM cards (Subscriber Identity Modules) which are memory cards or chips installed in the mobile device that can store data. The SIM card makes it easy to switch to a new phone by simply installing the SIM in the new device. The SIM generally stores personal identity information, cell phone number, contacts and phone book, text messages and other data. (Wise Geek, 2010) One significant risk of GSM networks is that it only provides robust access point security and therefore the data in transit can be at risk for capture, alteration, deletion and eavesdropping (Gadaix, 2001; Torrani, 2008).

EDGE, which stands for Enhanced Data rates for GSM Evolution, is a faster version of GSM. EDGE is a high-speed 3G technology that was built upon the GSM standard. It was launched in 2003 and was designed to deliver multimedia applications such as streaming television, audio and video to mobile phones. EDGE transmits data more than three times the capacity and performance of GSM (Fendleman, What is Edge?, 2010). AT&T currently uses the EDGE version of GSM. Although EDGE provide faster speeds and improved performance over GSM, the security risks associated with GSM remain with EDGE.

CDMA is a technique used for wireless communications that involves a combination of multiple frequencies and a transmission code. Currently, Sprint, Virgin Mobile and Verizon Wireless use the CDMA technology standard (Fendleman, 2010). Whereas GSM uses constant frequencies, CDMA uses multiple frequencies, or multiplexing, to transfer information. The use of the multiple transmission frequencies is referred to as spread spectrum technology. CDMA allows several transmitters to send information simultaneously over a single communication channel. This concept is called multiplexing. CDMA also requires that a certain code accompany the transmission. CDMA technology had its roots in military applications and cryptography (Nortel, 2005). CDMA technology is distinguished by two characteristics. First, the transmission signal occupies a bandwidth that is greater than necessary which results in immunity to interference and jamming. The transmission through the bandwidth is spread by means of a pseudo random code that is independent of data (Hendry, 1985). CDMA is considered secure due to its use of spread spectrum technology and the use of pseudo random codes. These spread spectrum techniques are used to form unique code channels for individual users in within the communication channel. "Because the signals of all calls in a coverage area are spread over the entire bandwidth, it creates a noise-like appearance to other mobiles or detectors in the network as a form of disguise, making the signal of any one call difficult to distinguish and decode" (Nortel, 2005). This protects the CDMA transmission from interference and interception and there have been no reports of highjacking or eavesdropping on CDMA transmissions in a commercial network. (Wise Geek, 2010; Nortel, 2004)

HISTORY OF MOBILE MALWARE

Unfortunately, all mobile devices and all mobile operating systems are subject to mobile malware attacks. The most popular mobile device operating systems are Nokia Symbian, Microsoft Mobile OS, Apple, iOS, and Google Android. As the Symbian operating system has been around the longest, it has been the most targeted platform. However, as new mobile operating systems emerged, such as the Apple iOs and Google Android, mobile malware soon followed.

Before the rise of the current smart phones, Palm Pilots and Windows Pocket PC's ruled the world of mobile computing. Soon after these devices were released, there was malware directed at their operating systems. In 2001, the first malware directed at the Palm Operating system was created. It was called *Liberty Crack,* and when executed, it deleted all applications from the PDA. This was followed shortly by a virus referred to as Phage. This virus infected mobile applications, and when executed, turned the application icons gray so that they were no longer visible to the user. (Chen, 2008).

The development release of smart phones began in the early years of the 21st century. Soon, the first malware program that infected smart phones occurred in 2004. It was a virus known as *Cabir* which attacked the Symbian mobile operating system and tried to replicate itself by attempting to establish a Bluetooth connection with other mobile devices. It did no damage other than running down the life of the battery (Hypponen, 2006). In 2004, mobile malware attacks against the Microsoft mobile operating system began occurring along with continued attacks against the Symbian operating system. During the years 2004 and 2006, the mobile threat landscape increased significantly and produced malware similar to those facing computer users: viruses, worms, Trojans, spyware, and backdoors. In 2004, mobile malware was capable of spreading viruses through SMS messaging and Bluetooth connection; infecting files stored on mobile devices; enabling remote control of the mobile device; modifying or replace icons or applications; combating antivirus programs, stealing data and locking memory cards. The potential damage of mobile malware has increased over the years and is now capable of causing even more damage by spreading malware via removable memory cards; damaging user data; stealing contact lists and other data; giving remote access to criminals; disabling operating system security measures; downloading files from the internet; and calling paid services (Gostev, 2010).

Malware has also kept current with the release of new phones and operating systems. For example, it did not take long after the release of the Apple iPhone for the development of mobile malware designed to attack that platform. In 2008, an 11 year old wrote the first iPhone Trojan horse malware which disguised itself as a "prep" program for the installation of an iPhone firmware update. Once installed, it did little damage, only displaying the word "shoes". Unfortunately, if you tried to remove the malware, it destroyed several iPhone legitimate programs and applications (Patrizio, 2008). In 2009, a more malicious malware attacked iPhones. This malware allowed the attacker to copy user information without knowledge of the users. (Mac Security Blog, 2009). In 2010, the first malware used against the Google Android Operating System was detected. Once installed, it sent SMS messages to premium numbers without the user's knowledge. Interestingly, the premium numbers only worked in Russia (Wells, 2010).

The pace of mobile malware has accelerated and, as of 2010, there are over 400 known mobile device viruses or malware. Cybercriminals consider mobile application downloads as a "fertile fields from which they can gain valuable information" (Culver, 2010, p. 1). Many of these attacks are now targeted toward gaining access to a mobile user's financial information. For example, several recent malware attacks are cybercriminals targeting users of mobile banking services. More than 50 fraudulent applications targeting mobile banking were identified and removed from the Google Android Application store in the early part of 2010 (McGlasson, 2010).

ENTERPRISE THREATS

Although security threats to personal users of mobile device can be disturbing, the threat to corporate enterprise systems can be devastating. The use of mobile devices such as smart phones is widespread in the business world, some of which are authorized by the enterprise, and some unauthorized. In fact, "More than 50% of enterprises have bowed to worker pressure and support personally-owned smart phone" (Lai, 2010, p. 1). Unfortunately, many corporate IT staffs continue to treat these devices as personal phones rather than portable computing devices rife with similar security concerns. The use of a smart phone or other intelligent mobile device can open a corporate network to several types of malicious attacks as discussed previously. However, the one unique aspect of enterprise mobile security is that the attacks and breaches can affect the entire corporate network rather than simply one individual's device. Enterprise networks also have the additional risk of destruction or access to confidential and proprietary data and information. "Unless actions are taken to secure this information, the mobile device represents a potentially severe security risk to the enterprise" (Good Technology, 2009, p. 1). Another concern related to mobile security is the cost of data breaches through mobile devices. "The average organizational cost of a data breach was US $3.4 million, but all countries in the study reported noticeably higher costs when the incidents involved mobile devices." (Lobel, 2010).

Therefore, it is essential for organizations to institute *Mobile Device Management (MDM)* policies and practices to protect the enterprise from security attacks. Mobile Device Management (MDM) includes technology, policies and procedures to manage mobile devices across the enterprise. The software and policies include the ability to remotely manage and wipe devices, password and PIN requirements, asset management and consistent application configuration settings (Joseph, 2006). Over-the-Air (OTA) ability is considered an essential component of MDM which allows the central IT department to remotely configure, lock or wipe a device Although MDM policies and practices have been in place for many years, the current emphasis of MDM is on security (Winthrop, 2010).

It is estimated that by the year 2011, there will be one billion mobile workers using mobile devices to perform their job functions and store enterprise data (Good Technology, 2009). Mobile security risks generally fall into two categories: lost or stolen mobile devices; devices that are connected to the enterprise and compromised by a security breach or malware. Most enterprise mobile devices store or have access to a myriad of enterprise data including emails, financial records, sales reports, customer information, and business plans. In many cases the use of passwords, PINs or encryption is not enforced by the enterprise. When we consider that over 250,000 mobile devices are left in U.S. airports everyone year, we can begin to see the security risks to corporate information (Ibid). Although the security malware attacks on enterprise mobile devices are the same as the ones discussed earlier such as viruses and worms, the impact on the enterprise is far greater. "Mobile device hackers target devices in order to launch larger attacks on corporate networks, with the intent of accessing business-critical information or hampering business activities." (Good Technology, 2009, p. 4).

Additionally, some enterprises have specific regulatory requirements that require protection of personal, financial and other confidential information. These enterprises include healthcare, financial services and the government. Each of these domains has regulations that require them to safeguard and protect customer data and have established several penalties for failure to comply with the regulations. These regulations have resulted in some industries, such as healthcare, to become a leader in mobile device security implement audit trails, device access management and encryption (Good Technology, 2009, p. 10).

PROTECTING MOBILE DEVICES

Enterprise Mobile Devices

As discussed above, it is important for enterprises to implement MDM technologies, policies and practices to protect the enterprise from data breaches and malware. These MDM challenges fall into several categories: authentication, data protection, application control and configuration, and encryption. One key to mobile security is being able to limit the use of the mobile devices to only the authorized user. This can be accomplished in several ways. All enterprise users of mobile devices should be required to make their devices password protected and should change those passwords on a regular basis. Additionally the devices should be configured to require a password after a relatively short period of idle time and to limit the number of password attempts before the data on the device is wiped clean. When access to enterprise data is involved, additional passwords should be required. These access rights should be integrated into the overall enterprise security management system that manages user roles and should have the ability to change mobile users' access rights when their role changes. (Good Technology, 2009). Mobile users should also limit their use of public WiFi networks as eavesdropping is a common threat. Eavesdropping involves the creation of a rogue access point to which the mobile user unknowingly connects. All data sent through that access point is available to review and capture. If an enterprise mobile user must use a public WiFi network for transmission, all data should be encrypted.

In order to protect the enterprise network, IT administrators require the ability to remotely deploy security policies and software. All enterprise mobile devices should be protected with anti-virus software and the enterprise IT department should have the ability to remotely update virus protection on a regular basis. It is also essential that the data maintained on mobile devices is backed up to an enterprise service so that it can be restored or accessed should a device be lost or compromised. Enterprise IT departments should configure their firewall and intrusion detection devices to identify and restrict access from unknown or compromised mobile devices (Trend Micro, 2007). Enterprise mobile users should also be required to use encryption when storing data on their mobile device. Encrypted data can be store on a memory card that when inserted into the mobile device, prompts for a password. Another recommended enterprise practice is to segment users into different categories of devices and access requirements and deploy group policies and practices specific to them (Lai, 2010). It is also important for the enterprise to restrict the download of applications onto mobile devices. As discussed earlier, one of the most common ways to distribute malware is through downloaded mobile applications. (Good Technology, 2009). It is also crucial that enterprise users of mobile devices receive training on mobile security threats and the policies and practices that have been implemented to prevent or minimize them. Compliance with enterprise polices require monitoring and enforcement to be effective and the enterprise must be willing to put these elements

into practice in order to protect their enterprise from mobile breaches and malware.

Personal Mobile Devices

"An increasing number of malware and spyware applications are targeting mobile users and are able to log every key typed, message sent or received and data within mobile banking or trading software. The scary part is almost none of these devices have anti-virus, encryption or other endpoint security tools installed," (Fendelman, 2010, p. 1). This threat becomes more challenging when the various mobile devices and operating systems are considered. This vast mobile landscape "imposes an extremely broad and challenging surface to defend." (Couture, 2010, p. 4).

There are two vital steps that every mobile user should implement to protect their device against malware and breaches. The first is the use of passwords and PINs. As discussed earlier, the most prevalent security risk is the lost mobile device. Requiring a password or PIN to access data stored on the mobile device will provide one level of protection. Establishing a relationship with a vendor who is able to wipe all the data from your device remotely if lost, is the next level of protection. The second essential step is the use of mobile anti-virus software. As mobile viruses become more prevalent, the use of mobile anti-virus software becomes essential. However, it is important to keep your anti-virus software updated by establishing an ongoing subscription with a mobile anti-virus software vendor.

Mobile users should also limit their Bluetooth and WiFi connections. "A simple defense against many forms of malware is to turn off Bluetooth, WiFi, infrared, and other wireless devices until they are needed." (Jansen, 2008, pp. 4-5). When using Bluetooth, users should only allow connections with devices that they know. One good piece of advice is to only activate the Bluetooth discoverability of the mobile device when a known and specific connection is desired. WiFi connections pose eavesdropping risks and mobile users should limit the types of data transmitted through public WiFi connections. If confidential or proprietary information must be transmitted, it should only be sent in an encrypted form. To protect against vishing, phishing, and smishing attacks, mobile users should remain wary of requests for personal or financial information through a message or voicemail. "Any messages or contacts received on a mobile phone from an unknown number or device should be treated with suspicion. Messages should be destroyed without opening and connections denied" (Jansen, 2008, pp. 4-4). Most legitimate companies will not request this type of information through messaging, email or voice calls (McGlasson, 2010).

In many ways users should treat their mobile devices similar to their computers. They should backup the data of their smartphone, install virus protection, and make sure that the current operating system updates are current. They should also be cautious about where they download their applications making sure it is from a verified source.

One of the more recent uses of mobile devices is mobile banking. It is imperative that mobile users understand the potential security risks of mobile banking. As mentioned previously, mobile applications are a primary source of security risks. Anyone can create and sell a mobile application and the vetting of these applications as to their security risks is often incomplete. As it relates to mobile banking, several banks have posted messages on their website informing its customers of risky mobile banking applications (McGlasson, 2010).

Although the risks related to mobile banking are quite serious, other mobile applications can cause serious damage to a mobile device. Mobile users need to be very cautious of the ringtones, wallpapers, and applications that they download. Only download applications from know sources, and even then make sure that your anti-virus software is up-to-date. Some mobile devices such as the Apple iPhone restrict the functionality of the

device and limit its ability to run only approved applications. In order to overcome these limits, some users attempt to *jailbreak* their devices. This involves getting access to the underlying operating system and changing its configurations to accept a broader selection of applications. Unfortunately, once the jailbreak occurs, many of the security protection features installed in the device are eliminated and the device becomes more susceptible to breaches and malware.

CONCLUSION

The three core pillars of information security are confidentiality, integrity and availability and these form the basis for developing and assessing all information security efforts. Threats against these security principles form the basis for malicious attacks and cyber criminal activity. Mobile devices are uniquely susceptible to security risks against confidentiality, integrity, availability and authenticity in that they are always on and accessible, and provide several means of communication and connectivity.

Mobile devices are clearly here to stay. Their use is increasing and has already exceeded the number of computers and laptops on a worldwide basis. Alongside this increasing use, the variations of mobile devices, operating systems, and applications continue to swell. Both the number of mobile devices in use, and the changing nature of the technology, makes these devices prime targets for malicious users. Mobile users find applications irresistible and as a result, these applications have become a primary means to spread malware. Mobile users also like the freedom to connect to other devices and communicate through public Wi-Fi networks. These services, although convenient, are also rife with security risks.

Mobile users want to enjoy the growing functionality of their devices and in many cases remain unaware or unconcerned about security risks. However, as the number of breaches and malware increases each year, mobile users will have to take steps to protect their devices and data. Some steps are relatively simple such as enabling passwords, installing anti-virus software, and disabling wireless connections until needed. Mobile users will also have to become more aware of smishing and vishing scams and the risk of downloaded applications. Although the risks to individual users are significant, the risks to enterprises can be devastating and must be controlled with a strong set of technology, policies and practices. As the number of workers using mobile devices swells to over one billion, there is an increasing need to implement Mobile Device Management (MDM). Mobile devices provide a gateway into the enterprise which must be guarded.

Mobile devices bring a tremendous amount of functionality and freedom to its users. However, as with all technology advances, misuse of that functionality and freedom soon follows. Together with the growing use of mobile devices, there must be an increasing awareness of security risks and the steps needed to mitigate those risks.

REFERENCES

Ananda, A. A. (2008). *M-banking: Security aspects.* Retrieved on December 27, 2010, from http://www.strathmore.edu/pdf/m-pesa.pdf

Ante, S. (2010). Dark side arises for phone apps. Retrieved on December 22, 2010, from http://online.wsj.com/article/SB10001424052748703340904575284532175834088.html

Best Security Tips. (2007). *Practical WiFi hacking.* Retrieved on December 23, 2010, from http://www.bestsecuritytips.com/news+article.storyid+192.htm

Bradley, T. (2010). *2011 security predictions: Web attacks, social engineering and mobile nightmare.* Retrieved on December 23, 2010, from http://www.itbusiness.ca/it/client/en/home/News.asp?id=60418

Brenner, B. (2010). *Mobile phone security dos and don'ts.* Retrieved on December 19, 2010, from http://www.csoonline.com/article/596163/mobile-phone-security-dos-and-don-ts

Brenner, B. (2010). *Mobile phone security dos and don'ts.* Retrieved on December 19, 2010, from http://www.csoonline.com/article/596163/mobile-phone-security-dos-and-don-ts

Chen, T. M. (2008). *Malicious software in mobile devices.* Retrieved October 3, 2010, from http://lyle.smu.edu/~tchen/papers/mobile-malware.pdf

Couture, E. (2010). *Mobile security.* Retrieved on December 22, 2010, from http://www.sans.org/reading_room/whitepapers/incident/wireless-mobile-security_33548

Culver, D. (2010). *Smartpones: The new hacker frontier.* Retrieved on December 23, 2010, from http://www.lightreading.com/document.asp?doc_id=196519

Fendelman, A. (2010). *Mobile security: Can you trust your bank account to your phone's mobile web?* Retrieved on December 19, 2010, from http://cellphones.about.com/od/mobilewebtips/a/mobilesecurity.htm

Fendleman, A. (2010). Cell phone glossary: What is GSM vs. EDGE vs. CDMA vs. TDMA? Retrieved on December 27, 2010, from http://cellphones.about.com/od/phoneglossary/tp/gsmcdmatdma.htm

Fendleman, A. (2010). *What is edge?* Retrieved on December 27, 2010, from http://cellphones.about.com/od/phoneglossary/g/edge.htm

Gadaix, E. (2001). *GSM and 3G security.* Retrieved on December 27, 2010, from ww.blackhat.com/presentations/bh-asia-01/gadiax.ppt.

Good Technology. (2009). *Mobile device security: Security the handheld, securing the enterprise.* Retrieved on December 24, 2010, from http://www.good.com/media/pdf/enterprise/mobile_device_security_wp.pdf

Gostev, A. (2010). *Mobile malware evolution: An overview, part 3.* Retrieved on December 22, 2010, from http://www.securelist.com/en/analysis/204792080/Mobile_Malware_Evolution_An_Overview_Part_3

Hendry, M. (1985). *Introduction to CDMA.* Retrieved on December 27, 2010, from http://www.bee.net/mhendry/vrml/library/cdma/cdma.htm

Higgins, K. (2010). *Smartphone weather app builds a mobile botnet.* Retrieved December 29, 2010, from http://www.darkreading.com/insider-threat/167801100/security/application-security/223200001/index.html

Hypponen, M. (2006). Malware goes mobile. *Scientific American*, (November): 2006.

Jain, A. A. (2010). *Mobile worms and viruses.* Retrieved on October 3, 2010, from http://www.it.iitb.ac.in/~JEEVAN/COURSES/SEM/Mobile_Viruses_AND_Worms_Report_IT653.pdf

Jansen, W. A. (2008). *Guidelines on cell phone and PDA security. National Institute of Standards and Technology.* US Department of Commerce.

Joseph, A. (2006). *Mobile device management – Brave new horizon or basic plumbing?* Retrieved on December 27, 2010, from http://www.device-management.org/content/view/20754/152/.

Kardos, J. (2010). *Mobile malware: How safe is your smartphone?* Retrieved on December 22, 2010, from http://bostinnovation.com/2010/04/29/mobile-malware-how-safe-is-your-smartphone/

Lai, E. (2010). *Forrester's top 20 mobile device management best practices for enterprises.* Retrieved on December 24, 2010, from http://www.zdnet.com/blog/sybase/forresters-top-20-mobile-device-management-best-practices-for-enterprises/743

Lobel, M. (2010). *Is it time to ban mobile devices?* Retrieved on December 24, 2010, from http://www.cnbc.com/id/38581570/Is_It_Time_to_Ban_Mobile_Devices

Mac Security Blog. (2009). *Intego security memo: Hacker tool copies personal information from iPhones.* Retrieved on December 23, 2010, from http://blog.intego.com/2009/11/11/intego-security-memo-hacker-tool-copies-personal-info-from-iphones/

Mahaffey, K. A. (2010). *App Attach: Surviving the explosive growth of mobile apps.* Retrieved December 29, 2010, from e https://media.blackhat.com/bh-us-10/presentations/Mahaffey_Hering/Blackhat-USA-2010-Mahaffey-Hering-Lookout-App-Genome-slides.pdf

McGlasson, L. (2010). *Fraudsters take aim at mobile banking.* Retrieved on December 23, 2010, from http://www.bankinfosecurity.com/articles.php?art_id=2085

Montalbano, E. (2010). *FBI warns of mobile cyber threats.* Retrieved on December 19, 2010 from http://www.informationweek.com/news/government/security/showArticle.jhtml?articleID=228400096&itc=ref-true

Nortel. (2005). *CDMA end-to-end security.* Retrieved on December 27, 2010, from http://www.nortel.com/solutions/wireless/collateral/nn_107760.09-15-04.pdf

Patrizio, A. (2008). *First iPhone malware found.* Retrieved on October 4, 2010, from http://www.internetnews.com/security/article.php/3721016/First-iPhone-Malware-Found.htm

Schwartz, M. (2010). *New mobile security threat: Fingerprint oil.* Retrieved on December 22, 2010, from http://www.darkreading.com/security/vulnerabilities/226700046/index.html

Security Watch. (2009). *Experts: More malware socializing in 2010.* Retrieved on December 23, 2010, from http://securitywatch.eweek.com/social_engineering/experts_more_malware_socializing_in_2010.html

Siciliano, R. (2010). *Mobile phone security under attack.* Retrieved on December 19, 2010, from http://blogs.mcafee.com/consumer/identity-theft/mobile-phone-security-under-attack

Stewart, J. (2008). *CISSP study guide.* John Wiley and Sons.

Symantec. (2008). *Internet security report.* Retrieved on December 22, 2010, from http://eval.symantec.com/mktginfo/enterprise/white_papers/b-whitepaper_exec_summary_internet_security_threat_report_xiii_04-2008.en-us.pdf

Torrani, M. a. (2008). *Solutions to GSM security weaknesses.* Retrieved on December 27, 2010, from http://www.slideshare.net/Garry54/solutions-to-the-gsm-security-weaknesses

Trend Micro. (2007). *Enterprise mobile security.* Retrieved on December 22, 2010, from http://emea.trendmicro.com/imperia/md/content/us/pdf/products/enterprise/mobilesecurity/wp03_tmms_071212us.pdf

Wells. (2010). *Lookout for Android issues.* Retrieved October 3, 2010, from http://www.androidtapp.com/lookout-for-android-issues-security-update-for-android-sms-trojan/

Winthrop, P. (2010). *Are we transitioning from mobile device management to mobile security?* Retrieved on December 24, 2010, from http://theemf.org/2010/11/17/are-we-transitioning-from-mobile-device-management-to-mobile-security/

Wise Geek. (2010). *What is a SIM card?* Retrieved on December 27, 2010, from http://www.wisegeek.com/what-is-a-sim-card.htm

Wise Geek. (2010). *What is mobile spam?* Retrieved on December 23, 2010, from http://www.wisegeek.com/what-is-mobile-phone-spam.htm

World, G. S. M. (2009). *Market data summary.* Retrieved on December 27, 2010, from http://www.gsmworld.com/newsroom/market-data/market_data_summary.htm

Compilation of References

ABI. (2009). *Dramatic growth for AR via smartphones*. Retrieved September 1, 2010, from http://www.abire-search.com/press/1516-ABI+Research+Anticipates+%93Dramatic+Growth%94+for+Augmented+Reality+via+Smartphones

Ahonen, T. (2010). *An inconceivable truth: MMS is a global success at 30B dollars*. Retrieved on December 16, 2010 from http://communities-dominate.blogs.com/brands/2010/06/an-inconceivable-truth-mms-is-a-global-success-at-30b-dollars.html

Ahonen, T. (2010). An inconceivable truth: MMS is a global success at 30B dollars. Retrieved on December 16, 2010, from http://communities-dominate.blogs.com/brands/2010/06/an-inconceivable-truth-mms-is-a-global-success-at-30b-dollars.html

Aizenman, N. C. (2010, July 20). Insurers tout disease management programs, but critics are wary. *The Washington Post.*

Alisi, T. M., Bimbo, A. D., & Valli, A. (2005, July). Natural interfaces to enhance visitors' experiences. *IEEE MultiMedia, 12*(3), 80–85. doi:10.1109/MMUL.2005.52

Allan, R. (2006). A brief history of telemedicine. *Electronic Design.* (July 2006).

Ally, M. (2009). *Mobile learning transforming the delivery of education and training*. Edmonton, Canada: AU Press.

Ambient Insights. (2008). *Marketing report.* The US market for eight learning technology products and services: 2008-2013 forecast and analysis.

Analytics, H. I. M. S. S. (2010). *Healthcare industry continues to overlook critical gaps in data security, according to new bi-annual report*. Retrieved on January 23, 2011, from http://www.himssanalytics.org/general/pr_20100421.asp

Ananda, A. A. (2008). *M-banking: Security aspects*. Retrieved on December 27, 2010, from http://www.strathmore.edu/pdf/m-pesa.pdf

Anonymous. (2009, June 13). *Mobile phones, Facebook, Youtube cut in Iran*. Retrieved from http://www.google.com/hostednews/afp/article/ALeqM5jSPlmVgh-SfeE-O9WhpOVG6Slnu0w

Ante, S. (2010). Dark side arises for phone apps. Retrieved on December 22, 2010, from http://online.wsj.com/article/SB10001424052748703340904575284532175834088.html

Army, U. S. (2010). *Top five apps for the army winners recognized at LandWarNet Conference*. Retrieved on January 4, 2011 from http://ciog6.army.mil/AppsfortheArmyChallengeBuilds53Appsin75D/tabid/67/Default.aspx

Ascari, A., & Bakshi, A. (2010). *McKinsey & Company - Global mobile healthcare opportunity*. Retrieved from August 23, 2010, from http://www.mckinsey.it/storage/first/uploadfile/attach/141765/file/global_mobile_health-care_opportunity.pdf

Association, G. S. M. (2003). *Location based services*. Retrieved on January 4, 2011 from http://www.gsmworld.com/documents/se23.pdf

Azuma, R. (1997). A survey of augmented reality. *Presence (Cambridge, Mass.)*, *6*(4), 355–385.

Baldwin, R. C. (2004). *Enabling an e-maintenance infrastructure*. Retrieved from www.mt-online.com

Ballantyne, N. (2010). Mobile learning: What is it and what's it got to do with me? *IATEFL Call Review, Spring/ Summer,* 8-11.

Barritt, K. (2010). *Wireless medical devices: Navigating government regulation in the new digital age*. BNA's Medical Devices, Law, & Industry Report.

Beaudoin, J. (2011). *Healthcare's increasing presence at CES harbinger of things to come*. Retrieved on January 9, 2011, from http://www.healthcareitnews.com/blog/healthcares-increasing-presence-ces-harbinger-things-come

Becker, M. A. (2010). *Mobile marketing for dummies*. Wiley Publishing, Inc.

Beier, K. P. (2004). *Virtual reality: A short introduction*. Retrieved September 2, 2010, from http://www-vrl.umich.edu/intro/

Bellina, L. (2010). *Mobile diagnosis*. Telecare Soapbox. Retrieved October 15, 2010, from http://www.telecareaware.com/index.php/telecare-soapbox-mobile-diagnosis.html

Benedict, K. (2008). Mobile applications for fighting crime, reporting potholes and birdwatching. Retrieved on January 4, 2011, from http://kevinbenedict.sys-con.com/node/1215961/mobile

Bennett, S. (2010). *GEOTREE: A tree inventory web application for citizen driven urban forestry asset management*. Big Rapids, MI: Master's of Information Systems Management Capstone Project, Ferris State University.

Bergman, C. (2010). *AR will be big in 5 years, says study*. Retrieved September 1, 2010, from http://www.lostremote.com/2010/01/06/augmented-reality-will-be-big-in-5-years-says-study/

Berry, D. McNeil & Parker. (2010). *Analysis of HITECH provisions in the American Recovery & Reinvestment Act*. Retrieved on January 23, 2011 from www.bdmp.com

Best Security Tips. (2007). *Practical WiFi hacking*. Retrieved on December 23, 2010, from http://www.bestsecuritytips.com/news+article.storyid+192.htm

Bhavnani, A., Won-Wai Chiu, R., Janakiram, S., & Silarszky, P. (2008). *The role of mobile phones in sustainable rural poverty reduction*. ICT Policy Division, Global Information and Communications Department (GICT). Washington, DC: World Bank. Retrieved from http://siteresources.worldbank.org/extinformationandcommunicationandtechnologies/Resources/

Bhutkar, G. K. (2009). *Major challenges with mobile healthcare applications*. Retrieved on December 28, 2010, from http://www.bjhcim.co.uk/features/2009/909004.htm

Bimber, O., & Raskar, R. (July, 2005). *Spatial augmented reality: Merging real and virtual worlds*. Wellesley, MA: A K Peters/CRC Press.

Boonstra, C., van der Klein, R., & Lens-Fitzgerald, M. (2010). *Contextual services in mobile*. Retrieved September 23, 2010, from http://www.slideshare.net/Thinkmobile/contextual-services-in-mobile-presentation

Booth, D., et al. (2004, February 11). *Web services architecture*. W3C working group note. Retrieved from http://www.w3.org/TR/2004/NOTE-ws-arch-20040211

Booz & Company. (2010). *App-Downloads generieren 2013 Umsatzvolumen von 17 Mill. Euro weltweit*. Retrieved September 7, 2010, from http://www.booz.com/de/home/Presse/Pressemitteilungen/pressemitteilung-detail/48203889

Bowser, M. (2009). *IVR - A marketer's dream*. Retrieved on December 18, 2010, from http://www.mobilemarketing-magazine.co.uk/content/ivr-marketers-dream-no-really

Bradley, T. (2010). *2011 security predictions: Web attacks, social engineering and mobile nightmare*. Retrieved on December 23, 2010, from http://www.itbusiness.ca/it/client/en/home/News.asp?id=60418

Brenner, B. (2010). *Mobile phone security dos and don'ts*. Retrieved on December 19, 2010, from http://www.csoonline.com/article/596163/mobile-phone-security-dos-and-don-ts

Brown, L. (2009). Using mobile learning to teach reading to ninth-grade students. *Journal for Computing Teachers, Fall*, 1-17. Retrieved September 1, 2010, from http://www.iste.org/jct

Bryson, S. (1996). Virtual reality in scientific visualization. *Communications of the ACM, 39*(5), 62–71. doi:10.1145/229459.229467

Bukheit, C. (2010, August 16). *How state political candidates are using mobile texting* [Web log message]. Retrieved from http://www.nonprofitmediaworks.com/2010/08/16/how-state-political-candidates-are-using-mobile-texting/

Bureau of Economics. (April, 2010). *Credit Card Accountability Responsibility And Disclosure Act Of 2009. Report on emergency technology for use with ATMs.* Federal Trade Commission.

Butcher, D. (2009, May 08). *Mobile software reinventing healthcare in developing world* [Web log message]. Retrieved from http://www.mobilemarketer.com/cms/news/software-technology/3204.html

Butcher, D. (2010). *7 key trends mobil marketers need to know.* Retrieved on December 15, 2010 from http://www.mobilemarketer.com/cms/news/research/7342.html

Butcher, D. (2010). *Snickers ties first branded mobile game to in-store marketing.* Retrieved on December 17, 2010 from http://www.mobilemarketer.com/cms/news/gaming/5468.html

Campbell, S. W. (2006). Perceptions of mobile phones in college classrooms: Ringing, cheating, and classroom policies. *Communication Education, 55*(3), 280–294. doi:10.1080/03634520600748573

Campbell, M. (2005). The impact of the mobile phone on young people. *Proceedings of the Social Change in the 21st Century Conference.* Brisbane, Australia: Queensland University of Technology-Centre for Social Change.

Cassavoy, L. (2007). *In pictures: A history of cell phones.* Retrieved on December 16, 2010, from http://www.pcworld.com/article/131450/in_pictures_a_history_of_cell_phones.html

Cassela, D. (2009). *What is augmented reality?* Retrieved on December 16, 2010, from http://www.digitaltrends.com/mobile/what-is-augmented-reality-iphone-apps-games-flash-yelp-android-ar-software-and-more/2/

CBS News. (2010). *Number of cell phones worldwide hits 4.6B.* Retrieved on December 15, 2010, from http://www.cbsnews.com/stories/2010/02/15/business/main6209772.shtml

Cebeci, Z., & Tekdal, M. (2006). Using podcasts as audio learning objects. *Interdisciplinary Journal of Knowledge and Learning Objects, 2*, 47–57.

Cellphone Advertising. (2007). *Product placement in mobile phone advertisement.* Retrieved on December 17, 2010 from http://www.cellphone-advertising.com/product-placement-in-mobile-phone-advertising/

Channel Insider. (2010). *Salesforce chatter social networking goes mobile.* Retrieved on December 16, 2010, from http://www.channelinsider.com/c/a/Cloud-Computing/Salesforce-Chatter-Social-Networking-Goes-Mobile-443229/

Chen, B. X. (2009). *How the iPhone could reboot education.* Retrieved September 1, 2010, from http://www.wired.com/gadgetlab/2009/12/iphone-university-abilene/

Chen, T. M. (2008). *Malicious software in mobile devices.* Retrieved October 3, 2010, from http://lyle.smu.edu/~tchen/papers/mobile-malware.pdf

Chong, C. Y., & Kumar, S. P. (2003). Sensor networks: Evolution, opportunities, and challenges. *Proceedings of the IEEE, 91*, 1247–1256. doi:10.1109/JPROC.2003.814918

Choudhary, V. (2007, January). Software as a service: Implications for investment in software development. *International Conference on System Sciences*, (p. 209)

Clark, R. (2007). *Intelligence analysis: A target-centric approach.* Washington, DC: CQ Press.

Clark, M., & Goodwin, N. (2010). *Sustaining innovation in telehealth and telecare.* Whole Systems Demonstrator Action Network Briefing Paper. The Kings Fund.

Collins, J. (May, 2006). *RFID's impact at Wal-Mart greater than expected.* Retrieved on November 28, 2010, from http://www.rfidjournal.com/article/print/2314

Commins, J. (2010). *Physicians: Mobile devices expedite decision making.* Retrieved on December 29, 2010, from http://www.healthleadersmedia.com/content/TEC-256203/Physicians-Mobile-Devices-Expedite-Decision-Making##

Consulting, F. (2009). *Managing and securing mobile healthcare data and devices.* A custom Tech Adoption Profile Commissioned by MaaS360 by Fiberlink. Retrieved on January 21, 2011 from http://www.informationweek.com/whitepaper/Mobility/Mobile-Business/managing-and-securing-mobile-healthcare-data-and--wp1269371115712

Conway, R. (2009). *The European mobile manifesto: How mobile will help achieve key European Union objectives. November 2009.* GSMA.

Conway-Smith, E. (2010). *Teaching with cell phones.* Retrieved September 1, 2010 from http://www.globalpost.com/dispatch/education/100720/south-africa-teaching-cell- phones?page=0,0

Copley, J. (2007). Audio and video podcasts of lectures for campus-based students: production and evaluation of student use. *Innovations in Education and Teaching International, 44*(4), 387–399. doi:10.1080/14703290701602805

Corbeil, J. R., & Valdes-Corbeil, M. E. (2007). Are you ready for mobile learning? *EDUCAUSE Quarterly, 30*(2). Retrieved September 1, 2010, from http://www.learning-centric.net/mobile.cc/relatedinfo.pdf.

Couture, E. (2010). *Mobile security.* Retrieved on December 22, 2010, from http://www.sans.org/reading_room/whitepapers/incident/wireless-mobile-security_33548

Cox, A. (2010). *Mobile healthcare opportunities: Monitoring, applications and mHealth strategies 2010 – 2014.* Juniper Research.

Cox, J. (2010, March 23). *Can the iphone save higher education?* [Web log message]. Retrieved from http://www.networkworld.com/news/2010/032310-iphone-higher-education.html

Credant. (2010). *Tips for securing healthcare data on mobile devices.* Retrieved on January 11, 2011, from http://www.executivehm.com/article/Tips-for-securing-healthcare-data-on-mobile-devices/

Crespo-Marquez, A., & Gupta, J. (2006). Contemporary maintenance management: Process, framework and supporting pillars. *Omega, 34*(3), 313–326. doi:10.1016/j.omega.2004.11.003

Crisp, M. S. (2010). *Modernizing school communication systems: Using text messaging to improve student academic performance.* Unpublished doctoral dissertation, Oregon State University, Oregon.

CTIA. (2011). *Basics of CSC FAQs.* Retrieved on December 16, 2010 from http://www.ctia.org/business_resources/short_code/index.cfm/AID/10341

Culver, D. (2010). *Smartpones: The new hacker frontier.* Retrieved on December 23, 2010, from http://www.lightreading.com/document.asp?doc_id=196519

Curthoys, P. C., Phillips, J. P., Aguilera, R. A., & Ochs, S. O. (2011, January). Apple's next big thing. *Mac Life, 5*(1), 22–31.

DARPA. (2006). *Spoken language communication and translation system for tactical use* (TRANSTAC). Retrieved January 17, 2011, from http://www.darpa.mil/i2o/solicit/baa/BAA-06-21_PIP.pdf

DARPA. (2010). *Mobile apps for the military.* Retrieved on January 4, 2011, from https://www.fbo.gov/index?s=opportunity&mode=form&id=6f438e64dcf6bd132987d9554442b851&tab=core&_cview=0

Davis, G. (2009). *Teaching mathematics with technology. Computational media: The universal acid of mathematics teaching,* vol. 4. Retrieved Sept 28, 2010 from http://republicofmath.wordpress.com/2009/12/29/computational-media-the-universal-acid-of-mathematics-teaching-4/

De Lorenzo, R. (2009). *The powerful combination of mobile devices and learning apps.* Retrieved on June 6, 2010, from http://themobilelearner.wordpress.com/2009/11/15/combining-mobile-devices-with-apps/

De Lorenzo, R. (2010). *Five difficulties in mobile learning implementation.* Retrieved on June 6, 2010, from http://themobilelearner.wordpress.com/2010/06/11/on-five-difficulties-in-mobile-learning-implementation/

De Waard, I. (2010). *Free mobile service getting to grips with bulling in schools: Bullyproof.* Retrieved September 1, 2010 from http://ignatiawebs.blogspot.com/2010/02/free-mobile-sms-service- getting-to.html

Dede, C. (2009). Immersive interfaces for engagement and learning. *Science, 323*(5910), 66–69. doi:10.1126/science.1167311

della Cava, M. (2010). *It's an app world, and it could swallow all computing.* Retrieved on February 16, 2011, from http://www.usatoday.com/tech/products/2010-03-31-1Aappworld31_CV_N.htm

Denison, D. (2009, October 14). Using cellphones to change the world. MIT project leads to programs that help health workers, farmers in developing countries. *Boston Globe.* Retrieved November 12, 2010, from http://www.boston.com/business/technology/articles/2009/10/14/mit_program_looks_at_ways_to_change_the_world_using_cellphones/

Department of Defense. (2010). *About the Department of Defense: Mission.* Retrieved on January 4, 2011, from http://www.defense.gov/about

Deubel, P. (2010). *Technology integration: Essential questions.* Retrieved October 1, 2010, from http://www.ct4me.net/technology_integr.htm

Deyo, R. (2004). *Gaps, tensions and conflicts in the FDA approval process: Medical devices.* Retrieved on January 17, 2011, from http://www.medscape.com/viewarticle/474285_4

Dini, P., De Wilde, P., Rowe, J., Petrou, M., & Heistracher, T. (2005). *The digital ecosystem research vision: 2010 and beyond.* Position paper FP7 Workshop, 2005. Retrieved from http://www.digital-ecosystems.org/events/2005.05/de_position_paper_vf.pdf

Dolan, B. (2010). Interview: The iPhone medical app denied 510(k). *mobihealthnews.* Retrieved August 11, 2010, from http://mobihealthnews.com/6932/interview-the-iphone-medical-app-denied-510k/2/

Dolan, B. (2010). *Investors pumped $233M into mobile health in 2010.* Retrieved February 2, 2011, from http://mobihealthnews.com/10087/investors-pumped-233-million-into-mobile-health-in-2010

Durrell, J. (2010). *Mobile game marketing.* Retrieved on December 16, 2010 from http://mmaglobal.com/articles/mobile-game-marketing-greystripe

Edwards, C. (2009). *Resilient nation.* London, UK: Demos.

Electronic Frontier Foundation. (n.d.). *National security letters.* Retrieved January 27, 2011, from https://ssd.eff.org/foreign/nsl

eMarketer. (2009). *Two-thirds of kids and teens now mobile.* Retrieved on December 19, 2010 froom http://www3.emarketer.com/Article.aspx?R=1007780

Entner, R. (2010). *Smartphones to overtake feature phones in U.S. by 2011.* Retrieved on December 19, 2010 from http://blog.nielsen.com/nielsenwire/consumer/smartphones-to-overtake-feature-phones-in-u-s-by-2011/

EPN. (2010). *Website.* Retrieved October 1, 2010, from http://epnweb.org/

Esfandiari, G. (2010). The Twitter devolution. *Foreign Policy.* Retrieved from http://www.foreignpolicy.com/articles/2010/06/07/the_twitter_revolution_that_wasnt

Evans, C. (2008). The effectiveness of m-learning in the form of podcast revision lectures in higher education. *Computers & Education, 50*(2), 491–498. doi:10.1016/j.compedu.2007.09.016

Evernote. (2011). *Learn more.* Retrieved from http://www.evernote.com/about/learn_more/

Execution of Saddam Hussein. (2010). *Wikipedia.* Retrieved September 23, 2010, from http://en.wikipedia.org/wiki/Execution_of_Saddam_Hussein#Execution_proceedings

Fayad, G. (2010). mHealth – Just what the doctor ordered. *Comm.* Retrieved from September 15, 2010 from http://comm.ae/2010/10/10/mhealth-just-what-the-doctor-ordered/

Federal Trade Commission. (2004). Sham site is a scam: There is no national do not e-mail registry. Retrieved on December 18, 2010, from http://www.ftc.gov/opa/2004/02/spamcam.shtm

Federal Trade Commission. (2010). *FTC testifies on do not track legislation.* Retrieved on December 18, 2010, from http://www.ftc.gov/opa/2010/12/dnttestimony.shtm

Fendelman, A. (2010). *Mobile security: Can you trust your bank account to your phone's mobile web?* Retrieved on December 19, 2010, from http://cellphones.about.com/od/mobilewebtips/a/mobilesecurity.htm

Fendleman, A. (2010). Cell phone glossary: What is GSM vs. EDGE vs. CDMA vs. TDMA? Retrieved on December 27, 2010, from http://cellphones.about.com/od/phoneglossary/tp/gsmcdmatdma.htm

Fendleman, A. (2010). *What is edge?* Retrieved on December 27, 2010, from http://cellphones.about.com/od/phoneglossary/g/edge.htm

Finley, K. (2011). *DARPA and Raytheon building new ad-hoc mobile network for the military.* Retrieved January 17, 2011, from http://www.readwriteweb.com/enterprise/2011/01/darpa-and-raytheon-building-ne.php

FlexiSpy. (2009). *How do I use SpyCall / remote listening?* Retrieved January 28, 2011, from http://support.flexispy.com/index.php?_m=knowledgebase&_a=viewarticle&kbarticleid=6

Foresman, C. (2009). *Apple responsible for 99.4% of mobile app sales in 2009 (Updated).* Retrieved on February 16, 2011, from http://arstechnica.com/apple/news/2010/01/apple-responsible-for-994-of-mobile-app-sales-in-2009.ars

Fout, T. (2001). *Universal plug and play in Windows XP.* Retrieved from http://technet.microsoft.com/en-us/library/bb457049.aspx

Free Trade Commission. (2004). *Sham site is a scam: There is no national do not e-mail registry.* Retrieved on December 18, 2010 from http://www.ftc.gov/opa/2004/02/spamcam.shtm

Free Trade Commission. (2010). *FTC testifies on do not track legislation.* Retrieved on December 18, 2010 from http://www.ftc.gov/opa/2010/12/dnttestimony.shtm

Fung, L. (2010). Marketing mobile games. Retrieved on December 17, 2010, from http://www.selfgrowth.com/articles/Marketing_Mobile_Games.html

Gadaix, E. (2001). *GSM and 3G security.* Retrieved on December 27, 2010, from ww.blackhat.com/presentations/bh-asia-01/gadiax.ppt.

Gaerdenfors, D., Haliburton, J., & Stark, P. (2010). *AR for the masses.* Retrieved September 1, 2010, from http://www.perey.com/MobileARSummit/TAT-Augmented-reality-for-the-masses.pdf

Ganapathy, K. (2010, July). *An mHealth perspective from India.* Paper presented at The World Congress 2nd Annual Leadership Summit on Mobile Health, Boston, MA.

Garten, M. (2010). *United Nations compendium of mHealth projects.*

Gartner. (2008). *Gartner identifies top ten disruptive technologies for 2008 to 2012.* Retrieved September 1, 2010, from http://www.ehomeupgrade.com/2008/05/28/gartner-identifies-top-ten-disruptive-technologies-for-2008-to-2012/

Gartner. (2009). *Gartner hype cycle 2009.* Retrieved September 1, 2010, from http://www.gartner.com/it/page.jsp?id=1124212

Gartner.com. (November, 2010). *Gartner outlines 10 mobile technologies to watch in 2010 and 2011.* Gartner Newsroom. Retrieved on November 7, 2010, from http://www.gartner.com/it/page.jsp?id=1328113

Gibb, K. (2010). *Military testing Nexus Ones for real-time translation in Afghanistan.* Retrieved on January 16, 2011, from http://www.androidcentral.com/military-testing-nexus-ones-real-time-translation-afganistan

Gogolin, G. (2010). The digital crime tsunami. [). Elsevier.]. *Digital Investigation, 7*(1-2), 3–8. doi:10.1016/j.diin.2010.07.001

Gogolin, G. & Jones, J. (2010). Law enforcement's ability to deal with digital crime and the implications for business. *Information Security Journal: A Global Perspective, 19*(3), 109-117. Doi:10.1080/19393555.2010.483931 Top of Form

Good Technology. (2009). *Mobile device security: Security the handheld, securing the enterprise.* Retrieved on December 24, 2010, from http://www.good.com/media/pdf/enterprise/mobile_device_security_wp.pdf

Google. (2010). *Google Translate.* Retrieved on January 17, 2011 from http://www.appbrain.com/app/com.google.android.apps.translate

Gostev, A. (2010). *Mobile malware evolution: An overview, part 3.* Retrieved on December 22, 2010, from http://www.securelist.com/en/analysis/204792080/Mobile_Malware_Evolution_An_Overview_Part_3

GPS. (2011). *Amazon search*. Retrieved from http://www.amazon.com/s/ref=nb_sb_noss?url=search-alias%3Daps&field-keywords=GPS&x=0&y=0

Grayson, M. S. (2011). *Building the mobile internet*. Indianoplis, IN: Cisco Press.

Gribbins, M. (2007). The perceived usefulness of podcasting in higher education: A survey of students' attitudes and intention to use. *Proceedings of the Second Midwest United States Association for Information Systems*, Springfield, IL May 18-19.

Griffith, A. (2010). *Revolutionizing healthcare delivery with mobile health (mHealth)*. Retrieved on December 29, 2010, from http://www.suite101.com/content/revolutionizing-healthcare-delivery-with-mobile-health-mhealth-a243935

Grossman, L. (2009, June 17). Iran protests: Twitter, the medium of the movement. *Time*. Retrieved from http://www.time.com/time/world/article/0,8599,1905125,00.html

GSMA. (2010). *European framework for safer mobile use by younger teenagers and children*. Retrieved on December 18, 2010 from http://www.eubusiness.com/topics/telecoms/gsma.10-06-09/

Gumpert, G. A. (2007). Mobile communication in the twenty-first century or everybody, everywhere, at any time. In Kleinman, S. (Ed.), *Displacing space* (pp. 7–20). New York, NY: Peter Lang Publishing.

Guy, R. (2009). *The evolution of mobile teaching and learning*. Santa Rosa, CA: Informing Science.

Hager, F. (2006). Mobile communications. Retrieved on December 18, 2010 from http://www.fredhager.com/index.asp?CategoryID=67&SubCategoryID=587&ContentID=1047

Han, T., & Yang, B.-S. (2006). Development of an e-maintenance system integrating advanced techniques. [Elsevier.]. *Computers in Industry*, *57*, 569–580. doi:10.1016/j.compind.2006.02.009

Harfoush, R. (2009). *Yes we did: An inside look at how social media built the Obama brand*. Berkeley, CA: New Riders.

Harvard Kennedy School. (2003). *311 system*. Retrieved on January 4, 2011 from http://www.innovations.harvard.edu/awards.html?id=3670

Hau, S. (2001). *Moving to mobile - Five key strategies for managing handheld devices in healthcare settings - Technology information*. Retrieved on December 20, 2011, from http://findarticles.com/p/articles/mi_m0DUD/is_7_22/ai_76548959/.

Hausladen, I., & Bechheim, C. (2004). E-maintenance platform as a basis for business process integration. *Proceedings of INDIN04, Second IEEE International Conference on Industrial Informatics*, (pp. 46–51). Berlin, Germany. i-Maintenance system. (2010). Retrieved September 30, 2010, from http://www.track-it-systems.com

Havenstein, M. (2008). *LinkedIn social networking goes mobile*. Retrieved on December 16, 2010 from http://www.cio.com/article/187401/LinkedIn_Social_Networking_Goes_Mobile

Heise News. (2010). *Telekom will Diensteanbieter zur Kasse bitten*. Retrieved September 7, 2010, from http://www.heise.de/newsticker/meldung/Telekom-will-Diensteanbieter-zur-Kasse-bitten-1042726.html

Hendry, M. (1985). *Introduction to CDMA*. Retrieved on December 27, 2010, from http://www.bee.net/mhendry/vrml/library/cdma/cdma.htm

Heuer, R. (1999). *Psychology of intelligence analysis*. Center for the Study of Intelligence, Central Intelligence Agency.

Hew, K. F., & Brush, T. (2007). Integrating technology into K-12 teaching and learning: Current knowledge gaps and recommendations for future research. *Educational Technology Research and Development*, *55*, 223–252. doi:10.1007/s11423-006-9022-5

Higgins, K. (2010). *Smartphone weather app builds a mobile botnet*. Retrieved December 29, 2010, from http://www.darkreading.com/insider-threat/167801100/security/application-security/223200001/index.html

Hockly, N. (2010). Mobile learning: What is it and what's it got to do with me? In *IATEFL*. *California Law Review*, (Spring/Summer): 5–8.

Hodge, N. (2010). *A combat zone iPhone? Soldiers have an app for that.* Retrieved on January 4, 2011 from http://www.wired.com/dangerroom/2010/03/a-combat-zone-iphone-soldiers-have-an-app-for-that/?utm_source=feedburner&utm_medium=feed&utm_campaign=Feed:+wired/index+(Wired:+Index+3+(Top+Stories+2))

Hoffman, M. (2009). *Fixes on the way for nonsecure UAV links.* Retrieved January 17, 2011, from http://www.navytimes.com/news/2009/12/airforce_uav_hack_121809w/

Hollerer, T. A. (2004). Mobile augmented reality . In Karimi, H., & Hammad, A. (Eds.), *Telegeoinformatics: Location-based computing and services.* Taylor and Francis Books Ltd.

Hoover, N. (2009). An Android app for the military. Retrieved on January 16, 2011 from http://www.informationweek.com/news/government/mobile/showArticle.jhtml?articleID=221200035

Huang, Y.-M., Jeng, Y.-L., & Huang, T.-C. (2009). An educational mobile blogging system for supporting collaborative learning. *Journal of Educational Technology & Society, 12*(2), 163–175.

Hypponen, M. (2006). Malware goes mobile. *Scientific American,* (November): 2006.

IBM. (2010). *Augmented reality projects at IBM.* Retrieved September 7, 2010, from http://www.perey.com/Mobile-ARSummit/IBM-Recent-AR-Project-Experiences.pdf

iBooks. (2011). *iPad features.* Retrieved from http://www.apple.com/ipad/features/ibooks.html

Imaging Technology News. (2011). *Mobile PACS application added for smart phone access.* Retrieved on January 17, 2011, from http://www.itnonline.net/node/38592/3/

Intel. (2010). *Intel and Nokia: A new lab for mobile 3D.* Retrieved September 7, 2010, from http://blogs.intel.com/research/2010/08/today_intel_along_with_nokia.php

Interactive Blend. (2010). *QR codes: The future of marketing.* Retrieved on December 16, 2010 from http://interactiveblend.com/blog/interactive/qr-codes/

International Telecommunications Union. (2005). Cellular standards for the third generation: The ITU's IMT-200 family. Retrieved on December 15, 2010, from http://www.itu.int/osg/spu/imt-2000/technology.html#Cellular%20Standards%20for%20the%20Third%20Generation

iPad. (2011). *Apple shop: iPad.* Retrieved from http://store.apple.com/us/browse/home/shop_ipad/family/ipad?mco=OTY2ODA0NQ

Iqbal Quadir. (2010). *Wikipedia.* Retrieved September 24, 2010, from http://en.wikipedia.org/wiki/Iqbal_Quadir

ITU. (2010). *Monitoring the WSIS targets: A mid-term review.* Geneva, Switzerland: International Telecommunications Union.

iTunes U. (2011). *What is iTunes U?* Retrieved from http://www.apple.com/education/itunes-u/what-is.html

Jain, A. A. (2010). *Mobile worms and viruses.* Retrieved on October 3, 2010, from http://www.it.iitb.ac.in/~JEEVAN/COURSES/SEM/Mobile_Viruses_AND_Worms_Report_IT653.pdf

Jansen, W. A. (2008). *Guidelines on cell phone and PDA security.* National Institute of Standards and Technology. US Department of Commerce.

Jobi, P. (2009). Cell phones in the (language) classroom: Recasting the debate. *EDUCAUSE Quarterly Magazine, 32*(4). Retrieved September 1, 2010, from http://www.educause.edu/EDUCAUSE+Quarterly/EDUCAUSEQuarterlyMagazineVolum/CellPhonesintheLanguageClassro/192995

Johansson, F., & Hagman, P. (2009). *MiniBands -A collaborative mobile music concept.* Unpublished Master's Thesis, Umea, Sweden.

Johnson, L., Levine, A., Smith, R., & Stone, S. (2010). *The 2010 horizon report.* Austin, TX: The New Media Consortium. Retrieved September 1, 2010, from http://wp.nmc.org/horizon2010

Joseph, A. (2006). *Mobile device management – Brave new horizon or basic plumbing?* Retrieved on December 27, 2010, from http://www.devicemanagement.org/content/view/20754/152/.

Kardos, J. (2010). *Mobile malware: How safe is your smartphone?* Retrieved on December 22, 2010, from http://bostinnovation.com/2010/04/29/mobile-malware-how-safe-is-your-smartphone/

Kawano, L. (2010). *DPD launches crime reporting app.* Retrieved on January 4, 2011 from http://www.myfoxdfw.com/dpp/news/100410-dpd-launches-crime-reporting-app.

Kay, R., & Mattern, F. (2004). The design space of wireless sensor networks. *IEEE Wireless Communications, 11*(6), 54–56. doi:10.1109/MWC.2004.1368897

Kee, T. (2010). *4 ways that 4G will impact mobile marketing in 2011.* Retrieved on December 16, 2010 from http://econsultancy.com/us/blog/6965-4g-or-not-4g-four-ways-it-will-impact-mobile-marketing-in-2011

Kim, W. (2009). Cloud computing: Today and tomorrow. *Journal of Object Technology, 8*(1), 65–72. doi:10.5381/jot.2009.8.1.c4

Kindle. (2011). *Amazon search.* Retrieved from http://www.amazon.com/dp/B002Y27P3M/?tag=googhydr-20&hvadid=6912512736&ref=pd_sl_98pmgzpmhj_e

King, R. (2009). Augmented reality goes mobile. *Businessweek.* Retrieved on November 22, 2010, from http://www.businessweek.com/technology/content/nov2009/tc2009112_434755.htm

King, R. C., & Buckner, F. (2004). Telemedicine. In ACLM (Ed.), *Legal medicine* - 6th edition, (pp. 424–431). Philadelphia, PA: Mosby, Inc.

Klimas, P. (2010). *City of Grand Rapids deploys MyGRCity311 mobile application for customer service.* Retrieved on January 4, 2011 from http://www.grandrapids.mi.us/download_upload/binary_object_cache/frontpage_311MEDIA%20RELEASE.pdf

Kloss, J. (2010). The role of standards for e-commerce in virtual worlds. In Ciaramitaro, B. (Ed.), *Virtual worlds and e-commerce.* Hershey, PA: IGI Global. doi:10.4018/978-1-61692-808-7.ch015

Koc, M., & Lee, J. (2001). A system framework for next-generation e-maintenance system. *Proceedings of the Second International Symposium on Environmentally Conscious Design and Inverse Manufacturing,* Tokyo, Japan, 2001.

Kolb, L. (2008). *Mobile learning is global: Examples of teachers using cell phones for learning!* Retrieved June 6, 2010, from http://www.cellphonesinlearning.com/

Kossen, J. S. (2001). *When e-learning becomes m-learning.* Retrieved June 6, 2010, from http://www.palmpower-enterprise.com/issuesprint/issue200106/elearning.html

Kozick, Z., & Gettliffe, C. (2010). *Why AR needs a reality check.* Retrieved September 21, 2010, from http://www.perey.com/MobileARSummit/Omniar-Augmented-Reality-Reality-Check.pdf

Krum, C. (2010). *Mobile marketing: Finding your customers no matter where they are.* Indianapolis, IN: Que.

Kukulska-Hulmes, A., & Traxler, J. (2005). *Mobile learning: A handbook for educators and trainers.* London, UK: Routledge.

Kukulska-Hulmes, A., & Pettit, J. (2009). Practitioners as innovators: Emergent practice in personal mobile teaching, learning, work, and leisure. In M. Ally (Ed.), *Mobile learning transforming the delivery of education and training.* Edmonton, Canada: AU Press. Retrieved from http://www.aupress.ca/books/120155/ebook/99Z_Mohamed_Ally_2009-MobileLearning.pdf

Labott, E. (2009, June 16). *State department to Twitter: Keep Iranian tweets coming* [Web log message]. Retrieved from http://ac360.blogs.cnn.com/2009/06/16/state-department-to-twitter-keep-iranian-tweets-coming/

Lai, E. (2010). *Forrester's top 20 mobile device management best practices for enterprises.* Retrieved on December 24, 2010, from http://www.zdnet.com/blog/sybase/forresters-top-20-mobile-device-management-best-practices-for-enterprises/743

Lamb, P. (2010). *Ever wider and deeper: Augmented reality in 2010.* Retrieved August 18, 2010, from http://www.perey.com/MobileARSummit/ARToolworks-Ever-Wider-and-Deeper-AR-in-2010.pdf

Lapeer, R., Rowland, R., & Chen, M. S. (2004). *PC-based volume rendering for medical visualisation and augmented reality based surgical navigation.* Eighth International Conference on Information Visualisation (IV'04).

Layar. (2010). *Website.* Retrieved August 14, 2010, from http://www.layar.com/

Layar. (2010b). *Layar reality browser adds 3D to its platform*. Retrieved August 18, 2010, from http://site.layar.com/company/blog/layar-reality-browser-adds-3d-to-its-platform/

Lechner, M., & Tripp, M. (2010). *ARML - An AR standard*. Retrieved September 23, 2010, from http://www.perey.com/MobileARSummit/Mobilizy-ARML.pdf

Lee, J. (2003). E-manufacturing: Fundamental, tools, and transformation. *Robotics and Computer-integrated Manufacturing*, *19*(6), 501–507. doi:10.1016/S0736-5845(03)00060-7

Lee, J., Ni, J., Djurdjanovi, D., Qiu, H., & Liao, H. (2006). Intelligent prognostics tools and e-maintenance. *Computers in Industry*, *57*, 476–489. doi:10.1016/j.compind.2006.02.014

Lee, J. (2001). A framework for next-generation e-maintenance system. *Proceedings of the Second International Symposium on Environmentally Conscious Design and Inverse Manufacturing*, Tokyo, Japan.

Lee, J., & Ni, J. (2004). *Infotronics-based intelligent maintenance system and its impacts to closed-loop product life cycle systems*. Invited keynote paper for IMS'2004—International Conference on Intelligent Maintenance Systems, Arles, France, 2004.

Lenhart, A. (2010). *Teens, cell phones and texting*. Retrieved on December 19, 2010, from http://pewresearch.org/pubs/1572/teens-cell-phones-text-messages

Lenhart, A., Ling, R., Campbell, S., & Purcell, K. (2010). *Teens and mobile phones*. Informally published manuscript. Washington, DC: Pew Internet and American Life Project, Pew Research Center. Retrieved from http://www.pewinternet.org/topics/Teens.aspx

Lens-Fitzgerald, M. (2010). *Was there movement on the AR hype cycle curve?* Retrieved September 1, 2010, from http://www.perey.com/MobileARSummit/Layar-Was-there-movement-on-the-AR-Hype-Cycle.pdf

Leonardi, M., Catterji, S., et al. (2008). Functioning and disability in ageing population in Europe: What policy for which interventions? *European Papers on the New Welfare, 9.*

Lewis, N. (2010). *Mobile devices to transform healthcare*. Retrieved on December 29, 2010, from http://www.informationweek.com/news/healthcare/mobile-wireless/showArticle.jhtml?articleID=227400122

Leyne, J. (2010, February 11). *How Iran's political battle is fought in cyberspace* [Web log message]. Retrieved from http://news.bbc.co.uk/2/hi/8505645.stm

Li, Y., Chun, L., Nee, A., & Ching, Y. (2005). An agent-based platform for web enabled equipment predictive maintenance. *Proceedings of IAT'05 IEEE/WIC/ACM International Conference on Intelligent Agent Technology*, Compiegne, France.

Limbo. (2008, February 08). *Limbo reports mobile advertising changes voters' attitudes and behaviors* [Web log message]. Retrieved from http://www.limbo.com/presscenter?pr=pr20080204.html

Linkous. (2010). *It's mHealth but will it be a revolution?* Retrieved December 12, 2010, from http://american-telemed.blogspot.com/2010/06/its-mhealth-but-will-it-be-revolution.html

Lobel, M. (2010). *Is it time to ban mobile devices?* Retrieved on December 24, 2010, from http://www.cnbc.com/id/38581570/Is_It_Time_to_Ban_Mobile_Devices

Lombardi. (2010). Mobile technology – Should medical smartphone applications be regulated? *Technology for Doctors*. Retrieved August 20, 2010 from http://www.canhealth.com/tfdnews0065.html

Mac Security Blog. (2009). *Intego security memo: Hacker tool copies personal information from iPhones*. Retrieved on December 23, 2010, from http://blog.intego.com/2009/11/11/intego-security-memo-hacker-tool-copies-personal-info-from-iphones/

MacCallum, K., & Jeffrey, L. (2009). Identifying discriminating variables that determine mobile learning adoption by educators: An initial study. In *Proceedings of ASCILITE 2009: Same Places, Different Spaces*. Retrieved October 1, 2010, from http://www.ascilite.org.au/conferences/auckland09/procs/maccallum.pdf

Maestri, N. (2010, January 16). *US texting raises $11 million for Haiti* [Web log message]. Retrieved from http://in.reuters.com/article/idINIndia-45435720100116

Mahaffey, K. A. (2010). *App Attach: Surviving the explosive growth of mobile apps.* Retrieved December 29, 2010, from e https://media.blackhat.com/bh-us-10/presentations/Mahaffey_Hering/Blackhat-USA-2010-Mahaffey-Hering-Lookout-App-Genome-slides.pdf

Mahoney, D. P. (2000, February). Innovative interfaces. *Computer Graphics World, 23*(2), 39-42, 44.

Management, M. (2010). *iPad is overwhelming tablet of choice in healthcare.* Retrieved on December 29, 2010, from http://www.visagemobile.com/news/news/managing-mobile-devices-news/5964/ipad-is-overwhelming-tablet-of-choice-in-healthcare/

Mankoff, J., Abowd, G. D., & Hudson, S. E. (2000). OOPS: A toolkit supporting mediation techniques for resolving ambiguity in recognition-based interfaces. *Computers & Graphics, 24*(6), 819–834. doi:10.1016/S0097-8493(00)00085-6

Marimon, D., et al. (2010). *Mobile visual recognition, the future of MAR.* Retrieved September 23, 2010, from http://www.perey.com/MobileARSummit/TelefonicaR&D-Mobile-Visual-Recognition.pdf

Martin, D. (2010). *Jibbigo: Your Star Trek universal translator for iPhone.* Retrieved January 17, 2011, from http://www.cultofmac.com/jibbigo-your-star-trek-universal-translator-for-iphone/44504

Martin, H. E. (2008). *Gartner's emerging technologies hype cycle 2008.* Retrieved September 1, 2010, from http://hemartin.blogspot.com/2008/08/gartners-emerging-technologies-hype.html

McGlasson, L. (2010). *Fraudsters take aim at mobile banking.* Retrieved on December 23, 2010, from http://www.bankinfosecurity.com/articles.php?art_id=2085

McKinney, D., Dyke, J., & Luber, E. S. (2009). iTunes university and the classroom: Can podcasts replace professors? *Computers & Education, 52,* 617–623. doi:10.1016/j.compedu.2008.11.004

MDG Gap Task Force (Ed.). (2010). *The global partnership for development at a critical juncture – Mdg Gap Task Force report 2010.* New York, NY: United Nations.

Meier, P. (2010). *Mobile augmented reality 2010.* Retrieved September 1, 2010, from http://www.perey.com/MobileARSummit/metaio-Mobile-AR-in-2010.pdf

Membridge. (2008). *History of cell phones.* Retrieved on December 16, 2010 from http://www.historyofcell-phones.net/

Merc Bank Michigan. (2010). *MercMobile® personal payments.* Retrieved on December 19, 2010, from https://www.mercbank.com/personal/electronic/p2p.asp

Merril, M. (2011). *Kalorama: Medical mobile app market worth $84.1M.* Retrieved on January 9, 2011, from http://www.healthcareitnews.com/news/kalorama-medical-mobile-app-market-worth-841m

Merrill, M. (2011b). *Report: 500M to use mHealth apps by 2015.* Retrieved on January 9, 2011, from http://www.healthcareitnews.com/news/report-500m-use-mhealth-apps-2015

Metaio. (2010). *Press release.* Retrieved August 18, 2010, from http://www.metaio.com/media-press/press-release/

Mignonneau, L., & Sommerer, C. (2005). Designing emotional, methaforic, natural and intuitive interfaces for interactive art, edutainment and mobile communications. *Computer Graphics, 29,* 837–851. doi:10.1016/j.cag.2005.09.001

Milian, M. (2011). *Reports say Egypt Web shutdown is coordinated, extensive.* Retrieved January 28, 2011, from http://www.cnn.com/2011/TECH/web/01/28/egypt.internet.shutdown/index.html?hpt=T2

Miller, B. (2009). *RFID technology being added to mobile marketing campaigns.* Retrieved on December 16, 2010, from http://blog.armoryideas.com/2009/06/11/rfid-technology-being-added-to-mobile-marketing-campaigns/

MJelly. (2009). *7 viral marketing tactics for mobile internet services.* Retrieved on December 18, 2010 from http://blog.mjelly.com/2009/01/viral-marketing-on-mobile.html

mobi Thinking. (2011). *Global mobile statistics 2011.* Retrieved on February 16, 2011 from http://mobithinking.com/mobile-marketing-tools/latest-mobile-stats

Mobile Augmented Reality. (2010). *The absolute latest in Android and iPhone augmented reality.* Retrieved on December 16, 2010, from http://www.mobileaugmentedreality.info/

Mobile Barcodes. (2010). *QR-code reader and software*. Retrieved on January 4, 2011 from http://www.mobile-barcodes.com/qr-code-software/

Mobile Behavior. (2010, February 16). Mobile in education: Mattel announces fisher-price ixl learning system [Web log message]. Retrieved from http://www.mobile-behavior.com/2010/02/16/mobile-in-education-mattel-announces-fisher-price-ixl-learning-system

Mobile Market Watch. (2010). *Research: Mobile proximity marketing to reach $750M by 2011 and nearly $6b by 2015*. Retrieved on December 16, 2010, from http://www.mobilemarketingwatch.com/research-mobile-proximity-marketing-to-reach-750m-by-2011-and-nearly-6b-by-2015-10252/

Mobile Marketing Association. (2008). *Code of conduct*. Retrieved on December 18, 2010 from http://www.mmaglobal.com/codeofconduct.pdf

Mobile Marketing Association. (2010). *Consumer best practices*. Retrieved on December 18, 2010 from http://www.mmaglobal.com/codeofconduct.pdf

Mobile Marketing Watch. (2010). *Research: Mobile proximity marketing to reach $750M by 2011 and nearly $6B by 2015*. Retrieved on December 16, 2010, from http://www.mobilemarketingwatch.com/research-mobile-proximity-marketing-to-reach-750m-by-2011-and-nearly-6b-by-2015-10252/

Mobilizy. (2010). *Press*. Retrieved August 18, 2010, from http://www.wikitude.org/category/07_press/press-press

Mohr, W. (2002). *Mobile communications beyond 3G in the global context*. Retrieved on December 15, 2010 from http://www.cu.ipv6tf.org/pdf/werner_mohr.pdf

Money, C. N. N. (2011). *Amazon sales pop as Kindle books overtake paperbacks*. Retrieved on February 16, 2011, from http://money.cnn.com/2011/01/27/technology/amazon_earnings/index.htm

Montagu, E. (1965). *The man who never was: World War II's boldest counter-intelligence operation*. Bantam Pathfinder.

Montalbano, E. (2010). *FBI warns of mobile cyber threats*. Retrieved on December 19, 2010 from http://www.informationweek.com/news/government/security/showArticle.jhtml?articleID=228400096&itc=ref-true

Monte, L. (2009). *The social pulpit – Barack Obama's social media toolkit*. New York, NY: Edelman.

Moore, W. J., & Starr, A. G. (2006). An intelligent maintenance system for continuous cost-based prioritisation of maintenance activities. *Computers in Industry, 57*(6), 595–606. doi:10.1016/j.compind.2006.02.008

Moore, A., Goulding, J., Brown, E., & Swan, J. (2009). AnswerTree – A hyperplace-based game for collaborative mobile learning. *Proceedings of mLearn*, Orlando, Florida.

Moore, D. (2003). *Species of competencies for intelligence analysis*. Retrieved January 27, 2011, from http://scip.cms-plus.com/files/Resources/Moore-Species-of-Competencies.pdf

Moore, J. (2010). *mHealth in the enterprise set to explode*. Retrieved on January 9, 2011, from http://www.healthcareitnews.com/blog/mhealth-enterprise-set-explode

Morris, J., & Koopman, P. (June, 2003). *Software defect masquerade fault in distributed embedded system*. DNS 2003. FastAbs.

Motion, X. GPS drive. (2011). *iTunes app*. Retrieved from http://itunes.apple.com/us/app/motionx-gps-drive/id328095974?mt=8

Muller, A., Crespo-Marquez, A., & Iung, B. (2008). On the concept of e-maintenance: Review and current research. *Reliability Engineering & System Safety, 93*(8), 1165–1187. doi:10.1016/j.ress.2007.08.006

Muller, A., Suhner, M.-C., & Iung, B. (2007). Maintenance alternative integration to prognosis process engineering. *Journal of Quality in Maintenance Engineering Journal of Quality in Maintenance Engineering, 13*(2).

Murakami, K., & Taguchi, H. (1991). Gesture recognition using recurrent neural networks. *Proceedings of the SIGCHI Conference on Human Factors in Computing Systems: Reaching Through Technology*, (pp. 237-242). New Orleans, Louisiana, United States.

Murphy, D. (2008). *It's as easy as IPTV*. Retrieved on December 18, 2010, from http://www.mobilemarketing-magazine.co.uk/content/its-easy-iptv

Nachira, F. (2005). *What is a digital ecosystem?* Retrieved from www.digital-ecosystems.org/events/2005.05/050518-fnak.pdf

Naish, J. (2009). *Mobile phones for children: A boon or a peril?* Retrieved on December 19, 2010, from http://women.timesonline.co.uk/tol/life_and_style/women/families/article6556283.ece

Nations, D. (2010). *A list of mobile Web browsers.* Retrieved on December 16, 2010, from http://webtrends.about.com/od/mobileweb20/tp/list_of_mobile_web_browsers.htm

Neustar. (2010). *CSC implementation: A mobile marketing plan.* Retrieved on December 16, 2010 from http://www.scribd.com/doc/21139025/CSC-Implementation-a-Mobile-Marketing-Plan

Newswire, P. R. (2010). *Blackboard launches mobile education platform in Mexico.* Retrieved October 1, 2010, from http://www.prnewswire.com/news-releases/blackboard-launches-mobile-education-platform-in-mexico-84415947.html

Nichols, R. (2010). *Arkansas.gov mobile apps put payment processing on smartphones.* Retrieved on January 4, 2011 from http://www.govtech.com/e-government/Arkansasgov-Mobile-Apps-Put-Payment-Processing.html

Nicol, J. R., Yechezkal, S. G., Paschetto, J., Rush, K. S., & Martin, C. (1999). How the Internet helps build collaborative multimedia applications. *Communications of the ACM, 42*(1), 79–85. doi:10.1145/291469.291474

Nielsen. (2010). *Nielsen unveils retail 2015 forecast.* Retrieved on December 15, 2010, from http://www.nielsen.com/us/en/insights/press-room/2010/nielsen_unveils_retail.html

NOAH. (2010). *Networked Organisms And Habitats home page.* Retrieved from http://www.networkedorganisms.com/about

Nokia. (2009). *Location, context, and mobile services.* Retrieved September 23, 2010, from http://research.nokia.com/files/insight/NTI_Location_&_Context_Jan_2009.pdf

Nook Color. (2011). *Features and tech specs.* Retrieved January 26, 2011, from http://www.barnesandnoble.com/nookcolor/features/techspecs/index.asp?cds2Pid=35607

Nook. (2011). *Features and tech specs.* Retrieved from http://www.barnesandnoble.com/nook/features/techspecs/index.asp?cds2Pid=35611

Nooriafshar, M., Williams, R., & Maraseni, T. N. (2004). *The use of virtual reality in education.* The American Society of Business and Behavioral Sciences (ASBBS) Seventh Annual International Conference, Cairns, Queensland, Australia, 6th-8th August.

Nortel. (2005). *CDMA end-to-end security.* Retrieved on December 27, 2010, from http://www.nortel.com/solutions/wireless/collateral/nn_107760.09-15-04.pdf

Oberman, J. (2006, March 08). *What some people do not get about the mobile buzz: Mobile at politics online day one* [Web log message]. Retrieved from http://personaldemocracy.com/content/what-some-people-do-not-get-about-mobile-buzz-mobile-politics-online-day-one

Och, F. (2009, June 18). *Google translates Persian* [Web log message]. Retrieved from http://googleblog.blogspot.com/2009/06/google-translates-persian.html

Office 2 HD. (2011). *iTunes app.* Retrieved from http://itunes.apple.com/us/app/id364361728?mt=8

Office of the Director of National Intelligence. (n.d.). *Members of the intelligence community.* Retrieved January 27, 2011, from http://www.odni.gov/members_IC.htm

Ogata, H., Hui, G. L., Yin, C., Ueda, T., Oishi, Y., & Yano, Y. (2008). LOCH: Supporting mobile language learning outside classrooms. *International Journal of Mobile Learning and Organisation, 2*(3), 271–282. doi:10.1504/IJMLO.2008.020319

Ogg, E. O. (2010, September 7). *iPad competitors lining up.* Retrieved from http://news.cnet.com/8301-31021_3-20015610-260.html

Oksman, V. (2009). Media content in mobiles. In Goggin, G. A. (Ed.), *Mobile technologies: From telecommuniciations to media* (pp. 118–130). New York, NY: Routledge.

Open Handset Alliance. (2010). *Home.* Retrieved August 14, 2010, from http://www.openhandsetalliance.com/index.html

OptumHealth. (2011). *NowClinic.* Retrieved January 25, 2011, from http://www.optumhealth.com/solutions-services/care-solutions/wellness/nowclinic/

O'Rourke, J. (2007). *US SMS penetration by age group.* Retrieved on December 19, 2010, from http://psmsus. blogspot.com/2007/01/us-sms-penetration-by-age-group. html

Parr, B. (2009, June 18). *Facebook releases Persian translation for #iranelection crisis* [Web log message]. Retrieved from http://mashable.com/2009/06/18/ facebook-persian/

Parviz, B. (2009). *Augmented reality in a contact lens.* Retrieved September 3, 2010, from http://spectrum.ieee. org/biomedical/bionics/augmented-reality-in-a-contact-lens/0

Patel, N. (March, 2008). *RFID credit cards easily hacked with an $8 reader.* Retrieved on November 28, 2010, from http://www.engadget.com/2008/03/19/rfid-credit-cards-easily-hacked-with-8-reader

Patrizio, A. (2008). *First iPhone malware found.* Retrieved on October 4, 2010, from http://www.internetnews.com/ security/article.php/3721016/First-iPhone-Malware-Found.htm

Pawlowski, A. (2010). *Paperless boarding takes off at United.* Retrieved on December 19, 2010, from http://articles.cnn.com/2010-03-15/travel/mobile.boarding.passes_1_boarding-airport-gates-passes?_s=PM:TRAVEL

Perey, C. (2010a). *Cross-platform AR: The market need and potential solutions.* Retrieved August 18, 2010, from http://www.perey.com/MobileARSummit/PEREY-CrossPlatform-AR.pdf

Perey, C. (2010b). *Clouds on our horizon? MAR risks and obstacles.* Retrieved September 3, 2010, from http://www.perey.com/MobileARSummit/PEREY-Clouds-on-the-Horizon.pdf

Perez, S. (2009). Opera reports explosive mobile web growth worldwide. Retrieved June 6, 2010, from http://www.readwriteweb.com

Perez, S. (2009, February 20). *Mobile phones to serve as doctors in developing countries* [Web log message]. Retrieved from http://www.readwriteweb.com/archives/mobile_phones_to_serve_as_doctors_in_developing_countries.php

Perez, S. (2010, September 22). *Android users crowd-source air pollution analysis* [Web log message]. Retrieved from http://www.readwriteweb.com/archives/android_users_crowd-source_air_pollution_analysis.php

Piike, J. (2010). *ECHELON.* Retrieved January 27, 2011, from http://www.fas.org/irp/program/process/echelon.htm

Pollard, S. (2008). *Mobile email marketing tips.* Retrieved on December 16, 2010 from http://www.lyris.com/resources/email-marketing/articles/mobile-email-marketing-tips/

Potts, H. (2010). *All bets are on m-health, but where to start to cash in on opportunities.* London: Mobile Healthcare Industry Summit.

Power, T., & Shrestha, P. (2010). *Mobile technologies for (English) language learning: An exploration in the context of Bangladesh.* In: IADIS International Conference: Mobile Learning, Porto, Portugal.

Purdue University. (2010). *Hotseat: Enabling collaborative micro-discussion in and out of the classroom.* Retrieved from http://www.itap.purdue.edu/tlt/hotseat/

PwC. (2010). Healthcare unwired: New business models delivering care anywhere.

Rajant. (2011). *Resilient tactical wireless broadband networks for military applications.* Retrieved January 17, 2011, from http://www.rajant.com/solutions/military

Ramirez, E. (2009). *Why teachers want to ban cellphone cameras from classrooms.* Retrieved September 1, 2010, from http://www.usnews.com/blogs/on-education/2009/03/23/why-teachers-want-to-ban-cellphone-cameras-from-classrooms.html

Rampell, C. (2010, April 6). Outsourced grades. *New York Times.* Retrieved September 28, 2010, from http://economix.blogs.nytimes.com/2010/04/06/outsourced-grades/

Rau, E. (2005, December). Combat science: The emergence of operational research in World War II. *Endeavour, 29*(4), 156–161. doi:10.1016/j.endeavour.2005.10.002

Ravuthula, S. (2002). *Web services applications and security: Part 1.* Retrieved from http://www.developer.com/services/article.php/1550461/Web-Services-Applications-and-Security-Part-1.htm

Raytheon. (2009). *System runs on Android mobile operating system*. Retrieved on January 14, 2011 from http://www.raytheon.com/newsroom/technology/rtn09_rats/

Reed. (2010). Sprint shows off 4G video apps. *IDG News*. Retrieved October 12, 2010 from http://www.oswmag.com/article/sprint-shows-4g-video-apps

Reisinger, D. (2010). *Mobile game revenue to top $11 billion by 2015*. Retrieved on Decemver 17, 2010 from http://news.cnet.com/8301-13506_3-20024103-17.html

Research, A. B. I. (2010). *Online social networking goes mobile: 140 million users by 2013*. Retrieved on December 16, 2010 from http://www.abiresearch.com/press/2998-Online+Social+Networking+Goes+Mobile%3A+140+Million+Users+by+2013

Rice, R. (2009). *Augmented vision and the decade of ubiquity*. Retrieved August 18, 2010, from http://curiousraven.squarespace.com/future-vision/2009/3/20/augmented-vision-and-the-decade-of-ubiquity.html

Richman, D. (2009, October 12). *Crowdsourcing digital signal strength* [Web log message]. Retrieved from http://www.msnbc.msn.com/id/33239992/ns/technology_and_science-wireless/

Roberti, F. (2005). *The history of RFID technology*. Retrieved on November 22, 2010, from http://www.rfidjournal.com/article/view/1338

Rogers, D. (2010). *BONDI augmented reality*. Retrieved September 23, 2010, from http://www.perey.com/MobileARSummit/OMTP-BONDI-Augmented-Reality-David%20Rogers.pdf

Ryu, J., & Lee, K. (2010). *How can we make the eyes for the MAR services*. Retrieved September 23, 2010, from http://www.perey.com/MobileARSummit/Olaworks-Eyes-for-MobileAR.pdf

Sacco, A. (2007). *A brief history of the mobile phone (1973-2007)*. Retrieved on December 16, 2010, from http://advice.cio.com/al_sacco/a_brief_history_of_the_mobile_phone_1973_2007?page=0%2C0

Sachoff, M. (2010). *Mobile social networking grows 240%*. Retrieved on December 16, 2010, from http://www.webpronews.com/topnews/2010/06/02/mobile-social-networking-grows-240

Sarigiannidis, G. (2006). *Localization for ad-hoc wireless sensor networks*. Master of Science Thesis, The Technical University of Delf, The Netherlands, August 2006.

Schneier, B. (2006). *Remotely eavesdropping on cell phone microphones*. Retrieved January 28, 2011, from http://www.schneier.com/blog/archives/2006/12/remotely_eavesd_1.html

Schroll, W., & Romescu, D. (2010). *Augmented citizen*. Retrieved September 21, 2010, from http://www.perey.com/MobileARSummit/Romescu-The-Augmented-Citizen.pdf

Schwartz, M. (2010). *New mobile security threat: Fingerprint oil*. Retrieved on December 22, 2010, from http://www.darkreading.com/security/vulnerabilities/226700046/index.html

SearchNetworking.com. (2011). *Telematics*. Retrieved on February 16, 2011, from http://searchnetworking.techtarget.com/definition/telematics

Security Watch. (2009). *Experts: More malware socializing in 2010*. Retrieved on December 23, 2010, from http://securitywatch.eweek.com/social_engineering/experts_more_malware_socializing_in_2010.html

Segall, L. (2011). *Mark Zuckerberg's Facebook page hacked*. Retrieved January 28, 2011, from http://money.cnn.com/2011/01/26/technology/facebook_hacked/index.htm

Sharda, N. (2006). Quality of service issues in mobile multimedia transmission . In Karmakar, G., & Dooley, L. (Eds.), *Mobile multimedia communications: Concepts, applications and challenges*. Hershey, PA: Idea Group Inc.

Sharda, N. (2009a). *A high-level model for and intelligent system to maximise the utilisation of harvested rainwater*. OZWATER 09 – Australia's National Water Conference & Exhibition, 16-18 March, 2009, Melbourne.

Sharda, N. (2009b). *Multimedia and communication technologies in digital ecosystems*. Seventh International Conference on Information, Communications and Signal Processing, Macau, 7-10 Dec 2009.

Sharda, N. (2011). Multimedia transmission over wireless sensor networks. In L.-M. Ang & K. P. Seng (Eds.), *Visual information processing in wireless sensor networks: Technology, trends and applications*. Hershey, PA: IGI Global. (Accepted for publication in 2011).

Sherwin, D. (2000). A review of overall models for maintenance management. [MCB University Press.]. *Journal of Quality in Maintenance Engineering, 6*(3), 138–164. doi:10.1108/13552510010341171

Shresta, S. (2007). Mobile web browsing: Usability study. In *Mobility '07: Proceedings of the 4th International Conference on Mobile Technology, Applications, and Systems and the 1st International Symposium on Computer Human Interaction in Mobile Technology*, (pp. 187-194).

Siciliano, R. (2010). *Mobile phone security under attack*. Retrieved on December 19, 2010, from http://blogs. mcafee.com/consumer/identity-theft/mobile-phone-security-under-attack

Singer, P. (2009). *Wired for war: The robotics revolution and conflict in the 21st century*. Penguin.

Sparkvue. (2010). *iTunes app*. Retrieved from http:// itunes.apple.com/us/app/sparkvue/id361907181?mt=8

Spyphone. (n.d.). *Top spyphone reviews*. Retrieved January 28, 2011, from http://www.topspyphonereviews.com /?gclid=CPGJkazV3aYCFY64KgodSB5e1Q

Stewart, J. (2008). *CISSP study guide*. John Wiley and Sons.

Svoboda, E. (2009, November 1). *Cellphonometry: Can kids really learn math from smartphones?* [Web log message]. Retrieved from http://www.fastcompany.com/ magazine/140/cellphonometry.htm

Symantec. (2008). *Internet security report*. Retrieved on December 22, 2010, from http://eval.symantec.com/ mktginfo/enterprise/white_papers/b-whitepaper_exec_ summary_internet_security_threat_report_xiii_04-2008. en-us.pdf

Synder, S. (2009). *The new world of wireless*. Wharton School Publishing.

Szymanski, J., Bangemann, T., Thomesse, J. P., Lang, C., Thron, M., Reboeuf, X., & Garcia, E. (2003). Proteus—A European initiative for e-maintenance platform development. *Proceedings of the Emerging Technologies and Factory Automation Conference, IEEE ETFA, 03*, 415–420.

Tagwhat. (2010). *Iryss*. Retrieved August 18, 2010, from http://tagwhat.com/

Talks, T. E. D. (Producer). (2006). *Iqbal Quadir says mobiles fight poverty*. Retrieved from http://www.ted.com/ talks/iqbal_quadir_says_mobiles_fight_poverty.html

Tao, B., Ding, H., & Xion, Y. L. (2003). IP sensor and its distributed networking application in e-maintenance. *Proceedings of the 2003 IEEE International Conference on Systems, Man and Cybernetics*, vol. 4, Washington, DC, USA, (pp. 3858–63).

Teo, L. (2010). *Why Facebook ad click through rates suck – And how to change that*. Retrieved on December 19, 2010, from http://www.ymarketing.com/blog/bid/49898/ Why-Facebook-Ad-Click-Through-Rates-Suck-And-How-To-Change-That

Terbilang, S. (2008). *Studying through SMS*. Retrieved September 1, 2010, from http://mjrevenge.blogspot. com/2009/12/studying-through-sms.html

The Mobile Health Crowd. (2010). *Introduction - The tipping point*. Retrieved on December 28, 2010, from http://www.themobilehealthcrowd.com/?q=node/7

The Mobile Health Crowd. (2010b). *101 applications - Engaging the mobile telecoms industry*. Retrieved on December 28, 2010, from http://www.themobilehealthcrowd. com/?q=node/9

The Mobile Health Crowd. (2010c). *PACS - Getting the picture at last*. Retrieved on December 28, 2011, from http://www.themobilehealthcrowd.com/q=node/13

The Mobile Health Crowd. (2010d). *Smart homes for the elderly*. Retrieved on December 28, 2010, from http:// www.themobilehealthcrown.com/?q=node/79

Théreaux, O. (2010). *Beyond the glorified tour guide and the dystopian future*. Retrieved September 1, 2010, from http://lab.pheromone.ca/2010/02/09/mobile-augmented-reality/

TMC. (2007). *The history of SMS messaging.* Retrieved on December 15, 2010, from http://www.tmcsms.com/sms-history.aspx

Tonchidot. (2010). *Home page.* Retrieved August 18, 2010, from http://www.tonchidot.com/

Topol, E. (2010). *Eric Topol: The wireless future of medicine.* TEDMED Conference 2009. Retrieved on January 17, 2011, from http://www.ted.com/talks/eric_topol_the_wireless_future_of_medicine.html

Torrani, M. a. (2008). *Solutions to GSM security weaknesses.* Retrieved on December 27, 2010, from http://www.slideshare.net/Garry54/solutions-to-the-gsm-security-weaknesses

Toure, H. (2010). *Monitoring the WSIS targets: A midterm review.* World Telecommunication/ICT Development Report 2010.

Traunmüler, R. (2010). Web 2.0 creates a new government. *Electronic Government and the Information Systems Perspective. Lecture Notes in Computer Science, 6267,* 77–83.

Traxler, J. (2009). Current state of mobile learning . In Ally, M. (Ed.), *Mobile learning transforming the delivery of education and training.* Edmonton, Canada: AU Press.

Trend Micro. (2007). *Enterprise mobile security.* Retrieved on December 22, 2010, from http://emea.trendmicro.com/imperia/md/content/us/pdf/products/enterprise/mobilesecurity/wp03_tmms_071212us.pdf

Tsang, A. (2002). Strategic dimensions of maintenance management. *Journal of Quality in Maintenance Engineering, 8*(1), 7–39. doi:10.1108/13552510210420577

Turner, M., Budgen, D., & Brereton, P. (2003, October). Turning software into a service. *Computer, 36*(10), 38–44. doi:10.1109/MC.2003.1236470

Txteagle. (2010). *Overview.* Retrieved from http://txteagle.com/?q=about/overview

Tzu, S. (1963). *The art of war* (Griffin, S. B., Trans.). Oxford University Press.

U.S. Department of Justice. (2010). *Amber alert website.* Retrieved on January 4, 2011, from http://www.amberalert.gov/

Ucar, M., & Qiu, R. G. (2005). E-maintenance in support of e-automated manufacturing systems. *Journal of the Chinese Institute of Industrial Engineering, 22*(1), 1–10. doi:10.1080/10170660509509271

United breaks guitars. (2010). *Wikipedia.* Retrieved September 06, 2010, from http://en.wikipedia.org/wiki/United_Breaks_Guitars

Ushahidi. (n.d.). Retrieved from http://www.ushahidi.com/media/Ushadidi_1-Pager.pdf

Valli, A. (2006). The design of natural interaction. *Multimedia Tools and Applications, 38*(3), 295–305. doi:10.1007/s11042-007-0190-z

Valli, A. (2004). *Notes on natural interaction.* Retrieved August 20, 2010, from http://www.citeulike.org/user/eckel/article/4324923

Van Rooyen, A. (2010). Effective integration of SMS communication into a distance education accounting module. *Meditari Accountancy Research, 18*(1), 47–57. doi:10.1108/10222529201000004

Vasdev, S. (2010). *The 21st-century mobile military.* Retrieved on January 16, 2011, from http://ndn.org/blog/2010/09/21st-century-mobile-military

Vassel, C. (2006, August). *Mobile learning: Using SMS to enhance education provision.* The 7th Annual Conference of the Higher Education Academy for Information and Computer Sciences, Dublin, Ireland.

Vaughan, T. (Ed.). (1993). *Multimedia: Making it work.* Berkeley, CA: Osborne/McGraw-Hill.

Vaughan-Nichols, S. V. (2010, March 12). *Here comes Linux's iPad clones.* Retrieved from http://blogs.computerworld.com/15742/here_comes_linuxs_iPad_clones

Vital Wave Consulting. (2008). *Sizing the business potential of mHealth in the global South: A practical approach.* Presented at The eHealth Connection. Bellagio, Italy.

Vital Wave Consulting. (2009). *Mhealth for development: The opportunity of mobile technology for healthcare in the developing world.* Washington, DC: UN Foundation and Vodafone Foundation.

W3. (2010). HTML5 differences from HTML 4. Retrieved on December 18, 2010 from http://www.w3.org/TR/html5-diff/

W3. (2010). *HTML5 differences from HTML 4*. Retrieved on December 18, 2010 from http://www.w3.org/TR/html5-diff/

Wajcman, J. B. (2009). Intimite connections: The impact of the mobile phone on work/life boundaries. In Goggin, G. A. (Ed.), *Mobile technologies: From telecommunications to media* (pp. 9–22). New York, NY: Routledge.

Wallace, L. (2009). *Blink-182 rocks augmented reality show in Doritos bag*. Retrieved on December 16, 2010 from http://www.wired.com/underwire/2009/07/blink-182-rocks-augmented-reality-show-in-doritos-bag/

Wang, M., Wang, M., Shen, R., Novak, D., & Pan, H. (2009). The impact of mobile learning on students' learning behaviours and performance: Report from a large blended classroom. *British Journal of Educational Technology*, *40*(4), 673–695. doi:10.1111/j.1467-8535.2008.00846.x

Wapner, J. (2010). iRegulate: Should medical apps face government oversight? *Scientific American*, (April): 2010.

Warren, C. (2010). *Mobile social networking usage soars*. Retrieved on December 16, 2010 from http://mashable.com/2010/03/03/comscore-mobile-stats/

Waters, A. (2010). *How barcodes and smartphones will rearchitect information*. Retrieved on January 4, 2011 from http://www.bukisa.com/articles/310287_how-barcodes-and-smartphones-will-rearchitect-information

Wauters, R. (2009). There's money in mobile dating. Retrieved on December 16, 2010, from http://techcrunch.com/2009/01/19/juniper-research-theres-money-in-mobile-dating-services/

Weld, D., Anderson, C., Domingos, P., Etzioni, O., Gajos, K., Lau, T., & Wolfman, S. (2003). Automatically personalizing user interfaces. *Proceedings of the International Joint Conference on Artificial Intelligence, (IJCAI03)*, Acapulco, México.

Wellman, B. (1997). An electronic group is virtually a social network. In Kielser, S. (Ed.), *Culture of the Internet* (pp. 179–205). Mahwah, NJ: Lawrence Erlbaum Associates.

Wells. (2010). *Lookout for Android issues*. Retrieved October 3, 2010, from http://www.androidtapp.com/lookout-for-android-issues-security-update-for-android-sms-trojan/

Wertime, K. A. (2008). *DigiMarketing: The essentail guide to new media and digital marketing*. John Wiley& Sons.

Winters, N., & Mor, Y. (2008). *CoMo: Supporting collaborative group work using mobile phones in distance education*. Retrieved September 1, 2010, from http://www.lkl.ac.uk/como/CoMo-Final-Report.pdf

Winthrop, P. (2010). *Are we transitioning from mobile device management to mobile security?* Retrieved on December 24, 2010, from http://theemf.org/2010/11/17/are-we-transitioning-from-mobile-device-management-to-mobile-security/

Wireless Internet. (2010). *What's this about 4G?* Retrieved on December 16, 2010 from http://www.wirelessinternet.org/4G-network.php

Wireshark (2011). *Wireshark*. Retrieved January 27, 2011, from http://www.wireshark.org/

Wise Geek. (2010). *What is a SIM card?* Retrieved on December 27, 2010, from http://www.wisegeek.com/what-is-a-sim-card.htm

Wise Geek. (2010). *What is mobile spam?* Retrieved on December 23, 2010, from http://www.wisegeek.com/what-is-mobile-phone-spam.htm

Wittstock, M. (2010, January 19). *Assignit - Mobile crowd-sourcing apps for hyper-local investigative journalism* [Web log message]. Retrieved from http://www.changemakers.com/node/68605

Wolfson, O. (2005). Mobi-dic: Mobile discovery of local resources in peer-to-peer wireless networks. *Bulletin of the IEEE Computer Society Technical Committee on Data Engineering*, *28*(3). Retrieved from http://cs.uic.edu/~boxu/mp2p/deb-wolfson.pdf.

Wolfson, O. (2009). *Foreword to the book. Mobile peer-to-peer computing for next generation distributed environments: Advancing conceptual and algorithmic applications* [Web log message]. Retrieved from http://www.cs.uic.edu/~boxu/mp2p/foreword.pdf

World, G. S. M. (2009). *Market data summary*. Retrieved on December 27, 2010, from http://www.gsmworld.com/newsroom/market-data/market_data_summary.htm

Yoshikawa, H. (1995). Manufacturing and the 21st century—Intelligent manufacturing systems and the renaissance of the manufacturing industry. *Technological Forecasting and Social Change, 49*(2), 165–213. doi:10.1016/0040-1625(95)00008-X

Yu, R., Iung, B., & Panetto, H. (2003). A mutli-agents based e-maintenance system with case-based reasoning decision support. *Engineering Applications of Artificial Intelligence, 16*, 321–333. doi:10.1016/S0952-1976(03)00079-4

Zapit, S. M. S. (2010). *SMS for education*. Retrieved October 1, 2010, from http://www.zapitsms.com/SMS-For-Education

Zetter, K. (2009, February 5). *TED: MIT students turn internet into a sixth human sense*. Retrieved from http://www.wired.com/epicenter/2009/02/ted-digital-six/

Zhang, W., Halang, W., & Diedrich, C. (2003). An agent-based platform for service integration in e-maintenance. *Proceedings of ICIT 2003, IEEE International Conference on Industrial Technology*, vol. 1, Maribor, Slovenia, (pp. 426–33).

Zhou, Y., Percival, G., Wang, X., Wang, Y., & Zhao, S. (2010). *MOGCLASS: A collaborative system of mobile devices for classroom music education*. Retrieved September 1, 2010, from www.comp.nus.edu.sg/~zhaosd/paper/mmshc05843-zhou.pdf

About the Contributors

Barbara L. Ciaramitaro, PhD, is an Assistant Professor with the College of Business at Ferris State University in Big Rapids, Michigan where she teaches graduate and undergraduate courses on a variety of topics including information security, business intelligence, social media and virtual worlds, and project management. Dr. Ciaramitaro entered the academic world after 30 years of experience in business, including 10 years as an Executive with General Motors. Dr. Ciaramitaro holds a Ph.D. from Nova Southeastern University in Information Systems with a graduate certificate in Information Security, a M.S. from Central Michigan University in Software Engineering Administration, and a B.A. in Psychology from Wayne State University. She is currently working on earning her MBA degree from Ferris State University. Dr. Ciaramitaro has earned several professional certifications including the CISSP (Certified Information System Security Professional), CSSLP (Certified Software System Lifecycle Professional), and PMP (Project Management Professional). Dr. Ciaramitaro is a frequent speaker and author on topics including social media, virtual worlds, security, and privacy. She edited the book and authored chapters in Virtual Worlds and E-Commerce: Technologies and Applications for Building Customer Relationships, published by IGI Global in August 2010. She is an active member of several professional organizations, including the Association of Virtual Worlds, Project Management Institute, ISACA, ISC2 and IEEE.

***

Ade Bamigboye is the Founder of Mobile Flow. Ade has been involved in the mobile industry for over 10 years. During that time he has been involved in early stage companies that have focused on remote patient monitoring and disease management using mobile and web technologies. His main area of interest is the commercialisation of digital healthcare. Previously, Ade was a Founding Investor and Chief Technology Officer at United Clearing from its inception until its UK stock market listing, and for over 20 years he has been creating innovative business solutions from emerging technology through positions that he has held at Shell, PA Consulting, and a number of software start-ups. Ade received an MBA from the Henley Management College and has a B.Sc (Hons) Applied Chemistry and M.Sc Artificial Intelligence. Other related publications include "The Consumerisation of Home Healthcare Technologies - Springer Handbook of Digital Homecare."

Douglas Blakemore is a Professor at Ferris State University in the United States. In that capacity, Dr. Blakemore has developed innovative courses in information systems management at both the graduate and undergraduate program for over 15 years. Prior to that, Dr. Blakemore co-managed a technology consulting and accounting firm for over 15 years with his wife Vivian. In addition to designing and managing computers, network systems, and client write-up systems, Dr. Blakemore worked with Michigan Dyslexia Institute as a certified reading consultant. Dr. Blakemore received his Ph.D. in Organization and Management from Capella University, a Master's in Information Systems Management from Ferris State University, and a B.A. from Spring Arbor College in Philosophy and Religion.

María del Puy Carretero studied Computer Science at the University of the Basque Country UPV/EHU (2004). She was an intern during the summer of 2002 and carried out her final year project at Vicomtech (July 2003-February 2004). Since October of 2004 she works as a researcher at Vicomtech, within the area of 3D Animation, Interactive Virtual Environments and HCI. She has a PhD in Computer Science from the University of Basque Country. She made the oral presentation of her PhD.-Thesis entitled 'Avatares Multidispositivo para Interación Multimodal' in June 2010. The thesis was graded Cum Laude A.

Greg Gogolin spent almost 20 years in IT before becoming a Professor at Ferris State University in 1999. He has worked as a programmer, database administrator, systems analyst, and project manager at small and multi-national corporations. Dr. Gogolin actively consults and is a licensed Private Investigator specializing in digital forensics cases. He has degrees in Arts, Computer Information Systems, Applied Biology, Computer Information Systems Management, and Administration, with his doctorate from Michigan State University. Dr. Gogolin has written many articles for journals such as *Information Security Journal: A Global Perspective, Digital Investigations,* and *Journal of Digital Forensic Practice,* and has previously collaborated with Dr. Barbara Ciaramitaro as a chapter contributor to Virtual Worlds and E-Commerce: Technologies and Applications for Building Customer Relationships. He holds CISSP, EnCE and PMP certifications, and is particularly thankful to have the opportunity to teach some of the brightest students in the world in the Bachelor of Science in Information Security & Intelligence and Master of Science in Information Systems Management programs at Ferris State University.

James Jones has worked in the information security field for the past 15 years with government, commercial, and research organizations. During this time he has led multiple security projects, directed a security operations center, led a digital forensics team, and served as the principle investigator for research projects with DARPA, DHS, and AFRL. Dr. Jones actively consults in the areas of digital forensics, information security, and intelligence analysis. Dr. Jones holds a B.S. in Industrial and Systems Engineering from Georgia Tech, a M.S. in Mathematical Sciences from Clemson University, and a Ph.D. in Computational Sciences and Informatics from George Mason University, and he holds CISSP, EnCE, and Security+ certifications. Dr. Jones is an Associate Professor at Ferris State University and a Principal Scientist for a defense contractor.

Joerg H. Kloss is one of the early pioneers of immersive technologies and interactive 3D graphics on the Internet. He is working in the field of Virtual Worlds, Virtual Reality and Augmented Reality since many years. Besides being a respected lecturer on many symposia, he published articles and reference books about 3D standards like VRML and X3D, as well as his latest book on Android programming. With a focus on Mobile Augmented Reality, Mr. Kloss currently fosters standardization efforts in the conversion of mobile and fixed network technologies and applications. He holds a Master's degree in Computational Linguistics, Psychology and Economics, and currently works as a telecommunications professional and consultant in Germany.

Juan Felipe Mogollón studied Technical Telecommunication Engineering, Electronic Systems specialty at Cantabria University. Formerly, he was a developer at Zitralia Security Solutions, developing security applications in GNU/Linux (October 2006 - June 2008). He has been working at Vicomtech as a research collaborator since July 2008.

Igor G. Olaizola received a six years degree in Electronic and Control Engineering from the University of Navarra, Spain (2001). He developed his master thesis at Fraunhofer Institut für Integrierte Schaltungen (IIS), Erlangen -Germany- 2001 where worked for a year as research assistant on several projects related to MPEG standard audio decoding. He is member of Vicomtech technological centre since 2002. Nowadays he is the head of the Digital Interactive TV and Multimedia Services department. In 2006 he also participated as a technology consultant in Vilau (www.vilau.net) company for one year. Moreover, he is working in his PhD based on automatic multimodal indexing and management of multimedia content at the Faculty of Informatics of San Sebastián, in the University of Basque Country.

David Oyarzun is the Head of the 3D Animation and Interactive Virtual Environments department at Vicomtech. He studied Computer Engineering in the University of the Basque Country. He received his Ph.D. in the field of 3D Virtual Environments standardization from the University of the Basque Country. From 2001 to 2003, he gained experience in topics such as biometric access control systems for networks or Artificial Intelligence. Since 2003, he has been working in Vicomtech, and he is coordinator of the Working Group V of Spanish eNEM platform, which is a working group focused in Virtual Reality and Virtual Worlds research composed by Spanish companies, universities, and research centres. He has several publications in topics related with 3D graphics in international conferences and journals.

Velislav Pavlov is the IT Coordinator for Ferris State University in Grand Rapids, MI. Velislav has nearly a decade of experience in information systems and security. He holds a Bachelor of Science in Information and Computer Security from Davenport University, and a Master of Science in Information System Management from Ferris State University. His professional certifications include CompTIA A+ and Security+. His research interests include penetration testing, digital forensics, IT compliance and governance, virtualization, and mobile security.

Shaunie Shammass is a phonetics-linguistics expert with over 20 years of experience in the application of linguistics to automatic speech recognition in the telephony and mobile environments. She is currently the linguistic expert at SpeakingPal, handling all aspects of content creation for innovative English language learning products on the mobile. Dr. Shammass has vast pedagogical expertise in e-learning, having worked at Edusoft and Digispeech, where she developed English language e-learning and pronunciation improvement programs. She has also lectured for many years at Bar-Ilan University and at several teachers colleges. Shaunie has worked in several high-tech firms, including NSC (Natural

Speech Communication), where she developed sophisticated linguistic resources in over 13 languages for automatic speech recognition, large lexica development and automatic speech-to-speech translation, and at Nice Systems, working on speech analytics. Shaunie holds a PhD in Linguistics, with distinction, from the University of Alberta, Canada.

Nalin Sharda gained B.Tech. and Ph.D. degrees from the Indian Institute of Technology, Delhi and he is an Associate Professor of Computer Science and Multimedia at Victoria University, Australia. Nalin's publications include the textbook entitled Multimedia Information Networking, Prentice Hall, and around 100 other publications. He has contributed to editing and publishing papers and articles on Multimedia Authoring and Networking in the Encyclopaedia of Multimedia, the Handbook of Multimedia Computing, the Handbook of Internet Computing, IEEE & ACM journals, the Communications of the AIS, and numerous international conferences. He is an editor for the *Informing Science Journal, the International Journal of Computers and Applications,* and the *Journal of Multimedia Tools and Applications.* Nalin's current research interests and related publications include applications of the Internet and multimedia systems to enhance communications and the development of tourism ICT systems. Nalin has developed a number of conceptual models for integrating the art, science, and technology of multimedia systems, including: Multimedia Design and Planning Pyramid (MUDPY), a metadesign framework for multimedia systems design; and Movement Oriented Design (MOD), a new paradigm for designing the temporal aspect of multimedia systems.

Marilyn K Skrocki is an Assistant Professor at Ferris State University in the College of Allied Health Science and teaches in the Health Care Systems Administration program. Dr. Skrocki worked in various clinical and administrative positions in healthcare and specialized in implementing health law toward assuring compliance with patient safety and quality. Dr. Skrocki was instrumental in developing a program that concentrates on medical informatics. Dr. Skrocki holds a Juris Doctorate from Thomas M. Cooley Law School, a Master's in Business Administration from Ferris State University, and a B.S. in Health Care Systems Administration.

David R Svacha is a student at Ferris State University in the United States. Mr. Svacha is currently pursuing a Bachelor's degree in the Information Security and Intelligence program along with a minor in Computer Information Systems. He actively involves himself in two student organizations, the Institute of Electrical/Electronic Engineers (IEEE) and Association of Information Technology Professionals (AITP). Upon graduation, he plans to obtain a job with the federal government.

Alejandro Ugarte obtained his Master's Degree in Telecommunication Engineering at the University of Deusto (Bilbao) in 2002. Since April 2003 he has been a researcher in the Digital Television and Multimedia Services department of Vicomtech. During this time he has participated in several projects in the fields of interactive TV applications and multimedia applications. In these projects he has collaborated in real scenarios with important consulting firms in engineering (Dominion) and TV broadcasters (ETB). It is worth pointing out his skills in the fields of t-government, multimedia, and augmented reality. In April 2007 he joined the quality committee of Vicomtech, taking part in the continuous improvement program of the quality system. Between 2007 and 2008 he completed an investigation stage in the Fraunhofer Institute for Computer Graphics (IGD) in Darmstadt, Germany, promoting the relations between the two institutes.

Ariel Velikovsky, with over 20 years in the telecom industry, has extensive experience in research and development of mobile products. Handling various marketing and technical aspects of mobile handset applications and communication systems, he specializes in mobile messaging and mobile learning solutions. Ariel is currently the CTO and co-founder of SpeakingPal, a mobile learning company that develops innovative English learning products. Prior to founding SpeakingPal, Ariel served as a consultant advisor for several mobile start-up companies. Ariel served as an R&D Manager, CTO of the Handset Line of Business and later as a Chief Architect in the CTO Office in Comverse, a world leading provider of software and systems for the mobile industry. He holds a BSc summa cum laude in Mathematics & Computer Science from Ben-Gurion University, an MA in Computer Science from Bar-Ilan University, and an MBA from the Kellogg Graduate School of Management at Northwestern University.

Mikel Zorrilla studied Telecommunication Engineering at the School of Engineering of Mondragon, Mondragon Unibertsitatea (2001-2007, www.mondragon.edu) and Technical Engineering in Telecommunication Systems at the same Engineering School (2001-2004). During his studies he worked half day in the area of communications of Ikerlan S. Coop. (www.ikerlan.es) as an area assistant (2002-2006). Afterwards, he developed the End of Degree Project also in Ikerlan S. Coop. about The Transport of Multimedia Traffic With an "Industrial Ethernet" Communication Bus (2006-2007). Since 2007 August he is working at Vicomtech (www.vicomtech.org), where he designs, develops, and leads projects of the digital television and multimedia services area.

238

Index

CPSIA information can be obtained at www.ICGtesting.com
Printed in the USA
BVOW060609221011

274084BV00010B/26/P